T0336010

An Informal Introduction to Stochastic Calculus with Applications

An Informal Introduction to Stochastic Calculus with Applications

Ovidiu Calin

Eastern Michigan University, USA

NEW JERSEY • LONDON • SINGAPORE • BEIJING • SHANGHAI • HONG KONG • TAIPEI • CHENNAI

Published by

World Scientific Publishing Co. Pte. Ltd.
5 Toh Tuck Link, Singapore 596224
USA office: 27 Warren Street, Suite 401-402, Hackensack, NJ 07601
UK office: 57 Shelton Street, Covent Garden, London WC2H 9HE

Library of Congress Cataloging-in-Publication Data
Calin, Ovidiu.
An informal introduction to stochastic calculus with applications / by Ovidiu Calin (Eastern Michigan University, USA).
pages cm
Includes bibliographical references and index.
ISBN 978-9814678933 (hardcover : alk. paper) -- ISBN 978-9814689915 (pbk : alk. paper)
1. Stochastic analysis. 2. Calculus. I. Title. II. Title: Introduction to stochastic calculus with applications.
QA274.2.C35 2015
519.2'2--dc23

2015014680

British Library Cataloguing-in-Publication Data
A catalogue record for this book is available from the British Library.

Printed in Singapore

Preface

Deterministic Calculus has been proved extremely useful in the last few hundred years for describing the dynamics laws for macro-objects, such as planets, projectiles, bullets, etc. However, at the micro-scale, the picture looks completely different, since at this level the classical laws of Newtonian mechanics cease to function "normally". Micro-particles behave differently, in the sense that their state cannot be determined accurately as in the case of macro-objects; their position or velocity can be described using probability densities rather than exact deterministic variables. Consequently, the study of nature at the micro-scale level has to be done with the help of a special tool, called Stochastic Calculus. The fact that nature at a small scale has a non-deterministic character makes Stochastic Calculus a useful and important tool for the study of Quantum Mechanics.

In fact, all branches of science involving random functions can be approached by Stochastic Calculus. These include, but they are not limited to, signal processing, noise filtering, stochastic control, optimal stopping, electrical circuits, financial markets, molecular chemistry, population evolution, etc.

However, all these applications assume a strong mathematical background, which takes a long time to develop. Stochastic Calculus is not an easy theory to grasp and, in general, requires acquaintance with probability, analysis and measure theory. This fact makes Stochastic Calculus almost always absent from the undergraduate curriculum. However, many other subjects studied at this level, such as biology, chemistry, economics, or electrical circuits, might be more completely understood if a minimum knowledge of Stochastic Calculus is assumed.

The attribute *informal*, present in the title of the book, refers to the fact that the approach is at an introductory level and not at its maximum mathematical detail. Many proofs are just sketched, or done "naively" without putting the reader through a theory with all the bells and whistles.

The goal of this work is to informally introduce elementary Stochastic Calculus to senior undergraduate students in Mathematics, Economics and Business majors. The author's goal was to capture as much as possible of the spirit of elementary Calculus, which the students have already been exposed to in the beginning of their majors. This assumes a presentation that mimics similar properties of deterministic Calculus as much as possible, which facilitates the understanding of more complicated concepts of Stochastic Calculus.

The reader of this text will get the idea that deterministic Calculus is just a particular case of Stochastic Calculus and that Ito's integral is not a too much harder concept than the Riemannian integral, while solving stochastic differential equations follows relatively similar steps as solving ordinary differential equations. Moreover, modeling real life phenomena with Stochastic Calculus rather than with deterministic Calculus brings more light, detail and significance to the picture.

The book can be used as a text for a one semester course in stochastic calculus and probabilities, or as an accompanying text for courses in other areas such as finance, economics, chemistry, physics, or engineering.

Since deterministic Calculus books usually start with a brief presentation of elementary functions, and then continue with limits, and other properties of functions, we employed here a similar approach, starting with elementary stochastic processes, different types of limits and pursuing with properties of stochastic processes. The chapters regarding differentiation and integration follow the same pattern. For instance, there is a product rule, a chain-type rule and an integration by parts in Stochastic Calculus, which are modifications of the well-known rules from elementary Calculus.

In order to make the book available to a wider audience, we sacrificed rigor and completeness for clarity and simplicity, emphasizing mainly on examples and exercises. Most of the time we assumed maximal regularity conditions for which the computations hold and the statements are valid. Many complicated proofs can be skipped at the first reading without affecting later understanding. This will be found attractive by both Business and Economics students, who might get lost otherwise in a very profound mathematical textbook where the forest's scenery is obscured by the sight of the trees. A flow chart indicating the possible order the reader can follow can be found at the end of this preface.

An important feature of this textbook is the large number of solved problems and examples which will benefit both the beginner as well as the advanced student.

This book grew from a series of lectures and courses given by the author at Eastern Michigan University (USA), Kuwait University (Kuwait) and Fu-Jen University (Taiwan). The student body was very varied. I had math, statistics, computer science, economics and business majors. At the initial stage, several students read the first draft of these notes and provided valuable feedback, supplying a list of corrections, which is far from exhaustive. Finding any typos or making comments regarding the present material are welcome.

Heartfelt thanks go to the reviewers who made numerous comments and observations contributing to the quality of this book, and whose time is very much appreciated.

Finally, I would like to express my gratitude to the World Scientific Publishing team, especially Rok-Ting Tan and Ying-Oi Chiew for making this endeavor possible.

O. Calin Michigan, January 2015

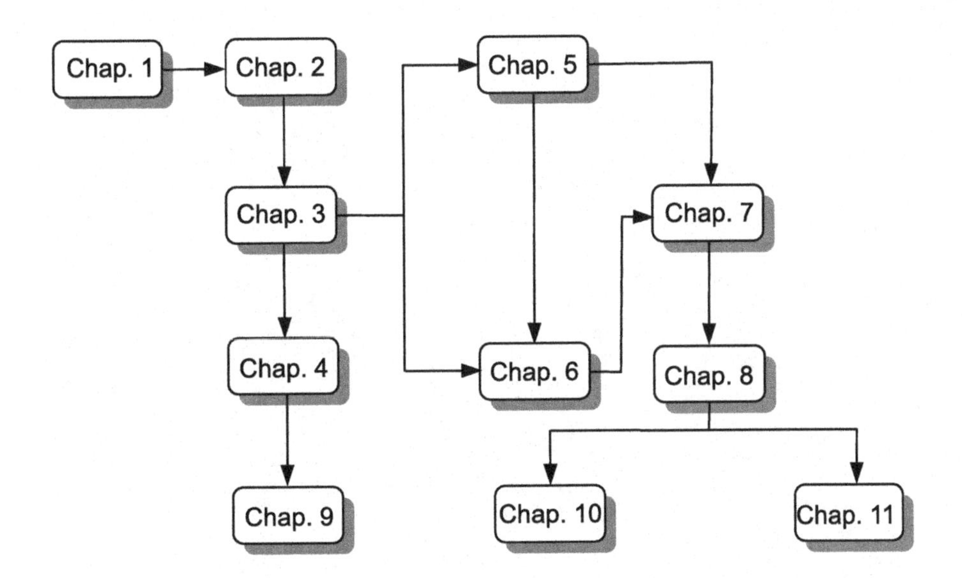

List of Notations and Symbols

The following notations have been frequently used in the text.

$(\Omega, \mathcal{F}, P)$	Probability space
Ω	Sample space
$\mathcal{F}$	σ-field
X	Random variable
X_t	Stochastic process
$\text{as} - \lim_{t\to\infty} X_t$	The almost sure limit of X_t
$\text{ms} - \lim_{t\to\infty} X_t$	The mean square limit of X_t
$\text{p} - \lim_{t\to\infty} X_t$	The limit in probability of X_t
$\mathcal{F}_t$	Filtration
$\mathcal{N}_t, \dot{W}_t, \frac{dW_t}{dt}$	White noise
W_t, B_t	Brownian motion
$\Delta W_t, \Delta B_t$	Jumps of the Brownian motion during time interval Δt
dW_t, dB_t	Infinitesimal jumps of the Brownian motion
$V(X_t)$	Total variation of X_t
$V^{(2)}(X_t), \langle X, X\rangle_t$	Quadratic variation of X_t
$F_X(x)$	Probability distribution function of X
$p_X(x)$	Probability density function of X
$p(x, y; t)$	Transition density function
$\mathbb{E}[\cdot]$	Expectation operator
$\mathbb{E}[X\|\mathcal{G}]$	Conditional expectation of X with respect to $\mathcal{G}$
$Var(X)$	Variance of the random variable X
$Cov(X, Y)$	Covariance of X and Y
$\rho(X, Y), Corr(X, Y)$	Correlation of X and Y
$\mathcal{A}_X, \mathcal{F}^X$	σ-algebras generated by X

$\Gamma(\,\cdot\,)$	Gamma function
$B(\,\cdot,\cdot\,)$	Beta function
N_t	Poisson process
S_n	Waiting time for Poisson process
T_n	Interarrival time for Poisson process
$\tau_1 \wedge \tau_2$	The minimum between τ_1 and τ_2 ($= \min\{\tau_1, \tau_2\}$)
$\tau_1 \vee \tau_2$	The maximum between τ_1 and τ_2 ($= \max\{\tau_1, \tau_2\}$)
$\bar{\tau}_n$	Sequence superior limit ($= \sup_{n\geq 1} \tau_n$)
$\underline{\tau}_n$	Sequence inferior limit ($= \inf_{n\geq 1} \tau_n$)
μ	Drift rate
σ	Volatility, standard deviation
$\partial_{x_k}, \dfrac{\partial}{\partial x_k}$	Partial derivative with respect to x_k
$\mathbb{R}^n$	n-dimensional Euclidean space
$\|x\|$	Euclidean norm ($= \sqrt{x_1^2 + \cdots + x_n^2}$)
Δf	Laplacian of f
$1_A, \chi_A$	The characteristic function of A
$\|f\|_{L^2}$	The L^2-norm ($= \sqrt{\int_a^b f(t)^2\,dt}$)
$L^2[0,T]$	Squared integrable functions on $[0,T]$
$C^2(\mathbb{R}^n)$	Functions twice differentiable with second derivative continuous
$C_0^2(\mathbb{R}^n)$	Functions with compact support of class C^2
R_t	Bessel process
$\widehat{\zeta}_t$	The mean square estimator of ζ_t

Contents

Chapter 1

A Few Introductory Problems

Even if deterministic Calculus is an excellent tool for modeling real life problems, however, when it comes to random exterior influences, Stochastic Calculus is the one which can allow for a more accurate modeling of the problem. In real life applications, involving trajectories, measurements, noisy signals, etc., the effects of many unpredictable factors can be averaged out, via the Central Limit Theorem, as a normal random variable. This is related to the Brownian motion, which was introduced to model the irregular movements of pollen grains in a liquid.

In the following we shall discuss a few problems involving random perturbations, which serve as motivation for the study of the Stochastic Calculus introduced in next chapters. We shall come back to some of these problems and solve them partially or completely in Chapter 11.

1.1 Stochastic Population Growth Models

Exponential growth model Let $P(t)$ denote the population at time t. In the time interval Δt the population increases by the amount $\Delta P(t) = P(t + \Delta t) - P(t)$. The classical model of population growth suggests that the relative percentage increase in population is proportional with the time interval, i.e.

$$\frac{\Delta P(t)}{P(t)} = r\Delta t,$$

where the constant $r > 0$ denotes the population growth. Allowing for infinitesimal time intervals, the aforementioned equation writes as

$$dP(t) = rP(t)dt.$$

This differential equation has the solution $P(t) = P_0 e^{rt}$, where P_0 is the initial population size. The evolution of the population is driven by its growth rate

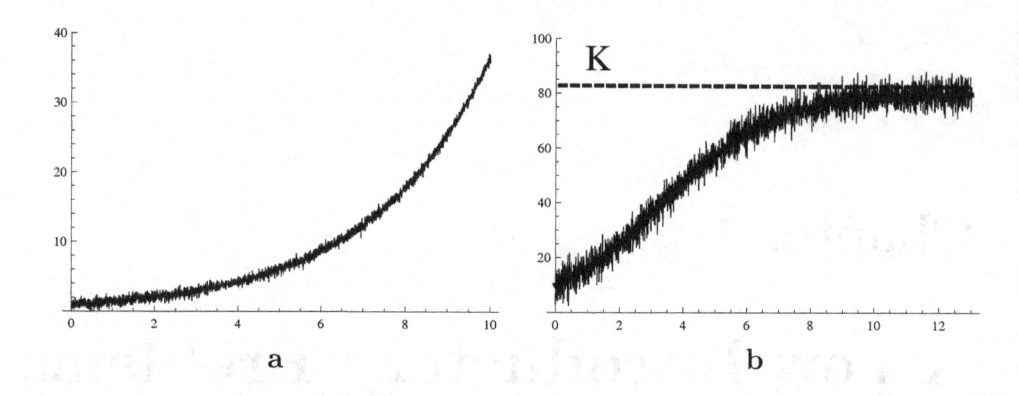

Figure 1.1: (a) *Noisy population with exponential growth.* (b) *Noisy population with logistic growth.*

r. In real life this rate is not constant. It might be a function of time t, or even more general, it might oscillate irregularly around some deterministic average function $a(t)$:

$$r_t = a(t) + \text{“noise”}.$$

In this case, r_t becomes a random variable indexed over time t. The associated equation becomes a stochastic differential equation

$$dP(t) = \big(a(t) + \text{“noise”}\big)P(t)dt. \tag{1.1.1}$$

Solving an equation of type (1.1.1) is a problem of Stochastic Calculus, see Fig. 1.1(a).

Logistic growth model The previous exponential growth model allows the population to increase indefinitely. However, due to competition, limited space and resources, the population will increase slower and slower. This model was introduced by P.F. Verhust in 1832 and rediscovered by R. Pearl in the twentieth century. The main assumption of the model is that the amount of competition is proportional with the number of encounters between the population members, which is proportional with the square of the population size

$$dP(t) = rP(t)dt - kP(t)^2dt. \tag{1.1.2}$$

The solution is given by the logistic function

$$P(t) = \frac{P_0K}{P_0 + (K - P_0)e^{-rt}},$$

where $K = r/k$ is the saturation level of the population. One of the stochastic variants of equation (1.1.2) is given by

$$dP(t) = rP(t)dt - kP(t)^2dt + \beta(\text{“noise”})P(t),$$

where $\beta \in \mathbb{R}$ is a measure of the size of the noise in the system. This equation is used to model the growth of a population in a stochastic, crowded environment, see Fig. 1.1(b).

1.2 Pricing Zero-coupon Bonds

A bond is a financial instrument which pays back at the end of its lifetime, T, an amount equal to B, and provides some periodical payments, called coupons. If the coupons are equal to zero, the bond is called a zero-coupon bond or a discount bond. Using the time value of money, the price of a bond at time t is $B(t) = Be^{-t(T-t)}$, where r is the risk-free interest rate. The bond satisfies the ordinary differential equation

$$dB(t) = rB(t)dt$$

with the final condition $B(T) = B$. In a "noisy" market the constant interest rate r is replaced by $r_t = r(t) +$ "noise", a fact that makes the bond pricing more complicated. This treatment can be achieved by Stochastic Calculus.

1.3 Noisy Pendulum

The free oscillations of a simple pendulum of unit mass can be described by the nonlinear equation $\ddot{\theta}(t) = -k^2 \sin\theta(t)$, where $\theta(t)$ is the angle between the string and the vertical direction. If the pendulum is moving under the influence of a time dependent exterior force $F = F(t)$, then the equation of the pendulum with forced oscillations is given by $\ddot{\theta}(t) + k^2 \sin\theta(t) = F(t)$. We may encounter the situation when the force is not deterministic and we have

$$F(t) = f(t) + (\text{"noise"}).$$

How does the noisy force influence the deviation angle $\theta(t)$? Stochastic Calculus can be used to answer this question.

1.4 Diffusion of Particles

Consider a flowing fluid with the velocity field $v(x)$. A particle that moves with the fluid has a trajectory $\phi(t)$ described by the equation $\phi'(t) = v\big(\phi(t)\big)$. A small particle, that is also subject to molecular bombardments, will be described by an equation of the type $\phi'(t) = v\big(\phi(t)\big) + \sigma(\text{"noise"})$, where the constant $\sigma > 0$ determines the size of the noise and controls the diffusion of the small particle in the fluid.

Now consider a drop of ink (which is made out of a very large number of tiny particles) left to diffuse in a liquid. Each ink particle performs a noisy trajectory in the liquid. Let $p(x, t)$ represent the density of particles that arrive about x at time t. After some diffusion time, the darker regions of the liquid represent the regions with higher density $p(x, t)$, while the lighter regions correspond to smaller density $p(x, t)$. Knowing the density $p(x, t)$ provides control over the dynamics of the diffusion process and can be used to find the probability that an ink particle reaches a certain region.

1.5 Cholesterol Level

The blood cholesterol level at time t is denoted by $C(t)$. This depends on the intaken food fat as well as organism absorption and individual production of cholesterol. The rate of change of the cholesterol level is given by

$$\frac{dC(t)}{dt} = a\big(C_0 - C(t)\big) + bE,$$

where C_0 is the natural level of cholesterol and E denotes the daily rate of intaken cholesterol; the constants a and b model the production and absorption of cholesterol in the organism. The solution of this linear differential equation is

$$C(t) = C_0 e^{-at} + \big(C_0 + \frac{b}{a}E\big)(1 - e^{-at}),$$

which in the long run tends to the saturation level of cholesterol $C_0 + \frac{b}{a}E$. Due to either observation errors or variations in the intake amount of food, the aforementioned equation will get the following noisy form

$$\frac{dC(t)}{dt} = a\big(C_0 - C(t)\big) + bE + \text{“noise”}.$$

This equation can be explicitly solved using Stochastic Calculus. Furthermore, we can also find the probability that the cholesterol level is over the allowed organism limit.

1.6 Electron Motion

Consider an electron situated at the initial distance $x(0)$ from the origin, which moves with a unit speed towards the origin. Its coordinate $x(t) \in \mathbb{R}^3$ is supposed to satisfy the equation

$$\frac{dx(t)}{dt} = -\frac{x(t)}{|x(t)|}.$$

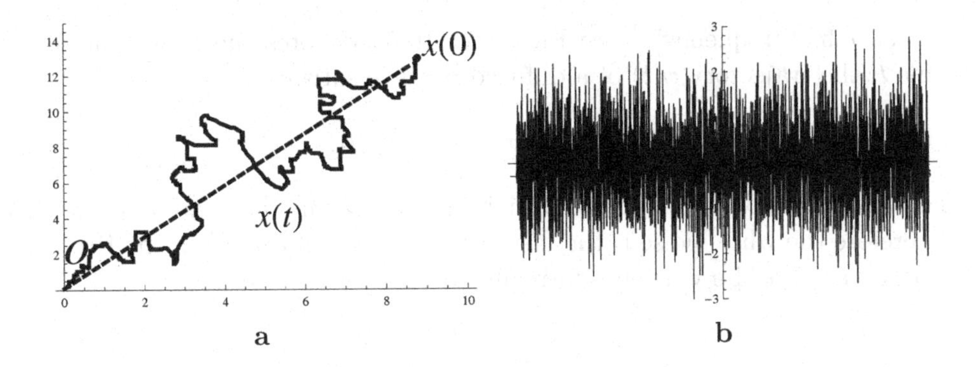

Figure 1.2: (a) *The trajectory of the electron $x(t)$ tends towards the origin.* (b) *White noise.*

Like in the case of the pollen grain, whose motion is agitated by the neighboring molecules, we assume that the electron is subject to bombardment by some "aether" particles, which makes its movement unpredictable, with constant tendency to go towards the origin, see Fig. 1.2 (a). Then its equation becomes

$$\frac{dx(t)}{dt} = -\frac{x(t)}{|x(t)|} + \text{"noise"}.$$

This type of description of electrons is usually seen in stochastic mechanics. This theory can be found in Fényes [18] and Nelson [36].

1.7 White Noise

All aforementioned problems involved a "noise" influence. This noise is directly related to the trajectory of a small particle which diffuses in a liquid due to the molecular bombardments (just consider the last example in the case of a static fluid, $v = 0$). This was observed first time by Brown [8] in 1828, and was called Brownian motion and it is customarily denoted by B_t. It has a very irregular, continuous trajectory, which from the mathematical point of view is nowhere differentiable. A satisfactory explanation of the Brownian motion was given by Einstein [17] in 1905. A different but likewise succesful decription of the Brownian motion was done by Langevin [32] in 1908.

The adjective "white" comes from signal processing, where it refers to the fact that the noise is completely unpredictable, i.e. it is not biased towards any specific "frequency",[1] see Fig. 1.2. If $\mathcal{N}_t$ denotes the "white noise" at time t, the trajectory $\phi(t)$ of a diffused particle satisfies

$$\phi'(t) = \sigma \mathcal{N}_t, \qquad t \geq 0.$$

The solution depends on the Brownian motion starting at x, i.e $\phi(t) = x + B_t$. Therefore the white noise is the instantaneous rate of change of the Brownian motion, and can be written informally as

$$\mathcal{N}_t = \frac{dB_t}{dt}.$$

This looks contradictory, since B_t is not differentiable. However, there is a way of making sense of the previous formula by considering the derivative in the following "generalized sense":

$$\int_{\mathbb{R}} \mathcal{N}_t f(t)\, dt = -\int_{\mathbb{R}} B_t f'(t)\, dt,$$

for any compact supported, smooth function f. From this point of view, the white noise $\mathcal{N}_t$ is a *generalized function* or a *distribution.* We shall get back to the notion of white noise in section 11.1.

1.8 Bounded and Quadratic Variation

The graph of a C^1-differentiable function, defined on a compact interval, has finite length. Unlike the case of differentiable functions, trajectories of Brownian motions, or other stochastic processes, are not of finite length. We may say that to a certain extent, the role of the "length" in this case is played by the "quadratic variation". This is actually a measure of the roughness of the process, see Fig. 1.3. This section will introduce these notions in an informal way. We shall cover these topics in more detail later in sections 4.11 and 4.12.

Let $f : [a, b] \to \mathbb{R}$ be a continuously differentiable function, and consider the partition $a = x_0 < x_1 < \cdots < x_n = b$ of the interval $[a, b]$. A smooth curve $y = f(x)$, $a \leq x \leq b$, is rectifiable (has length) if the sum of the lengths of the line segments with vertices at $P_0(x_0, f(x_0)), \cdots, P_n(x_n, f(x_n))$ is bounded by a given constant, which is independent of the number n and the choice of the division points x_i. Assuming the division is equidistant, $\Delta x = (b - a)/n$, the curve length becomes

[1] The white light is an equal mixture of radiations of all visible frequencies.

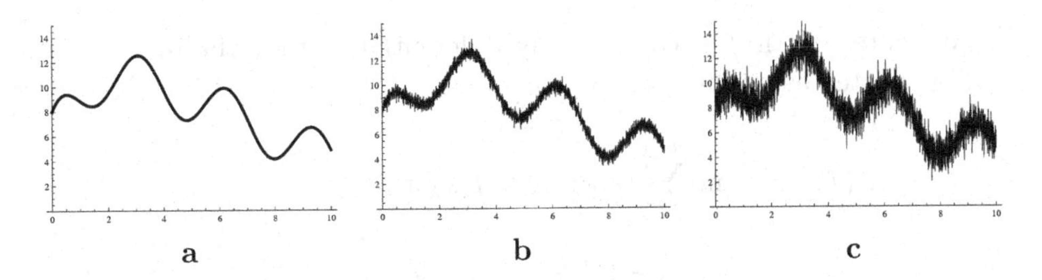

Figure 1.3: (a) *Smooth.* (b) *Rough.* (c) *Very rough.*

$$\ell = \sup_{x_i} \sum_{k=0}^{n-1} |P_k P_{k+1}| = \sup_{x_i} \sum_{k=0}^{n-1} \sqrt{(\Delta x)^2 + (\Delta f)^2}.$$

Furthermore, if f is continuously differentiable, the computation can be continued as

$$\ell = \lim_{n\to\infty} \sum_{k=0}^{n-1} \sqrt{1 + \left(\frac{\Delta f}{\Delta x}\right)^2}\, \Delta x = \int_a^b \sqrt{1 + f'(x)^2}\, dx,$$

where we used that the limit of an increasing sequence is equal to its superior limit.

Definition 1.8.1 *The function $f(x)$ has bounded variation on the interval $[a, b]$ if for any division $a = x_0 < x_1 < \cdots < x_n = b$ the sum*

$$\sum_{k=0}^{n-1} |f(x_{k+1}) - f(x_k)|$$

is bounded above by a given constant.

The *total variation* of f on $[a, b]$ is defined by

$$V(f) = \sup_{x_i} \sum_{k=0}^{n-1} |f(x_{k+1}) - f(x_k)|. \qquad (1.8.3)$$

The amount $V(f)$ measures in a certain sense the "roughness" of the function. If f is a constant function, then $V(f) = 0$. If f is a stair-type function, then $V(f)$ is the sum of the absolute value of its jumps.

We note that if f is continuously differentiable, then the total variation can be written as an integral

$$\begin{aligned} V(f) &= \sup_{x_i} \sum_{k=0}^{n-1} |f(x_{k+1}) - f(x_k)| \\ &= \lim_{n \to \infty} \sum_{k=0}^{n-1} \frac{|f(x_{k+1}) - f(x_k)|}{x_{k+1} - x_k} \Delta x = \int_a^b |f'(x)|\, dx. \end{aligned}$$

The next result states a relation between the length of the graph and the total variation of a function f, which is not differentiable.

Proposition 1.8.2 *Let $f : [a, b] \to \mathbb{R}$ be a function. Then the graph $y = f(x)$ has length if and only if $V(f) < \infty$.*

Proof: Consider the simplifying notations $(\Delta f)_k = f(x_{k+1}) - f(x_k)$ and $\Delta x = x_{k+1} - x_k$. Taking the summation in the double inequality

$$|(\Delta f)_k| \leq \sqrt{(\Delta x)^2 + |(\Delta f)_k|^2} \leq \Delta x + |(\Delta f)_k|$$

and then applying the "sup" yields

$$V(f) \leq \ell \leq (b - a) + V(f),$$

which implies the desired conclusion. ∎

By virtue of the previous result, the functions with infinite total variations have graphs of infinite lengths.

The following informal computation shows that the Brownian motion has infinite total variation (a real proof of this fact is given in section 4.12)

$$V(B_t) = \sup_{t_k} \sum_{k=0}^{n-1} |B_{t_{k+1}} - B_{t_k}| = \int_a^b \left| \frac{dB_t}{dt} \right| dt = \int_a^b |\mathcal{N}_t|\, dt = \infty,$$

since the area under the curve $t \to |\mathcal{N}_t|$ is infinite.

We can try to model a finer "roughness" of the function using the quadratic variation of f on the interval $[a, b]$

$$V^{(2)}(f) = \sup_{x_i} \sum_{k=0}^{n-1} |f(x_{k+1}) - f(x_k)|^2. \tag{1.8.4}$$

where the "sup" is taken over all divisions $a = x_0 < x_1 < \cdots < x_n = b$.

It is worth noting that if f has bounded total variation, $V(f) < \infty$, then $V^{(2)}(f) = 0$. This comes from the following inequality

$$V^{(2)}(f) \leq \max_{x_k} |f(x_{k+1}) - f(x_k)| V(f) \to 0, \quad \text{as } |\Delta x| \to 0.$$

The total variation for the Brownian motion does not provide much information. It turns out that the correct measure for the roughness of the Brownian motion is the quadratic variation

$$V^{(2)}(B_t) = \sup_{t_i} \sum_{k=0}^{n-1} |B(t_{k+1}) - B(t_k)|^2. \tag{1.8.5}$$

It will be shown that $V^{(2)}(B_t)$ is equal to the time interval $b - a$.

Chapter 2

Basic Notions

2.1 Probability Space

The modern theory of probability stems from the work of Kolmogorov [28], published in 1933. Kolmogorov associates a random experiment with a probability space, which is a triplet, $(\Omega, \mathcal{F}, P)$, consisting of the set of outcomes, Ω, a σ-field, $\mathcal{F}$, with Boolean algebra properties, and a probability measure, P. In the following sections, each of these elements will be discussed in more detail.

2.2 Sample Space

A *random experiment* in the theory of probability is an experiment whose outcomes cannot be determined in advance. When an experiment is performed, all possible outcomes form a set called the *sample space*, which will be denoted by Ω.

For instance, flipping a coin produces the sample space with two states $\Omega = \{H, T\}$, while rolling a die yields a sample space with six states $\Omega = \{1, \cdots, 6\}$. Choosing randomly a number between 0 and 1 corresponds to a sample space, which is the entire segment $\Omega = (0, 1)$.

In financial markets one can regard Ω as the *states of the world*, by this, we mean all possible states the world might have. The number of states of the world that affect the stock market is huge. These would contain all possible values for the vector parameters that describe the world, which is practically infinite.

All subsets of the sample space Ω form a set denoted by 2^{Ω}. The reason for this notation is that the set of parts of Ω can be put into a bijective correspondence with the set of binary functions $f : \Omega \to \{0, 1\}$. The number of elements of this set is $2^{|\Omega|}$, where $|\Omega|$ denotes the cardinal of Ω. If the set is finite, $|\Omega| = n$, then 2^Ω has 2^n elements. If Ω is infinitely countable (i.e. can be put into a bijective correspondence with the set of natural numbers), then $2^{|\Omega|}$ is infinite and its cardinal is the same as that of the real number set $\mathbb{R}$.

Remark 2.2.1 Pick a natural number at random. Any subset of the sample space corresponds to a sequence formed with 0 and 1. For instance, the subset $\{1, 3, 5, 6\}$ corresponds to the sequence $10101100000\dots$ having 1 on the 1st, 3rd, 5th and 6th places and 0 in rest. It is known that the number of these sequences is infinite and can be put into a bijective correspondence with the real number set $\mathbb{R}$. This can also be written as $|2^{\mathbb{N}}| = |\mathbb{R}|$, and stated by saying that the set of all subsets of natural numbers $\mathbb{N}$ has the same cardinal as the real numbers set $\mathbb{R}$.

2.3 Events and Probability

The set of parts 2^Ω satisfies the following properties:

1. It contains the empty set $\emptyset$;
2. If it contains a set A, then it also contains its complement $\bar{A} = \Omega \backslash A$;
3. It is closed with regard to unions, i.e., if $A_1, A_2, \dots$ is a sequence of sets, then their union $A_1 \cup A_2 \cup \cdots$ also belongs to 2^Ω.

Any subset $\mathcal{F}$ of 2^Ω that satisfies the previous three properties is called a *σ-field*. The sets belonging to $\mathcal{F}$ are called *events*. This way, the complement of an event, or the union of events is also an event. We say that an event occurs if the outcome of the experiment is an element of that subset.

The chance of occurrence of an event is measured by a probability function $P : \mathcal{F} \to [0, 1]$ which satisfies the following two properties:

1. $P(\Omega) = 1$;
2. For any mutually disjoint events $A_1, A_2, \dots \in \mathcal{F}$,

$$P(A_1 \cup A_2 \cup \cdots) = P(A_1) + P(A_2) + \cdots .$$

The triplet $(\Omega, \mathcal{F}, P)$ is called a *probability space*. This is the main setup in which the probability theory works.

Example 2.3.1 *In the case of a coin flipping, the probability space has the following elements:* $\Omega = \{H, T\}$, $\mathcal{F} = \{\emptyset, \{H\}, \{T\}, \{H, T\}\}$ *and* P *is defined by* $P(\emptyset) = 0$, $P(\{H\}) = \frac{1}{2}$, $P(\{T\}) = \frac{1}{2}$, $P(\{H, T\}) = 1$.

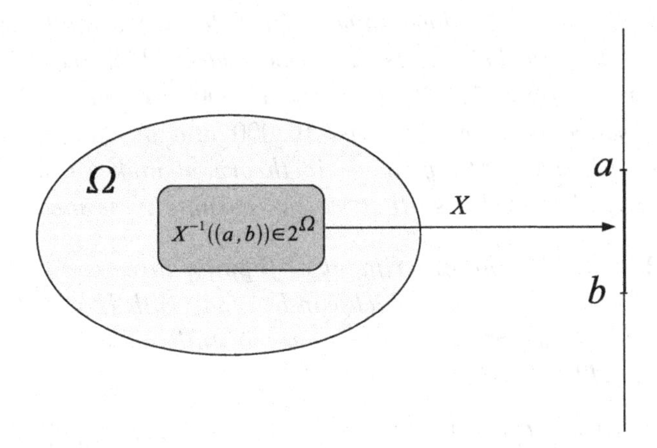

Figure 2.1: *If any set $X^{-1}\big((a,b)\big)$ is "known", then the random variable $X : \Omega \to \mathbb{R}$ is 2^{Ω}-measurable.*

Example 2.3.2 *Consider a finite sample space $\Omega = \{s_1, \dots, s_n\}$, with the σ-field $\mathcal{F} = 2^{\Omega}$, and probability given by $P(A) = |A|/n$, $\forall A \in \mathcal{F}$. Then $(\Omega, \mathcal{F}, P)$ is a probability space.*

Example 2.3.3 *Let $\Omega = [0,1]$ and consider the σ-field $\mathcal{B}([0,1])$ given by the set of all open or closed intervals on $[0,1]$, or any unions, intersections, and complementary sets. Define $P(A) = \lambda(A)$, where λ stands for the Lebesgue measure (in particular, if $A = (a,b)$, then $P(A) = b - a$ is the length of the interval). It can be shown that $(\Omega, \mathcal{B}([0,1]), P)$ is a probability space.*

2.4 Random Variables

Since the σ-field $\mathcal{F}$ provides the knowledge about which events are possible on the considered probability space, then $\mathcal{F}$ can be regarded as the information component of the probability space $(\Omega, \mathcal{F}, P)$. A *random variable* X is a function that assigns a numerical value to each state of the world, $X : \Omega \to \mathbb{R}$, such that the values taken by X are known to someone who has access to the information $\mathcal{F}$. More precisely, given any two numbers $a, b \in \mathbb{R}$, then all the states of the world for which X takes values between a and b forms a set that is an event (an element of $\mathcal{F}$), i.e.

$$\{\omega \in \Omega; a < X(\omega) < b\} \in \mathcal{F}.$$

Another way of saying this is that X is an *$\mathcal{F}$-measurable function.*

Example 2.4.1 *Let $X(\omega)$ be the number of people who want to buy houses, given the state of the market ω. Is X measurable? This would mean that given two numbers, say $a = 10,000$ and $b = 50,000$, we know all the market situations ω for which there are at least $10,000$ and at most $50,000$ people willing to purchase houses. Many times, in theory, it makes sense to assume that we have enough knowledge so that we can assume X is measurable.*

Example 2.4.2 *Consider the experiment of flipping three coins. In this case Ω is the set of all possible triplets, which can be made with H and T. Consider the random variable X which gives the number of tails obtained. For instance $X(HHH) = 0$, $X(HHT) = 1$, etc. The sets*

$$\{\omega; X(\omega) = 0\} = \{HHH\}, \quad \{\omega; X(\omega) = 1\} = \{HHT, HTH, THH\},$$
$$\{\omega; X(\omega) = 3\} = \{TTT\}, \quad \{\omega; X(\omega) = 2\} = \{HTT, THT, TTH\}$$

belong to 2^Ω, and hence X is a random variable.

2.5 Integration in Probability Measure

The notion of expectation is based on integration on measure spaces. In this section we recall briefly the definition of an integral with respect to the probability measure P. For more insight on measurable functions and integration theory the reader is referred to the classical text of Halmos [21].

Let $X : \Omega \to \mathbb{R}$ be a random variable on the probability space $(\Omega, \mathcal{F}, P)$. A partition $(\Omega_i)_{1\leq i\leq n}$ of Ω is a family of subsets $\Omega_i \subset \Omega$, with $\Omega_i \in \mathcal{F}$, satisfying

1. $\Omega_i \cap \Omega_j = \emptyset$, for $i \neq j$;
2. $\displaystyle\bigcup_i^n \Omega_i = \Omega$.

Each Ω_i is an event and its associated probability is $P(\Omega_i)$. Consider the *characteristic function* of a set $A \subset \Omega$ defined by $\chi_A(\omega) = \begin{cases} 1, & \text{if } \omega \in A \\ 0, & \text{if } \omega \notin A \end{cases}$. More properties of χ_A can be found in Exercise 2.12.9. The integral will be defined in the following three steps:

(*i*) A *simple function* is a sum of characteristic functions $f = \sum_i^n c_i \chi_{\Omega_i}$, $c_i \in \mathbb{R}$. This means $f(\omega) = c_k$ for $\omega \in \Omega_k$. The integral of the simple function f is defined by

$$\boxed{\int_\Omega f\, dP = \sum_i^n c_i P(\Omega_i).}$$

(*ii*) If $X : \Omega \to \mathbb{R}$ is a random variable, then from the measure theory it is known that there is a sequence of simple functions $(f_n)_{n\geq 1}$ satisfying

$\lim_{n\to\infty} f_n(\omega) = X(\omega)$. Furthermore, if $X \geq 0$, then we may assume that $f_n \leq f_{n+1}$. Then we define

$$\boxed{\int_\Omega X\,dP = \lim_{n\to\infty} \int_\Omega f_n\,dP.}$$

(*iii*) If X is not non-negative, we can write $X = X^+ - X^-$ with $X^+ = \sup\{X, 0\} \geq 0$ and $X^- = \sup\{-X, 0\} \geq 0$. Define

$$\boxed{\int_\Omega X\,dP = \lim_{n\to\infty} \int_\Omega X^+\,dP - \lim_{n\to\infty} \int_\Omega X^-\,dP,}$$

where we assume that at least one of the integrals is finite.

From now on, the integral notations $\int_\Omega X\,dP$ or $\int_\Omega X(\omega)\,dP(\omega)$ will be used interchangeably. In the rest of the chapter the integral notation will be used informally, without requiring a direct use of the previous definition.

Two widely used properties of the integral defined above are:

Linearity: For any two random variables X and Y and $a, b \in \mathbb{R}$

$$\int_\Omega (aX + bY)\,dP = a \int_\Omega X\,dP + b \int_\Omega Y\,dP;$$

Positivity: If $X \leq 0$ then

$$\int_\Omega X\,dP \geq 0.$$

2.6 Two Convergence Theorems

During future computations we shall often need to swap the limit symbol with the integral. There are two basic measure theory results that allow doing this. We shall state these results below and use them whenever needed.

Theorem 2.6.1 (The monotone convergence theorem) *Let $(\Omega, \mathcal{F}, P)$ be a probability space and $(f_n)_{n\geq 1}$ a sequence of measurable functions, $f_n : \Omega \to [0, \infty)$ such that:*

(*i*) $0 \leq f_k(\omega) \leq f_{k+1}(\omega)$, $\forall \omega \in \Omega$, $k \geq 1$;

(*ii*) *the sequence is pointwise convergent*

$$f(\omega) = \lim_{n\to\infty} f_n(\omega), \quad \forall \omega \in \Omega.$$

Then

(1) *f is measurable;*

(2) $\displaystyle \lim_{n\to\infty} \int_\Omega f_n\,dP = \int_\Omega f\,dP.$

Theorem 2.6.2 (The dominated convergence theorem) *Let $(\Omega, \mathcal{F}, P)$ be a probability space and $(f_n)_{n\geq 1}$ a sequence of measurable functions, $f_n : \Omega \to \mathbb{R}$. Assume that:*

(i) (f_n) is pointwise convergent

$$f(\omega) = \lim_{n\to\infty} f_n(\omega), \quad \forall \omega \in \Omega;$$

(ii) there is an integrable function g, (i.e. $\int_\Omega |g|\, dP < \infty$) such that

$$|f_n(\omega)| \leq g(\omega), \quad \forall \omega \in \Omega, n \geq 1.$$

Then f is integrable and

$$\lim_{n\to\infty} \int_\Omega f_n\, dP = \int_\Omega f\, dP.$$

2.7 Distribution Functions

Let X be a random variable on the probability space $(\Omega, \mathcal{F}, P)$. The *distribution function* of X is the function $F_X : \mathbb{R} \to [0, 1]$ defined by

$$F_X(x) = P(\omega; X(\omega) \leq x).$$

It is worth observing that since X is a random variable, then the set $\{\omega; X(\omega) \leq x\}$ belongs to the information set $\mathcal{F}$.

The distribution function is non-decreasing and satisfies the limits

$$\lim_{x\to-\infty} F_X(x) = 0, \qquad \lim_{x\to+\infty} F_X(x) = 1.$$

If we have

$$\frac{d}{dx} F_X(x) = p(x),$$

then we say that $p(x)$ is the *probability density function* of X.

It is important to note the following relation among distribution function, probability and probability density function of the random variable X

$$F_X(x) = P(X \leq x) = \int_{\{X\leq x\}} dP(\omega) = \int_{-\infty}^{x} p(u)\, du. \tag{2.7.1}$$

The probability density function $p(x)$ has the following properties:

(*i*) $p(x) \geq 0$

(*ii*) $\int_{-\infty}^{\infty} p(u)\, du = 1.$

The first one is a consequence of the fact that the distribution function $F_X(x)$ is non-decreasing. The second follows from (2.7.1) by making $x \to \infty$

$$\int_{-\infty}^{\infty} p(u)\,du = \int_{\Omega} dP = P(\Omega) = 1.$$

As an extension of formula (2.7.1) we have for any $\mathcal{F}$-measurable function h

$$\int_{\Omega} h(X(\omega))\,dP(\omega) = \int_{\mathbb{R}} h(x)p(x)\,dx. \qquad (2.7.2)$$

Another useful propety, which follows from the Fundamental Theorem of Calculus is

$$P(a < X < b) = P(\omega; a < X(\omega) < b) = \int_a^b p(x)\,dx.$$

In the case of discrete random variables the aforementioned integral is replaced by the following sum

$$P(a < X < b) = \sum_{a<x<b} P(X = x).$$

For more details the reader is referred to a traditional probability book, such as Wackerly et al. [13].

2.8 Independence

Roughly speaking, two random variables X and Y are independent if the occurrence of one of them does not change the probability density of the other. More precisely, if for any two open intervals $A, B \subset \mathbb{R}$, the events

$$E = \{\omega; X(\omega) \in A\}, \qquad F = \{\omega; Y(\omega) \in B\}$$

are independent, i.e., $P(E \cap F) = P(E)P(F)$, then X and Y are called *independent* random variables.

Proposition 2.8.1 *Let X and Y be independent random variables with density functions $p_X(x)$ and $p_Y(y)$. Then the joint density function of (X, Y) is given by $p_{X,Y}(x, y) = p_X(x)\, p_Y(y)$.*

Proof: Using the independence of sets, we have[1]

$$\begin{aligned} p_{X,Y}(x,y)\,dxdy &= P(x<X<x+dx,\, y<Y<y+dy) \\ &= P(x<X<x+dx)P(y<Y<y+dy) \\ &= p_X(x)\,dx\; p_Y(y)\,dy \\ &= p_X(x)p_Y(y)\,dxdy. \end{aligned}$$

Dropping the factor $dxdy$ yields the desired result. We note that the converse also holds true. ■

The *σ-algebra generated by a random variable* $X:\Omega\to\mathbb{R}$ is the σ-algebra generated by the unions, intersections and complements of events of the form $\{\omega; X(\omega)\in(a,b)\}$, with $a<b$ real numbers. This will be denoted by $\mathcal{A}_X$.

Two σ-fields $\mathcal{G}$ and $\mathcal{H}$ included in $\mathcal{F}$ are called *independent* if

$$P(G\cap H) = P(G)P(H), \qquad \forall G\in\mathcal{G}, H\in\mathcal{H}.$$

The random variable X and the σ-field $\mathcal{G}$ are called *independent* if the algebras $\mathcal{A}_X$ and $\mathcal{G}$ are independent.

2.9 Expectation

A random variable $X:\Omega\to\mathbb{R}$ is called *integrable* if

$$\int_\Omega |X(\omega)|\,dP(\omega) = \int_{\mathbb{R}} |x|p(x)\,dx < \infty,$$

where $p(x)$ denotes the probability density function of X. The previous identity is based on changing the domain of integration from Ω to $\mathbb{R}$.

The *expectation* of an integrable random variable X is defined by

$$\mathbb{E}[X] = \int_\Omega X(\omega)\,dP(\omega) = \int_{\mathbb{R}} x\,p(x)\,dx.$$

Customarily, the expectation of X is denoted by μ and is called the *mean*. In general, for any measurable function $h:\mathbb{R}\to\mathbb{R}$, we have

$$\mathbb{E}[h(X)] = \int_\Omega h\big(X(\omega)\big)\,dP(\omega) = \int_{\mathbb{R}} h(x)p(x)\,dx.$$

In the case of a discrete random variable X the expectation is defined as

$$\mathbb{E}[X] = \sum_{k\geq 1} x_k P(X=x_k).$$

[1]We are using the useful approximation $P(x<X<x+dx) = \int_x^{x+dx} p(u)\,du = p(x)dx$.

Proposition 2.9.1 *The expectation operator* $\mathbb{E}$ *is linear, i.e. for any integrable random variables* X *and* Y

1. $\mathbb{E}[cX] = c\mathbb{E}[X], \qquad \forall c \in \mathbb{R}$;
2. $\mathbb{E}[X+Y] = \mathbb{E}[X] + \mathbb{E}[Y]$.

Proof: It follows from the fact that the integral is a linear operator. ■

Proposition 2.9.2 *Let* X *and* Y *be two independent integrable random variables. Then*

$$\mathbb{E}[XY] = \mathbb{E}[X]\mathbb{E}[Y].$$

Proof: This is a variant of Fubini's theorem, which in this case states that a double integral is a product of two simple integrals. Let p_X, p_Y, $p_{X,Y}$ denote the probability densities of X, Y and (X,Y), respectively. Since X and Y are independent, by Proposition 2.8.1 we have

$$\mathbb{E}[XY] = \iint xyp_{X,Y}(x,y)\,dxdy = \int xp_X(x)\,dx \int yp_Y(y)\,dy = \mathbb{E}[X]\mathbb{E}[Y].$$

■

Definition 2.9.3 *The covariance of two random variables is defined by*

$$Cov(X,Y) = \mathbb{E}[XY] - \mathbb{E}[X]\mathbb{E}[Y].$$

The variance of X *is given by*

$$Var(X) = Cov(X,X).$$

Proposition 2.9.2 states that if X and Y are independent, then $Cov(X,Y) = 0$. It is worth to note that the converse in not necessarily true, see Exercise 2.9.6. However, the converse holds true if both X and Y are assumed normally distributed.

Exercise 2.9.4 *Show that*

(*a*) $Cov(X,Y) = \mathbb{E}[(X-\mu_X)(Y-\mu_Y)]$, *where* $\mu_X = \mathbb{E}[X]$ *and* $\mu_Y = \mathbb{E}[Y]$;

(*b*) $Var(X) = \mathbb{E}[(X-\mu_X)^2]$;

From Exercise 2.9.4 (*b*), we have $Var(X) \geq 0$, so, there is a real number $\sigma > 0$ such that $Var(X) = \sigma^2$. The number σ is called *standard deviation.*

Exercise 2.9.5 *Let μ and σ denote the mean and the standard deviation of the random variable X. Show that*

$$\mathbb{E}[X^2] = \mu^2 + \sigma^2.$$

Exercise 2.9.6 *Consider two random variables with the following table of joint probabilities:*

$Y \backslash X$	-1	0	1
-1	$1/16$	$3/16$	$1/16$
0	$3/16$	0	$3/16$
1	$1/16$	$3/16$	$1/16$

Show the following:

(a) $\mathbb{E}[X] = \mathbb{E}[Y] = \mathbb{E}[XY] = 0$;

(b) $Cov(X, Y) = 0$;

(c) $P(0, 0) \neq P_X(0)P_Y(0)$;

(d) X *and* Y *are not independent.*

The covariance can be standardized in the following way. Let σ_X and σ_Y be the standard deviations of X and Y, respectively. The *correlation coefficient* of X and Y is defined as

$$\rho(X, Y) = \frac{Cov(X, Y)}{\sigma_X \sigma_Y}.$$

Exercise 2.9.7 *(a) Prove that for any random variables A and B we have*

$$\mathbb{E}[AB]^2 \leq \mathbb{E}[A^2]\mathbb{E}[B^2].$$

(b) Use part (a) to show that for any random variables X and Y we have $-1 \leq \rho(X, Y) \leq 1$.

(c) What can you say about the random variables X and Y if $\rho(X, Y) = 1$?

2.9.1 The best approximation of a random variable

Let X be a random variable. We would like to approximate X by a single (nonrandom) number x. The "best" value of x is chosen in the sense of the "least squares", i.e. x is picked such that the expectation of the error square $(X - x)^2$ is minimum. Denote $\mu = \mathbb{E}[X]$ and $\sigma^2 = Var(X)$. Since

$$\begin{aligned}
\mathbb{E}[(X - x)^2] &= \mathbb{E}[X^2] - 2x\mathbb{E}[X] + x^2 \\
&= \sigma^2 + \mu^2 - 2x\mu + x^2 \\
&= \sigma^2 + (x - \mu)^2,
\end{aligned}$$

the minimum is obtained for $x = \mu$, and in this case

$$\min_x \mathbb{E}[(X - x)^2] = \sigma^2.$$

It follows that the mean, μ, is the best approximation of the random variable X in the least squares sense.

2.9.2 Change of measure in an expectation

Let $P, Q : \mathcal{F} \to \mathbb{R}$ be two probability measures on Ω, such that there is an integrable random variable $f : \Omega \to \mathbb{R}$, such that $dQ = f dP$. This means

$$Q(A) = \int_A dQ = \int_A f(\omega)\, dP(\omega), \qquad \forall A \in \mathcal{F}.$$

Denote by $\mathbb{E}^P$ and $\mathbb{E}^Q$ the expectations with respect to the measures P and Q, respectively. Then we have

$$\mathbb{E}^Q[X] = \int_\Omega X(\omega)\, dQ(\omega) = \int_\Omega X(\omega) f(\omega)\, dP(\omega) = \mathbb{E}^P[fX].$$

Exercise 2.9.8 *Let $g : [0, 1] \to [0, \infty)$ be a integrable function with*

$$\int_0^1 g(x)\, dx = 1.$$

Consider $Q : \mathcal{B}([0, 1]) \to \mathbb{R}$, given by $Q(A) = \int_A g(x)\, dx$. Show that Q is a probability measure on $(\Omega = [0, 1], \mathcal{B}([0, 1]))$.

2.10 Basic Distributions

We shall recall a few basic distributions, which are most often seen in applications.

Normal distribution A random variable X is said to have a *normal distribution* if its probability density function is given by

$$p(x) = \frac{1}{\sigma\sqrt{2\pi}} e^{-(x-\mu)^2/(2\sigma^2)},$$

with μ and $\sigma > 0$ constant parameters, see Fig. 2.2(a). The mean and variance are given by

$$\mathbb{E}[X] = \mu, \qquad Var[X] = \sigma^2.$$

If X has a normal distribution with mean μ and variance σ^2, we shall write

$$X \sim N(\mu, \sigma^2).$$

Exercise 2.10.1 *Let $\alpha, \beta \in \mathbb{R}$. Show that if X is normal distributed, with $X \sim N(\mu, \sigma^2)$, then $Y = \alpha X + \beta$ is also normal distributed, with*

$$Y \sim N(\alpha\mu + \beta, \alpha^2\sigma^2).$$

Log-normal distribution Let X be normally distributed with mean μ and variance σ^2. Then the random variable $Y = e^X$ is said to be *log-normal distributed.* The mean and variance of Y are given by

$$\begin{aligned} \mathbb{E}[Y] &= e^{\mu+\frac{\sigma^2}{2}} \\ Var[Y] &= e^{2\mu+\sigma^2}(e^{\sigma^2} - 1). \end{aligned}$$

The density function of the log-normal distributed random variable Y is given by

$$p(x) = \frac{1}{x\sigma\sqrt{2\pi}} e^{-\frac{(\ln x - \mu)^2}{2\sigma^2}}, \quad x > 0,$$

see Fig. 2.2(b).

Definition 2.10.2 *The moment generating function of a random variable X is the function $m_X(t) = \mathbb{E}[e^{tX}] = \int e^{tx} p(x)\, dx$, where $p(x)$ is the probability density function of X, provided the integral exists.*

The name comes from the fact that the nth moments of X, given by $\mu_n = \mathbb{E}[X^n]$, are generated by the derivatives of $m_X(t)$

$$\frac{d^n m_X(t)}{dt^n}_{|t=0} = \mu_n.$$

It is worth noting the relation between the Laplace transform and the moment generating function, in the case $x \geq 0$, $\mathcal{L}(p(x))(t) = \int_0^\infty e^{-tx} p(x)\, dx = m_X(-t)$.

Exercise 2.10.3 *Find the moment generating function for the exponential distribution $p(x) = \lambda e^{-x\lambda}$, $x \geq 0$, $\lambda > 0$.*

Exercise 2.10.4 *Show that if X and Y are two independent random variables, then $m_{X+Y}(t) = m_X(t) m_Y(t)$.*

Exercise 2.10.5 *Given that the moment generating function of a normally distributed random variable $X \sim N(\mu, \sigma^2)$ is $m(t) = \mathbb{E}[e^{tX}] = e^{\mu t + t^2\sigma^2/2}$, show that*

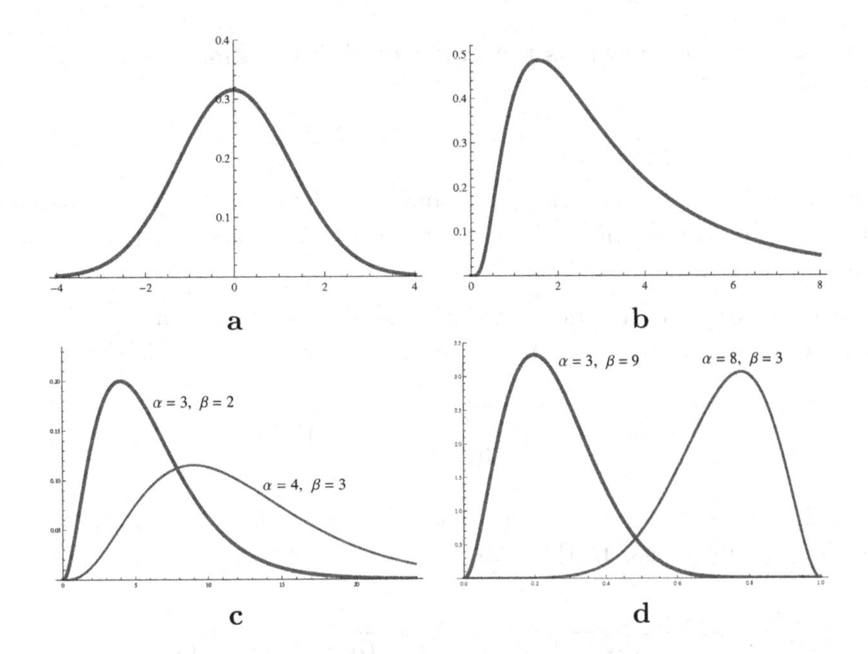

Figure 2.2: (a) *Normal distribution.* (b) *Log-normal distribution.* (c) *Gamma distributions.* (d) *Beta distributions.*

(*a*) $\mathbb{E}[Y^n] = e^{n\mu + n^2\sigma^2/2}$, *where* $Y = e^X$.

(*b*) *Show that the mean and variance of the log-normal random variable* $Y = e^X$ *are*

$$\mathbb{E}[Y] = e^{\mu + \sigma^2/2}, \quad Var[Y] = e^{2\mu + \sigma^2}(e^{\sigma^2} - 1).$$

Gamma distribution A random variable X is said to have a *gamma distribution* with parameters $\alpha > 0$, $\beta > 0$ if its density function is given by

$$p(x) = \frac{x^{\alpha-1}e^{-x/\beta}}{\beta^\alpha \Gamma(\alpha)}, \quad x \geq 0,$$

where $\Gamma(\alpha)$ denotes the gamma function

$$\Gamma(\alpha) = \int_0^\infty y^{\alpha-1}e^{-y}\, dy.$$

It is worth noting that for $\alpha = n$, integer, we have $\Gamma(n) = (n-1)!$. The gamma distribution is provided in Fig. 2.2(c). The mean and variance are given by

$$\mathbb{E}[X] = \alpha\beta, \qquad Var[X] = \alpha\beta^2.$$

The case $\alpha = 1$ is known as the *exponential distribution*, see Fig. 2.3(a). In this case

$$p(x) = \frac{1}{\beta} e^{-x/\beta}, \quad x \geq 0.$$

The particular case when $\alpha = n/2$ and $\beta = 2$ becomes the χ^2-*distribution* with n degrees of freedom. This characterizes also a sum of n independent standard normal distributions.

Beta distribution A random variable X is said to have a *beta distribution* with parameters $\alpha > 0$, $\beta > 0$ if its probability density function is of the form

$$p(x) = \frac{x^{\alpha-1}(1-x)^{\beta-1}}{B(\alpha,\beta)}, \quad 0 \leq x \leq 1,$$

where $B(\alpha, \beta)$ denotes the beta function.[2] See see Fig. 2.2(d) for two particular density functions. In this case

$$\mathbb{E}[X] = \frac{\alpha}{\alpha+\beta}, \quad Var[X] = \frac{\alpha\beta}{(\alpha+\beta)^2(\alpha+\beta+1)}.$$

Poisson distribution A discrete random variable X is said to have a *Poisson probability distribution* if

$$P(X = k) = \frac{\lambda^k}{k!} e^{-\lambda}, \qquad k = 0, 1, 2, \ldots,$$

with $\lambda > 0$ parameter, see Fig. 2.3(b). In this case $\mathbb{E}[X] = \lambda$ and $Var[X] = \lambda$.

Pearson 5 distribution Let $\alpha, \beta > 0$. A random variable X with the density function

$$p(x) = \frac{1}{\beta\Gamma(\alpha)} \frac{e^{-\beta/x}}{(x/\beta)^{\alpha+1}}, \qquad x \geq 0$$

is said to have a Pearson 5 distribution[3] with positive parameters α and β. It can be shown that

$$\mathbb{E}[X] = \begin{cases} \dfrac{\beta}{\alpha-1}, & \text{if } \alpha > 1 \\ \infty, & \text{otherwise,} \end{cases} \qquad Var(X) = \begin{cases} \dfrac{\beta^2}{(\alpha-1)^2(\alpha-2)}, & \text{if } \alpha > 2 \\ \infty, & \text{otherwise.} \end{cases}$$

---

[2]Two definition formulas for the beta functions are $B(\alpha, \beta) = \frac{\Gamma(\alpha)\Gamma(\beta)}{\Gamma(\alpha+\beta)}$ and $B(\alpha, \beta) = \int_0^1 y^{\alpha-1}(1-y)^{\beta-1}\, dy$.

[3]The Pearson family of distributions was designed by Pearson between 1890 and 1895. There are several Pearson distributions, this one being distinguished by the number 5.

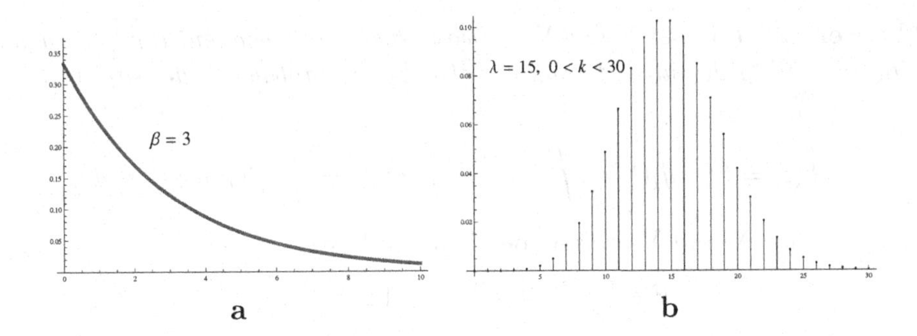

Figure 2.3: (a) *Exponential distribution.* (b) *Poisson distribution.*

The mode of this distribution is equal to $\frac{\beta}{\alpha + 1}$.

The Inverse Gaussian distribution Let $\mu, \lambda > 0$. A random variable X has an inverse Gaussian distribution with parameters μ and λ if its density function is given by

$$p(x) = \frac{\lambda}{2\pi x^3} e^{-\frac{\lambda(x-\mu)^2}{2\mu^2 x}}, \qquad x > 0. \tag{2.10.3}$$

We shall write $X \sim IG(\mu, \lambda)$. Its mean, variance and mode are given by

$$\mathbb{E}[X] = \mu, \qquad Var(X) = \frac{\mu^3}{\lambda}, \qquad Mode(X) = \mu\Big(\sqrt{1 + \frac{9\mu^2}{4\lambda^2}} - \frac{3\mu}{2\lambda}\Big),$$

where the mode denotes the value x_m for which $p(x)$ is maximum, i.e., $p(x_0) = \max_x p(x)$. This distribution will be used to model the time instance when a Brownian motion with drift exceeds a certain barrier for the first time.

2.11 Sums of Random Variables

Let X be a positive random variable with probability density f. We note first that for any $s > 0$

$$\mathbb{E}[e^{-sX}] = \int_0^\infty e^{-sx} f(x)\, dx = \mathcal{L}(f(x))(s), \tag{2.11.4}$$

where $\mathcal{L}$ denotes the Laplace transform.

The following result provides the relation between the convolution and the probability density of a sum of two random variables.

Theorem 2.11.1 *Let X and Y be two positive, independent random variables with probability densities f and g. Let h be the probability density of the sum $X+Y$. Then*

$$h(t) = (f*g)(t) = \int_0^t f(t-\tau)g(\tau)\,d\tau = \int_0^t f(\tau)g(t-\tau)\,d\tau.$$

Proof: Since X and Y are independent, we have

$$\mathbb{E}[e^{-s(X+Y)}] = \mathbb{E}[e^{-sX}]\,\mathbb{E}[e^{-sY}].$$

Using (2.11.4) this can be written in terms of Laplace transforms as

$$\mathcal{L}(h)(s) = \mathcal{L}(f)(s)\mathcal{L}(g)(s).$$

Using Exercise 2.11.2, the density h can be written as the desired convolution $h(t) = (f*g)(t)$. ■

Exercise 2.11.2 *If $F(s) = \mathcal{L}(f(t))(s)$, $G(s) = \mathcal{L}(g(t))(s)$ both exist for $s > a \geq 0$, then*

$$H(s) = F(s)G(s) = \mathcal{L}(h(t))(s),$$

for

$$h(t) = (f*g)(t) = \int_0^t f(t-\tau)g(\tau)\,d\tau = \int_0^t f(\tau)g(t-\tau)\,d\tau.$$

Using the associativity of the convolution

$$(f*g)*k = f*(g*k) = f*g*k$$

we obtain that if f, g and k are the probability densities of the positive, independent random variables X, Y and Z, respectively, then $f*g*k$ is the probability density of the sum $X+Y+Z$. The aforementioned result can be easily extended to the sum of n random variables.

Example 2.11.3 *Consider two independent, exponentially distributed random variables X and Y. We shall investigate the distribution of the sum $X+Y$. Consider $f(t) = g(t) = \lambda e^{-\lambda t}$ in Theorem 2.11.1 and obtain the probability density of the sum*

$$h(t) = \int_0^t \lambda e^{-\lambda(t-\tau)}\lambda e^{-\lambda\tau}\,d\tau = \lambda^2 t e^{-\lambda t}, \qquad t \geq 0,$$

which is Gamma distributed, with parameters $\alpha = 2$ and $\beta = 1/\lambda$.

Exercise 2.11.4 *Consider the independent, exponentially distributed random variables $X \sim \lambda_1 e^{-\lambda_1 t}$ and $Y \sim \lambda_2 e^{-\lambda_2 t}$, with $\lambda_1 \neq \lambda_2$. Show that the sum is distributed as*

$$X+Y \sim \frac{\lambda_1\lambda_2}{\lambda_1-\lambda_2}(e^{-\lambda_2 t} - e^{-\lambda_1 t}), \qquad t \geq 0.$$

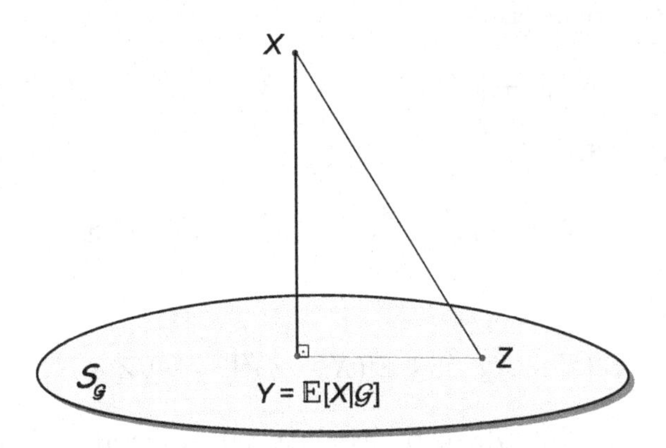

Figure 2.4: *The orthogonal projection of the random variable X on the space $S_{\mathcal{G}}$ is the conditional expectation $Y = \mathbb{E}[X|\mathcal{G}]$.*

2.12 Conditional Expectations

Let X be a random variable on the probability space $(\Omega, \mathcal{F}, P)$, and $\mathcal{G}$ be a σ-field contained in $\mathcal{F}$. Since X is $\mathcal{F}$-measurable, the expectation of X, given the information $\mathcal{F}$ must be X itself, a fact that can be written as $\mathbb{E}[X|\mathcal{F}] = X$ (for details see Example 2.12.5).

On the other hand, the information $\mathcal{G}$ does not completely determine X. The random variable that makes a prediction for X based on the information $\mathcal{G}$ is denoted by $\mathbb{E}[X|\mathcal{G}]$, and is called the *conditional expectation* of X given $\mathcal{G}$. This is defined as the random variable $Y = \mathbb{E}[X|\mathcal{G}]$, which is the best approximation of X in the least squares sense, i.e.

$$\mathbb{E}[(X - Y)^2] \leq \mathbb{E}[(X - Z)^2], \tag{2.12.5}$$

for any $\mathcal{G}$-measurable random variable Z.

The set of all square integrable random variables on Ω forms a Hilbert space with the inner product

$$\langle X, Y \rangle = \mathbb{E}[XY],$$

see Exercise 2.12.10. This defines the norm $\|X\|^2 = \mathbb{E}[X^2]$, which induces the distance $d(X, Y) = \|X - Y\|$. Denote by $S_{\mathcal{G}}$ the set of all $\mathcal{G}$-measurable random variables on Ω. We shall show that the element of $S_{\mathcal{G}}$ that is the closest to X in the aforementioned distance is the conditional expectation $Y = \mathbb{E}[X|\mathcal{G}]$, see Fig. 2.4. Let $X_{\perp}$ denote the orthogonal projection of X on the space $S_{\mathcal{G}}$. This satisfies

$$\mathbb{E}[(X - X_{\perp})(Z - X_{\perp})] = 0, \qquad \forall Z \in S_{\mathcal{G}}. \tag{2.12.6}$$

The Pythagorean relation

$$\|X - X_\perp\|^2 + \|Z - X_\perp\|^2 = \|X - Z\|^2, \qquad \forall Z \in S_\mathcal{G}$$

implies the inequality

$$\|X - X_\perp\|^2 \le \|X - Z\|^2, \qquad \forall Z \in S_\mathcal{G},$$

which is equivalent to

$$\mathbb{E}[(X - X_\perp)^2] \le \mathbb{E}[(X - Z)^2], \qquad \forall Z \in S_\mathcal{G},$$

which yields $X_\perp = \mathbb{E}[X|\mathcal{G}]$, so the conditional expectation is the orthogonal projection of X on the space $S_\mathcal{G}$. The uniqueness of this projection is a consequence of the Pythagorean relation. The orthogonality relation (2.12.6) can be written equivalently as

$$\mathbb{E}[(X - Y)U] = 0, \qquad \forall U \in S_\mathcal{G}.$$

Therefore, the conditional expectation Y satisfies the identity

$$\mathbb{E}[XU] = \mathbb{E}[YU], \qquad \forall U \in S_\mathcal{G}.$$

In particular, if we choose $U = \chi_A$, the characteristic function of a set $A \in \mathcal{G}$, then the foregoing relation yields

$$\int_A X\,dP = \int_A Y\,dP, \qquad \forall A \in \mathcal{G}.$$

We arrive at the following equivalent characterization of the conditional expectations.

The conditional expectation of X given $\mathcal{G}$ is a random variable satisfying:

1. $\mathbb{E}[X|\mathcal{G}]$ *is* $\mathcal{G}$*-measurable;*
2. $\int_A \mathbb{E}[X|\mathcal{G}]\,dP = \int_A X\,dP, \qquad \forall A \in \mathcal{G}.$

Exercise 2.12.1 *Consider the probability space $(\Omega, \mathcal{F}, P)$, and let $\mathcal{G}$ be a σ-field included in $\mathcal{F}$. If X is a $\mathcal{G}$-measurable random variable such that*

$$\int_A X\,dP = 0 \qquad \forall A \in \mathcal{G},$$

then $X = 0$ a.s.

It is worth mentioning here an equivalent famous result, which relates to conditional expectations:

Theorem 2.12.2 (Radon-Nikodym) *Let $(\Omega, \mathcal{F}, P)$ be a probability space and $\mathcal{G}$ be a σ-field included in $\mathcal{F}$. Then for any random variable X there is a $\mathcal{G}$-measurable random variable Y such that*

$$\int_A X\,dP = \int_A Y\,dP, \qquad \forall A \in \mathcal{G}. \tag{2.12.7}$$

Radon-Nikodym's theorem states the existence of Y. In fact this is unique almost surely by the application of Exercise 2.12.1.

Example 2.12.3 *Show that if $\mathcal{G} = \{\emptyset, \Omega\}$, then $\mathbb{E}[X|\mathcal{G}] = \mathbb{E}[X]$.*

Proof: We need to show that $\mathbb{E}[X]$ satisfies conditions 1 and 2. The first one is obviously satisfied since any constant is $\mathcal{G}$-measurable. The latter condition is checked on each set of $\mathcal{G}$. We have

$$\begin{aligned}\int_\Omega X\,dP &= \mathbb{E}[X] = \mathbb{E}[X]\int_\Omega dP = \int_\Omega \mathbb{E}[X]dP \\ \int_\emptyset X\,dP &= \int_\emptyset \mathbb{E}[X]dP.\end{aligned}$$

■

Example 2.12.4 *Show that $\mathbb{E}[\mathbb{E}[X|\mathcal{G}]] = \mathbb{E}[X]$, i.e. all conditional expectations have the same mean, which is the mean of X.*

Proof: Using the definition of expectation and taking $A = \Omega$ in the second relation of the aforementioned definition, yields

$$\mathbb{E}[\mathbb{E}[X|\mathcal{G}]] = \int_\Omega \mathbb{E}[X|\mathcal{G}]\,dP = \int_\Omega X dP = \mathbb{E}[X],$$

which ends the proof. ■

Example 2.12.5 *The conditional expectation of X given the total information $\mathcal{F}$ is the random variable X itself, i.e.*

$$\mathbb{E}[X|\mathcal{F}] = X.$$

Proof: The random variables X and $\mathbb{E}[X|\mathcal{F}]$ are both $\mathcal{F}$-measurable (from the definition of the random variable). From the definition of the conditional expectation we have

$$\int_A \mathbb{E}[X|\mathcal{F}]\,dP = \int_A X\,dP, \quad \forall A \in \mathcal{F}.$$

Exercise 2.12.1 implies that $\mathbb{E}[X|\mathcal{F}] = X$ almost surely. ∎

General properties of the conditional expectation are stated below without proof. The proof involves more or less simple manipulations of integrals and can be taken as an exercise for the reader.

Proposition 2.12.6 *Let X and Y be two random variables on the probability space $(\Omega, \mathcal{F}, P)$. We have*
1. Linearity:

$$\mathbb{E}[aX + bY|\mathcal{G}] = a\mathbb{E}[X|\mathcal{G}] + b\mathbb{E}[Y|\mathcal{G}], \qquad \forall a, b \in \mathbb{R};$$

2. Factoring out the measurable part:

$$\mathbb{E}[XY|\mathcal{G}] = X\mathbb{E}[Y|\mathcal{G}]$$

if X is $\mathcal{G}$-measurable. In particular, $\mathbb{E}[X|\mathcal{G}] = X$.
3. Tower property ("the least information wins"):

$$\mathbb{E}[\mathbb{E}[X|\mathcal{G}]|\mathcal{H}] = \mathbb{E}[\mathbb{E}[X|\mathcal{H}]|\mathcal{G}] = \mathbb{E}[X|\mathcal{H}], \text{ if } \mathcal{H} \subset \mathcal{G};$$

4. Positivity:

$$\mathbb{E}[X|\mathcal{G}] \geq 0, \text{ if } X \geq 0;$$

5. Expectation of a constant is a constant:

$$\mathbb{E}[c|\mathcal{G}] = c.$$

6. An independent condition drops out:

$$\mathbb{E}[X|\mathcal{G}] = \mathbb{E}[X],$$

if X is independent of $\mathcal{G}$.

Exercise 2.12.7 *Prove the property 3 (tower property) given in the previous proposition.*

Exercise 2.12.8 *Toss a fair coin 4 times. Each toss yields either H (heads) or T (tails) with equal probability.*

(a) How many elements does the sample space Ω have?

(b) Consider the events $A = \{$Two of the 4 tosses are H$\}$, $B = \{$The first toss is H$\}$, and $C = \{$3 of the 4 tosses are H$\}$. Compute $P(A)$, $P(B)$ and $P(C)$.

(c) Compute $P(A \cap B)$ and $P(B \cap C)$.

(d) Are the events A and B independent?

(e) Are the events B and C independent? Find $P(B|C)$.

(f) Consider the following information sets (σ-algebras)

$$\mathcal{F} = \{\text{we know the outcomes of the first two tosses}\}$$

$$\mathcal{G} = \{\text{we know the outcomes of the tosses but not the order}\}.$$

How can you state in words the information set $\mathcal{F} \cap \mathcal{G}$?

(g) Prove or disprove: (i) $A \in \mathcal{G}$, (ii) $B \in \mathcal{F}$, and (iii) $C \in \mathcal{G}$.

(h) Define the random variables

$$X = \text{number of } H - \text{number of } T$$

$$Y = \text{number of } T \text{ before the first } H.$$

Show that X is $\mathcal{G}$-measurable while Y is not $\mathcal{G}$-measurable.

(i) Find the expectations $\mathbb{E}[X]$, $\mathbb{E}[Y]$ and $\mathbb{E}[X|\mathcal{G}]$.

Exercise 2.12.9 *Let X be a random variable on the probability space $(\Omega, \mathcal{F}, P)$, which is independent of the σ-field $\mathcal{G} \subset \mathcal{F}$. Consider the characteristic function of a set $A \subset \Omega$ defined by* $\chi_A(\omega) = \begin{cases} 1, & \text{if } \omega \in A \\ 0, & \text{if } \omega \notin A. \end{cases}$

Show the following:

(a) χ_A is $\mathcal{G}$-measurable for any $A \in \mathcal{G}$;

(b) $P(A) = \mathbb{E}[\chi_A]$;

(c) X and χ_A are independent random variables;

(d) $\mathbb{E}[\chi_A X] = \mathbb{E}[X]P(A)$ for any $A \in \mathcal{G}$;

(e) $\mathbb{E}[X|\mathcal{G}] = \mathbb{E}[X]$.

Exercise 2.12.10 *Let $L^2(\Omega, \mathcal{F}, P)$ be the space of square integrable random variables on the probability space $(\Omega, \mathcal{F}, P)$. Define the following scalar product on $L^2(\Omega, \mathcal{F}, P)$*

$$\langle X, Y \rangle = \mathbb{E}[XY].$$

(a) Show that $L^2(\Omega, \mathcal{F}, P)$ becomes a Hilbert space;

(b) Show that if ξ is a random variable in $L^2(\Omega, \mathcal{F}, P)$ and $\mathcal{G}$ is a σ-field contained in $\mathcal{F}$, then $\mathbb{E}[\xi|\mathcal{G}]$ is the orthogonal projection of ξ onto the subspace of $L^2(\Omega, \mathcal{F}, P)$ consisting of $\mathcal{G}$-measurable random variables.

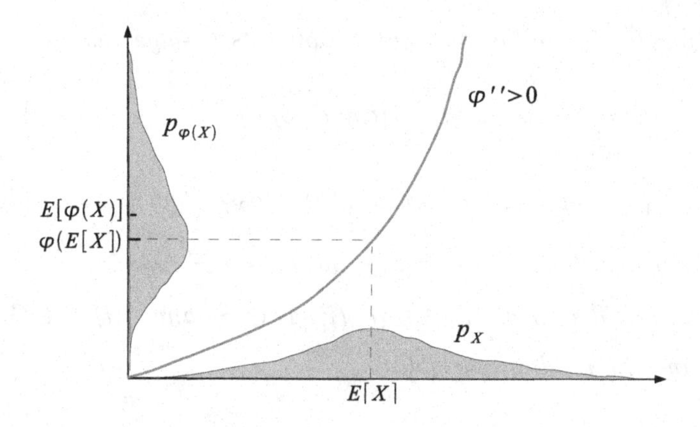

Figure 2.5: *Jensen's inequality* $\varphi(\mathbb{E}[X]) < \mathbb{E}[\varphi(X)]$ *for a convex function* φ.

2.13 Inequalities of Random Variables

This section prepares the reader for the limits of sequences of random variables and limits of stochastic processes. We recall first that an infinite differentiable function $f(x)$, has a *Taylor series* at a if

$$f(x) = f(a) + \frac{f'(a)}{1!}(x-a) + \frac{f''(a)}{2!}(x-a)^2 + \frac{f'''(a)}{3!}(x-a)^3 + \dots,$$

where x belongs to an interval neighborhood of a.

Exercise 2.13.1 *Let $f(x)$ be a function that is $n+1$ times differentiable on an interval I, containing a. Show that there is a $\xi \in I$ such that for any $x \in I$*

$$f(x) = f(a) + \frac{f'(a)}{1!}(x-a) + \cdots + \frac{f^{(n)}(a)}{n!}(x-a)^n + \frac{f^{(n+1)}(a)}{(n+1)!}(\xi - a)^{n+1}.$$

We shall start with a classical inequality result regarding expectations.

Theorem 2.13.2 (Jensen's inequality) *Let $\varphi : \mathbb{R} \to \mathbb{R}$ be a convex function and let X be an integrable random variable on the probability space $(\Omega, \mathcal{F}, P)$. If $\varphi(X)$ is integrable, then*

$$\varphi(\mathbb{E}[X]) \le \mathbb{E}[\varphi(X)].$$

Proof: We shall assume φ twice differentiable with φ'' continuous. Let $\mu = \mathbb{E}[X]$. Expand φ in a Taylor series about μ, see Exercise 2.13.1, and get

$$\varphi(x) = \varphi(\mu) + \varphi'(\mu)(x-\mu) + \frac{1}{2}\varphi''(\xi)(\xi-\mu)^2,$$

with ξ in between x and μ. Since φ is convex, $\varphi'' \geq 0$, and hence

$$\varphi(x) \geq \varphi(\mu) + \varphi'(\mu)(x - \mu),$$

which means the graph of $\varphi(x)$ is above the tangent line at $\big(x, \varphi(x)\big)$. Replacing x by the random variable X, and taking the expectation yields

$$\begin{aligned} \mathbb{E}[\varphi(X)] &\geq \mathbb{E}[\varphi(\mu) + \varphi'(\mu)(X - \mu)] = \varphi(\mu) + \varphi'(\mu)(\mathbb{E}[X] - \mu) \\ &= \varphi(\mu) = \varphi(\mathbb{E}[X]), \end{aligned}$$

which proves the result. ■

Fig. 2.5 provides a graphical interpretation of Jensen's inequality. If the distribution of X is symmetric, then the distribution of $\varphi(X)$ is skewed, with $\varphi(\mathbb{E}[X]) < \mathbb{E}[\varphi(X)]$.

It is worth noting that the inequality is reversed for φ concave. We shall present next a couple of applications.

A random variable $X : \Omega \to \mathbb{R}$ is called *square integrable* if

$$\mathbb{E}[X^2] = \int_\Omega |X(\omega)|^2 \, dP(\omega) = \int_\mathbb{R} x^2 p(x) \, dx < \infty.$$

Application 2.13.3 *If X is a square integrable random variable, then it is integrable.*

Proof: Jensen's inequality with $\varphi(x) = x^2$ becomes

$$\mathbb{E}[X]^2 \leq \mathbb{E}[X^2].$$

Since the right side is finite, it follows that $\mathbb{E}[X] < \infty$, so X is integrable. ■

Application 2.13.4 *If $m_X(t)$ denotes the moment generating function of the random variable X with mean μ, then*

$$m_X(t) \geq e^{t\mu}.$$

Proof: Applying Jensen's inequality with the convex function $\varphi(x) = e^x$ yields

$$e^{\mathbb{E}[X]} \leq \mathbb{E}[e^X].$$

Substituting tX for X implies that

$$e^{\mathbb{E}[tX]} \leq \mathbb{E}[e^{tX}]. \tag{2.13.8}$$

Using the definition of the moment generating function $m_X(t) = \mathbb{E}[e^{tX}]$ and that $\mathbb{E}[tX] = t\mathbb{E}[X] = t\mu$, then (2.13.8) leads to the desired inequality. ■

The *variance* of a square integrable random variable X is defined by

$$Var(X) = \mathbb{E}[X^2] - \mathbb{E}[X]^2.$$

By Application 2.13.3 we have $Var(X) \geq 0$, so that there is a constant $\sigma_X > 0$, called *standard deviation*, such that

$$\sigma_X^2 = Var(X).$$

Exercise 2.13.5 *Prove the following identity:*

$$Var[X] = \mathbb{E}[(X - \mathbb{E}[X])^2].$$

Exercise 2.13.6 *Prove that a non-constant random variable has a nonzero standard deviation.*

Exercise 2.13.7 *Prove the following extension of Jensen's inequality: If φ is a convex function, then for any σ-field $\mathcal{G} \subset \mathcal{F}$ we have*

$$\varphi(\mathbb{E}[X|\mathcal{G}]) \leq \mathbb{E}[\varphi(X)|\mathcal{G}].$$

Exercise 2.13.8 *Show the following:*

(a) $|\mathbb{E}[X]| \leq \mathbb{E}[|X|]$;

(b) $|\mathbb{E}[X|\mathcal{G}]| \leq \mathbb{E}[|X|\ |\mathcal{G}]$, *for any σ-field $\mathcal{G} \subset \mathcal{F}$*;

(c) $|\mathbb{E}[X]|^r \leq \mathbb{E}[|X|^r]$, *for $r \geq 1$*;

(d) $|\mathbb{E}[X|\mathcal{G}]|^r \leq \mathbb{E}[|X|^r\ |\mathcal{G}]$, *for any σ-field $\mathcal{G} \subset \mathcal{F}$ and $r \geq 1$.*

Theorem 2.13.9 (Markov's inequality) *For any $\lambda, p > 0$, we have the following inequality:*

$$P(\omega; |X(\omega)| \geq \lambda) \leq \frac{1}{\lambda^p}\mathbb{E}[|X|^p].$$

Proof: Let $A = \{\omega; |X(\omega)| \geq \lambda\}$. Then

$$\begin{aligned}
\mathbb{E}[|X|^p] &= \int_\Omega |X(\omega)|^p\, dP(\omega) \geq \int_A |X(\omega)|^p\, dP(\omega) \geq \int_A \lambda^p\, dP(\omega) \\
&= \lambda^p \int_A dP(\omega) = \lambda^p P(A) = \lambda^p P(|X| \geq \lambda).
\end{aligned}$$

Dividing by λ^p leads to the desired result. ■

Theorem 2.13.10 (Tchebychev's inequality) *If X is a random variable with mean μ and variance σ^2, then*

$$P(\omega; |X(\omega) - \mu| \geq \lambda) \leq \frac{\sigma^2}{\lambda^2}.$$

Proof: Let $A = \{\omega; |X(\omega) - \mu| \geq \lambda\}$. Then

$$\begin{aligned} \sigma^2 &= Var(X) = \mathbb{E}[(X-\mu)^2] = \int_\Omega (X-\mu)^2\, dP \geq \int_A (X-\mu)^2\, dP \\ &\geq \lambda^2 \int_A dP = \lambda^2 P(A) = \lambda^2 P(\omega; |X(\omega) - \mu| \geq \lambda). \end{aligned}$$

Dividing by λ^2 leads to the desired inequality. ∎

The next result deals with exponentially decreasing bounds on tail distributions.

Theorem 2.13.11 (Chernoff bounds) *Let X be a random variable. Then for any $\lambda > 0$ we have*

1. $P(X \geq \lambda) \leq \dfrac{\mathbb{E}[e^{tX}]}{e^{\lambda t}}, \forall t > 0;$
2. $P(X \leq \lambda) \leq \dfrac{\mathbb{E}[e^{tX}]}{e^{\lambda t}}, \forall t < 0.$

Proof: 1. Let $t > 0$ and denote $Y = e^{tX}$. By Markov's inequality

$$P(Y \geq e^{\lambda t}) \leq \frac{\mathbb{E}[Y]}{e^{\lambda t}}.$$

Then we have

$$\begin{aligned} P(X \geq \lambda) &= P(tX \geq \lambda t) = P(e^{tX} \geq e^{\lambda t}) \\ &= P(Y \geq e^{\lambda t}) \leq \frac{\mathbb{E}[Y]}{e^{\lambda t}} = \frac{\mathbb{E}[e^{tX}]}{e^{\lambda t}}. \end{aligned}$$

2. The case $t < 0$ is similar. ∎

In the following we shall present an application of the Chernoff bounds for the normal distributed random variables.

Let X be a random variable normally distributed with mean μ and variance σ^2. It is known that its moment generating function is given by

$$m(t) = \mathbb{E}[e^{tX}] = e^{\mu t + \frac{1}{2}t^2\sigma^2}.$$

Using the first Chernoff bound we obtain

$$P(X \geq \lambda) \leq \frac{m(t)}{e^{\lambda t}} = e^{(\mu-\lambda)t + \frac{1}{2}t^2\sigma^2}, \forall t > 0,$$

which implies

$$P(X \geq \lambda) \;\; \leq \;\; e^{\min\limits_{t>0}[(\mu-\lambda)t + \frac{1}{2}t^2\sigma^2]}.$$

It is easy to see that the quadratic function $f(t) = (\mu - \lambda)t + \frac{1}{2}t^2\sigma^2$ has the minimum value reached for $t = \dfrac{\lambda - \mu}{\sigma^2}$. Since $t > 0$, λ needs to satisfy $\lambda > \mu$. Then

$$\min_{t>0} f(t) = f\Big(\frac{\lambda-\mu}{\sigma^2}\Big) = -\frac{(\lambda-\mu)^2}{2\sigma^2}.$$

Substituting into the previous formula, we obtain the following result:

Proposition 2.13.12 *If X is a normally distributed variable, $X \sim N(\mu, \sigma^2)$, then for any $\lambda > \mu$*

$$P(X \geq \lambda) \leq e^{-\frac{(\lambda-\mu)^2}{2\sigma^2}}.$$

Exercise 2.13.13 *Let X be a Poisson random variable with mean $\lambda > 0$.*
(a) Show that the moment generating function of X is $m(t) = e^{\lambda(e^t-1)}$;
(b) Use a Chernoff bound to show that

$$P(X \geq k) \leq e^{\lambda(e^t-1)-tk}, \qquad t > 0.$$

Markov's, Tchebychev's and Chernoff's inequalities will be useful later when computing limits of random variables.

Proposition 2.13.14 *Let X be a random variable and f and g be two functions, both increasing or decreasing. Then*

$$\mathbb{E}[f(X)g(X)] \geq \mathbb{E}[f(X)]\mathbb{E}[g(X)]. \tag{2.13.9}$$

Proof: For any two independent random variables X and Y, we have

$$\Big(f(X) - f(Y)\Big)\Big(g(X) - g(Y)\Big) \geq 0.$$

Applying expectation yields

$$\mathbb{E}[f(X)g(X)] + \mathbb{E}[f(Y)g(Y)] \geq \mathbb{E}[f(X)]\mathbb{E}[f(Y)] + \mathbb{E}[f(Y)]\mathbb{E}[f(X)].$$

Considering Y as an independent copy of X we obtain

$$2\mathbb{E}[f(X)g(X)] \geq 2\mathbb{E}[f(X)]\mathbb{E}[g(X)].$$

■

Exercise 2.13.15 *Show the following inequalities:*

(*a*) $\mathbb{E}[X^2] \geq \mathbb{E}[X]^2$;

(*b*) $\mathbb{E}[X \sinh(X)] \geq \mathbb{E}[X]\mathbb{E}[\sinh(X)]$;

(*c*) $\mathbb{E}[X^6] \geq \mathbb{E}[X]\mathbb{E}[X^5]$;

(*d*) $\mathbb{E}[X^6] \geq \mathbb{E}[X^3]^2$.

Exercise 2.13.16 *For any* $n, k \geq 1$, *show that*

$$\mathbb{E}[X^{2(n+k+1)}] \geq \mathbb{E}[X^{2k+1}]\mathbb{E}[X^{2n+1}].$$

2.14 Limits of Sequences of Random Variables

Consider a sequence $(X_n)_{n\geq 1}$ of random variables defined on the probability space $(\Omega, \mathcal{F}, P)$. There are several ways of making sense of the limit expression $X = \lim\limits_{n\to\infty} X_n$. This is the subject treated in the following sections.

Almost Sure Limit The sequence X_n converges *almost surely* to X, if for all states of the world ω, except a set of probability zero, we have

$$\lim_{n\to\infty} X_n(\omega) = X(\omega).$$

More precisely, this means

$$P\Big(\omega; \lim_{n\to\infty} X_n(\omega) = X(\omega)\Big) = 1,$$

and we shall write $\text{as-}\lim\limits_{n\to\infty} X_n = X$. An important example where this type of limit occurs is the Strong Law of Large Numbers:

If X_n is a sequence of independent and identically distributed random variables with the same mean μ, then $\text{as-}\lim\limits_{n\to\infty} \dfrac{X_1 + \cdots + X_n}{n} = \mu$.

This result ensures that the sample mean tends to the (unknown) population mean μ almost surely as $n \to \infty$, a fact that makes it very useful in statistics.

Mean Square Limit Another possibility of convergence is to look at the mean square deviation of X_n from X. We say that X_n converges to X in the *mean square* if

$$\lim_{n\to\infty} \mathbb{E}[(X_n - X)^2] = 0.$$

More precisely, this should be interpreted as

$$\lim_{n\to\infty} \int_\Omega \big(X_n(\omega) - X(\omega)\big)^2 \, dP(\omega) = 0.$$

This limit will be abbreviated by $\text{ms-}\lim\limits_{n\to\infty} X_n = X$. The mean square convergence is useful when defining the Ito integral.

Proposition 2.14.1 *Consider a sequence X_n of random variables such that there is a constant k with $\mathbb{E}[X_n] \to k$ and $Var(X_n) \to 0$ as $n \to \infty$. Show that* $\text{ms-}\lim_{n\to\infty} X_n = k$.

Proof: Since we have

$$\begin{aligned}\mathbb{E}[|X_n - k|^2] &= \mathbb{E}[X_n^2 - 2kX_n + k^2] = \mathbb{E}[X_n^2] - 2k\mathbb{E}[X_n] + k^2 \\ &= \big(\mathbb{E}[X_n^2] - \mathbb{E}[X_n]^2\big) + \big(\mathbb{E}[X_n]^2 - 2k\mathbb{E}[X_n] + k^2\big) \\ &= Var(X_n) + \big(\mathbb{E}[X_n] - k\big)^2,\end{aligned}$$

the right side tends to 0 when taking the limit $n \to \infty$. ■

Exercise 2.14.2 *Show the following relation*

$$\mathbb{E}[(X-Y)^2] = Var[X] + Var[Y] + \big(\mathbb{E}[X] - \mathbb{E}[Y]\big)^2 - 2Cov(X,Y).$$

Exercise 2.14.3 *If X_n tends to X in mean square, with $\mathbb{E}[X^2] < \infty$, show that:*

(a) $\mathbb{E}[X_n] \to \mathbb{E}[X]$ *as* $n \to \infty$;

(b) $\mathbb{E}[X_n^2] \to \mathbb{E}[X^2]$ *as* $n \to \infty$;

(c) $Var[X_n] \to Var[X]$ *as* $n \to \infty$;

(d) $Cov(X_n, X) \to Var[X]$ *as* $n \to \infty$.

Exercise 2.14.4 *If X_n tends to X in mean square, show that $\mathbb{E}[X_n|\mathcal{H}]$ tends to $\mathbb{E}[X|\mathcal{H}]$ in mean square.*

Limit in Probability The random variable X is the *limit in probability* of X_n if for n large enough the probability of deviation from X can be made smaller than any arbitrary ϵ. More precisely, for any $\epsilon > 0$

$$\lim_{n\to\infty} P\big(\omega; |X_n(\omega) - X(\omega)| \le \epsilon\big) = 1.$$

This can also be written as

$$\lim_{n\to\infty} P\big(\omega; |X_n(\omega) - X(\omega)| > \epsilon\big) = 0.$$

This limit is denoted by $\text{p-}\lim_{n\to\infty} X_n = X$.

It is worth noting that both almost certain convergence and convergence in mean square imply the convergence in probability.

Proposition 2.14.5 *The convergence in mean square implies the convergence in probability.*

Proof: Let $\text{ms-}\lim\limits_{n\to\infty} Y_n = Y$. Let $\epsilon > 0$ be arbitrarily fixed. Applying Markov's inequality with $X = Y_n - Y$, $p = 2$ and $\lambda = \epsilon$, yields

$$0 \le P(|Y_n - Y| \ge \epsilon) \le \frac{1}{\epsilon^2}\mathbb{E}[|Y_n - Y|^2].$$

The right side tends to 0 as $n \to \infty$. Applying the Squeeze Theorem we obtain

$$\lim_{n\to\infty} P(|Y_n - Y| \ge \epsilon) = 0,$$

which means that Y_n converges stochastically to Y. ∎

Example 2.14.6 *Let X_n be a sequence of random variables such that $\mathbb{E}[|X_n|] \to 0$ as $n \to \infty$. Prove that* $\text{p-}\lim\limits_{n\to\infty} X_n = 0$.

Proof: Let $\epsilon > 0$ be arbitrarily fixed. We need to show

$$\lim_{n\to\infty} P\big(\omega; |X_n(\omega)| \ge \epsilon\big) = 0. \tag{2.14.10}$$

From Markov's inequality (see Theorem 2.13.9) we have

$$0 \le P\big(\omega; |X_n(\omega)| \ge \epsilon\big) \le \frac{\mathbb{E}[|X_n|]}{\epsilon}.$$

Using the Squeeze Theorem we obtain (2.14.10). ∎

Remark 2.14.7 The conclusion still holds true even in the case when there is a $p > 0$ such that $\mathbb{E}[|X_n|^p] \to 0$ as $n \to \infty$.

Limit in Distribution We say the sequence X_n converges *in distribution* to X if for any continuous bounded function $\varphi(x)$ we have

$$\lim_{n\to\infty} \mathbb{E}[\varphi(X_n)] = \mathbb{E}[\varphi(X)].$$

We make the remark that this type of limit is even weaker than the stochastic convergence, i.e. it is implied by it.

An application of the limit in distribution is obtained if we consider $\varphi(x) = e^{itx}$. In this case the expectation becomes the Fourier transform of the probability density

$$\mathbb{E}[\varphi(X)] = \int e^{itx} p(x)\, dx = \hat{p}(t),$$

and it is called the *characteristic function* of the random variable X. It follows that if X_n converges in distribution to X, then the characteristic function of X_n converges to the characteristic function of X. From the properties of the

Fourier transform, the probability density of X_n approaches the probability density of X.

It can be shown that the convergence in distribution is equivalent to

$$\lim_{n\to\infty} F_n(x) = F(x),$$

whenever F is continuous at x, where F_n and F denote the distribution functions of X_n and X, respectively. This is the reason that this convergence bears its name.

Exercise 2.14.8 *Consider the probability space $(\Omega, \mathcal{F}, P)$, with $\Omega = [0,1]$, $\mathcal{F}$ the σ-algebra of Borel sets, and P the Lebesgue measure (see Example 2.3.3). Define the sequence X_n by*

$$X_{2n} = \begin{cases} 0, & \text{if } \omega < 1/2 \\ 1, & \text{if } \omega \geq 1/2, \end{cases} \qquad X_{2n+1} = \begin{cases} 1, & \text{if } \omega < 1/2 \\ 0, & \text{if } \omega \geq 1/2. \end{cases}$$

Show that X_n converges in distribution, but does not converge in probability.

2.15 Properties of Mean-Square Limit

This section deals with the main properties of the mean-square limit, which will be useful in later applications regarding the Ito integral.

Lemma 2.15.1 *If* $\text{ms-}\lim_{n\to\infty} X_n = 0$ *and* $\text{ms-}\lim_{n\to\infty} Y_n = 0$, *then*

$$\text{ms-}\lim_{n\to\infty} (X_n + Y_n) = 0.$$

Proof: It follows from the inequality

$$(x+y)^2 \leq 2x^2 + 2y^2.$$

The details are left to the reader. ■

Proposition 2.15.2 *If the sequences of random variables X_n and Y_n converge in the mean square, then*

1. $\text{ms-}\lim_{n\to\infty} (X_n + Y_n) = \text{ms-}\lim_{n\to\infty} X_n + \text{ms-}\lim_{n\to\infty} Y_n$
2. $\text{ms-}\lim_{n\to\infty} (cX_n) = c \cdot \text{ms-}\lim_{n\to\infty} X_n, \quad \forall c \in \mathbb{R}.$

Proof: 1. Let $\text{ms-}\lim_{n\to\infty} X_n = X$ and $\text{ms-}\lim_{n\to\infty} Y_n = Y$. Consider the sequences $X'_n = X_n - X$ and $Y'_n = Y_n - Y$. Then $\text{ms-}\lim_{n\to\infty} X'_n = 0$ and $\text{ms-}\lim_{n\to\infty} Y'_n = 0$. Applying Lemma 2.15.1 yields

$$\text{ms-}\lim_{n\to\infty}(X_n' + Y_n') = 0.$$

This is equivalent to

$$\text{ms-}\lim_{n\to\infty}(X_n - X + Y_n - Y) = 0,$$

which becomes

$$\text{ms-}\lim_{n\to\infty}(X_n + Y_n) = X + Y.$$

2. The second relation can be proved in a similar way and is left as an exercise to the reader. ■

Remark 2.15.3 It is worthy to note that

$$\text{ms-}\lim_{n\to\infty}(X_nY_n) \neq \text{ms-}\lim_{n\to\infty}(X_n) \cdot \text{ms-}\lim_{n\to\infty}(Y_n).$$

Counter-examples can be found, see Exercise 2.15.5.

Exercise 2.15.4 *Use a computer algebra system to show the following:*

(a) $\displaystyle\int_0^\infty \frac{x^2}{x^5+1}\,dx = \frac{\pi\sqrt{2}}{4}\sqrt{1-5^{-1/2}}$;

(b) $\displaystyle\int_0^\infty \frac{x^4}{x^5+1}\,dx = \infty$;

(c) $\displaystyle\int_0^\infty \frac{1}{x^5+1}\,dx = \frac{1}{5}\Gamma\Big(\frac{1}{5}\Big)\Gamma\Big(\frac{4}{5}\Big)$.

Exercise 2.15.5 *Let X be a random variable with the probability density function*

$$p(x) = \frac{5}{\Gamma(1/5)\Gamma(4/5)}\frac{1}{x^5+1}, \qquad x \geq 0.$$

(a) *Show that* $\mathbb{E}[X^2] < \infty$ *and* $\mathbb{E}[X^4] = \infty$;

(b) *Construct the sequences of random variables* $X_n = Y_n = \frac{1}{n}X$. *Show that* $\text{ms-}\lim_{n\to\infty} X_n = 0$, $\text{ms-}\lim_{n\to\infty} Y_n = 0$, *but* $\text{ms-}\lim_{n\to\infty}(X_nY_n) = \infty$.

2.16 Stochastic Processes

A *stochastic process* on the probability space $(\Omega, \mathcal{F}, P)$ is a family of random variables X_t parameterized by $t \in \mathbf{T}$, where $\mathbf{T} \subset \mathbb{R}$. If $\mathbf{T}$ is an interval we say that X_t is a stochastic process in *continuous time.* If $\mathbf{T} = \{1, 2, 3, \dots\}$ we shall say that X_t is a stochastic process in *discrete time.* The latter case describes a sequence of random variables. The reader interested in these type of processes can consult Brzezniak and Zastawniak [9].

The aforementioned types of convergence can be easily extended to continuous time. For instance, X_t converges almost surely to X as $t \to \infty$ if

$$P\Big(\omega; \lim_{t\to\infty} X_t(\omega) = X(\omega)\Big) = 1.$$

The evolution in time of a given state of the world $\omega \in \Omega$ given by the function $t \longmapsto X_t(\omega)$ is called a *path* or *realization* of X_t. The study of stochastic processes using computer simulations is based on retrieving information about the process X_t given a large number of its realizations.

Next we shall structure the information field $\mathcal{F}$ with an order relation parameterized by the time t. Consider that all the information accumulated until time t is contained by the σ-field $\mathcal{F}_t$. This means that $\mathcal{F}_t$ contains the information containing events that have already occurred until time t, and which did not. Since the information is growing in time, we have

$$\mathcal{F}_s \subset \mathcal{F}_t \subset \mathcal{F}$$

for any $s, t \in \mathbf{T}$ with $s \leq t$. The family $\mathcal{F}_t$ is called a *filtration.*

A stochastic process X_t is said to be *adapted* to the filtration $\mathcal{F}_t$ if X_t is $\mathcal{F}_t$- measurable, for any $t \in \mathbf{T}$. This means that the information at time t determines the value of the random variable X_t.

Example 2.16.1 *Here there are a few examples of filtrations:*

1. $\mathcal{F}_t$ represents the information about the evolution of a stock until time t, with $t > 0$.

2. $\mathcal{F}_t$ represents the information about the evolution of a Black-Jack game until time t, with $t > 0$.

3. $\mathcal{F}_t$ represents the medical information of a patient until time t.

Example 2.16.2 *If X is a random variable, consider the conditional expectation*

$$X_t = \mathbb{E}[X|\mathcal{F}_t].$$

From the definition of conditional expectation, the random variable X_t is $\mathcal{F}_t$-measurable, and can be regarded as the measurement of X at time t using the information $\mathcal{F}_t$. If the accumulated knowledge $\mathcal{F}_t$ increases and eventually equals the σ-field $\mathcal{F}$, then $X = \mathbb{E}[X|\mathcal{F}]$, i.e. we obtain the entire random variable. The process X_t is adapted to $\mathcal{F}_t$.

Example 2.16.3 *Don Joe goes to a doctor to get an estimation of how long he still has to live. The age at which he will pass away is a random variable, denoted by X. Given his medical condition today, which is contained in $\mathcal{F}_t$, the doctor can infer an average age, which is the average of all random instances that agree with the information to date; this is given by the conditional expectation $X_t = \mathbb{E}[X|\mathcal{F}_t]$. The stochastic process X_t is adapted to the medical knowledge $\mathcal{F}_t$.*

We shall define next an important type of stochastic process.[4]

Definition 2.16.4 *A process X_t, $t \in \mathbf{T}$, is called a martingale with respect to the filtration $\mathcal{F}_t$ if*

1. *X_t is integrable for each $t \in \mathbf{T}$;*
2. *X_t is adapted to the filtration $\mathcal{F}_t$;*
3. *$X_s = \mathbb{E}[X_t|\mathcal{F}_s]$, $\forall s < t$.*

Remark 2.16.5 The first condition states that the unconditional forecast is finite $\mathbb{E}[|X_t|] = \int_\Omega |X_t| \, dP < \infty$. Condition 2 says that the value X_t is known, given the information set $\mathcal{F}_t$. This can also be stated by saying that X_t is $\mathcal{F}_t$-measurable. The third relation asserts that the best forecast of unobserved future values is the last observation on X_t.

Example 2.16.6 *Let X_t denote Mr. Li Zhu's salary after t years of work at the same company. Since X_t is known at time t and it is bounded above, as all salaries are, then the first two conditions hold. Being honest, Mr. Zhu expects today that his future salary will be the same as today's, i.e. $X_s = \mathbb{E}[X_t|\mathcal{F}_s]$, for $s < t$. This means that X_t is a martingale.*

Exercise 2.16.7 *If X is an integrable random variable on $(\Omega, \mathcal{F}, P)$, and $\mathcal{F}_t$ is a filtration. Prove that $X_t = \mathbb{E}[X|\mathcal{F}_t]$ is a martingale.*

Exercise 2.16.8 *Let X_t and Y_t be martingales with respect to the filtration $\mathcal{F}_t$. Show that for any $a, b, c \in \mathbb{R}$ the process $Z_t = aX_t + bY_t + c$ is an $\mathcal{F}_t$-martingale.*

Exercise 2.16.9 *Let X_t and Y_t be martingales with respect to the filtration $\mathcal{F}_t$.*

(a) *Is the process X_tY_t always a martingale with respect to $\mathcal{F}_t$?*

(b) *What about the processes X_t^2 and Y_t^2?*

Exercise 2.16.10 *Two processes X_t and Y_t are called conditionally uncorrelated, given $\mathcal{F}_t$, if*

$$\mathbb{E}[(X_t - X_s)(Y_t - Y_s)|\mathcal{F}_s] = 0, \qquad \forall 0 \leq s < t < \infty.$$

Let X_t and Y_t be martingale processes. Show that the process $Z_t = X_tY_t$ is a martingale if and only if X_t and Y_t are conditionally uncorrelated. Assume that X_t, Y_t and Z_t are integrable.

---

[4] The concept of martingale was introduced by Lévy in 1934.

In the following, if X_t is a stochastic process, the minimum amount of information resulted from knowing the process X_s until time t is denoted by $\mathcal{F}_t = \sigma(X_s; s \le t)$. This is the σ-algebra generated by the events $\{\omega; X_s(\omega) \in (a, b)\}$, for any real numbers $a < b$ and $s \le t$.

In the case of a discrete process, the minimum amount of information resulted from knowing the process X_k until time n is $\mathcal{F}_n = \sigma(X_k; k \le n)$, the σ-algebra generated by the events $\{\omega; X_k(\omega) \in (a, b)\}$, for any real numbers $a < b$ and $k \le n$.

Exercise 2.16.11 *Let X_n, $n \ge 0$ be a sequence of integrable independent random variables, with $\mathbb{E}[X_n] < \infty$, for all $n \ge 0$. Let $S_0 = X_0$, $S_n = X_0 + \cdots + X_n$. Show the following:*

(a) $S_n - \mathbb{E}[S_n]$ is an $\mathcal{F}_n$-martingale.

(b) If $\mathbb{E}[X_n] = 0$ and $\mathbb{E}[X_n^2] < \infty$, $\forall n \ge 0$, then $S_n^2 - Var(S_n)$ is an $\mathcal{F}_n$-martingale.

Exercise 2.16.12 *Let X_n, $n \ge 0$ be a sequence of independent, integrable random variables such that $\mathbb{E}[X_n] = 1$ for $n \ge 0$. Prove that $P_n = X_0 \cdot X_1 \cdots \cdot X_n$ is an $\mathcal{F}_n$-martingale.*

Exercise 2.16.13 *(a) Let X be a normally distributed random variable with mean $\mu \neq 0$ and variance σ^2. Prove that there is a unique $\theta \neq 0$ such that $\mathbb{E}[e^{\theta X}] = 1$.*

(b) Let $(X_i)_{i \ge 0}$ be a sequence of identically normally distributed random variables with mean $\mu \neq 0$. Consider the sum $S_n = \sum_{j=0}^{n} X_j$. Show that $Z_n = e^{\theta S_n}$ is a martingale, with θ defined in part (a).

In section 10.1 we shall encounter several processes which are martingales.

Chapter 3

Useful Stochastic Processes

This chapter deals with the most common used stochastic processes and their basic properties. The two main basic processes are the Brownian motion and the Poisson process. The other processes described in this chapter are derived from the previous two. For more advanced topics on the Brownian motion, the reader may consult Freedman [19], Hida [22], Knight [27], Karatzas and Shreve [26], or Mörters and Peres [34].

3.1 The Brownian Motion

The observation first made by the botanist Robert Brown in 1827, that small pollen grains suspended in water have a very irregular and unpredictable state of motion, led to the definition of the Brownian motion, which is formalized in the following.

Definition 3.1.1 *A Brownian motion process is a stochastic process B_t, $t \geq 0$, which satisfies*

1. The process starts at the origin, $B_0 = 0$;

2. B_t has independent increments;

3. The process B_t is continuous in t;

4. The increments $B_t - B_s$ are normally distributed with mean zero and variance $|t - s|$,

$$B_t - B_s \sim N(0, |t - s|).$$

The process $X_t = x + B_t$ has all the properties of a Brownian motion that starts at x. Condition 4 states that the increments of a Brownian motion are stationary, i.e. the distribution of $B_t - B_s$ depends only on the time interval $t - s$

$$P(B_{t+s} - B_s \leq a) = P(B_t - B_0 \leq a) = P(B_t \leq a).$$

It is worth noting that even if B_t is continuous, it is nowhere differentiable. From condition 4 we get that B_t is normally distributed with mean $\mathbb{E}[B_t] = 0$ and $Var[B_t] = t$

$$B_t \sim N(0, t).$$

This implies also that the second moment is $\mathbb{E}[B_t^2] = t$. Let $0 < s < t$. Since the increments are independent, we can write

$$\mathbb{E}[B_s B_t] = \mathbb{E}[(B_s - B_0)(B_t - B_s) + B_s^2] = \mathbb{E}[B_s - B_0]\mathbb{E}[B_t - B_s] + \mathbb{E}[B_s^2] = s.$$

Consequently, B_s and B_t are not independent.

Condition 4 also has a physical explanation. A pollen grain suspended in water is kicked about by a very large number of water molecules. The influence of each molecule on the grain is independent of the other molecules. These effects are averaged out into a resultant increment of the grain coordinate. According to the Central Limit Theorem, this increment has to be normally distributed.

If the exterior stochastic activity on the pollen grain is represented at time t by the noise $\mathcal{N}_t$, then the cummulative effect on the grain during the time interval $[0, t]$ is represented by the integral $W_t = \int_0^t \mathcal{N}_s \, ds$, which is the Brownian motion.

There are three distinct classical constructions of the Brownian motion, due to Wiener [47], Kolmogorov [28] and Lévy [33]. However, the existence of the Brownian motion process is beyond the goal of this book.

It is worth noting that the processes with stationary and independent increments form a special class of stochastic processes, called *Lévy processes*; so, in particular, Brownian motions are Lévy processes.

Proposition 3.1.2 *A Brownian motion process B_t is a martingale with respect to the information set $\mathcal{F}_t = \sigma(B_s; s \le t)$.*

Proof: The integrability of B_t follows from Jensen's inequality

$$\mathbb{E}[|B_t|]^2 \le \mathbb{E}[B_t^2] = Var(B_t) = |t| < \infty.$$

B_t is obviously $\mathcal{F}_t$-measurable. Let $s < t$ and write $B_t = B_s + (B_t - B_s)$. Then

$$\begin{aligned}
\mathbb{E}[B_t|\mathcal{F}_s] &= \mathbb{E}[B_s + (B_t - B_s)|\mathcal{F}_s] \\
&= \mathbb{E}[B_s|\mathcal{F}_s] + \mathbb{E}[B_t - B_s|\mathcal{F}_s] \\
&= B_s + \mathbb{E}[B_t - B_s] = B_s + \mathbb{E}[B_{t-s} - B_0] = B_s,
\end{aligned}$$

where we used that B_s is $\mathcal{F}_s$-predictable (from where $\mathbb{E}[B_s|\mathcal{F}_s] = B_s$) and that the increment $B_t - B_s$ is independent of previous values of B_t contained in the information set $\mathcal{F}_t = \sigma(B_s; s \le t)$. ∎

A process with similar properties as the Brownian motion was introduced by Wiener.

Definition 3.1.3 *A Wiener process W_t is a process adapted to a filtration $\mathcal{F}_t$ such that*

1. *The process starts at the origin, $W_0 = 0$;*
2. *W_t is a squared integrable $\mathcal{F}_t$-martingale with*

$$\mathbb{E}[(W_t - W_s)^2] = t - s, \qquad s \leq t;$$

3. *The process W_t is continuous in t.*

Since W_t is a martingale, its increments satisfy

$$\mathbb{E}[W_t - W_s] = \mathbb{E}[W_t - W_s|\mathcal{F}_s] = \mathbb{E}[W_t|\mathcal{F}_s] - W_s = W_s - W_s = 0,$$

and hence $\mathbb{E}[W_t] = 0$. It is easy to show that

$$Var[W_t - W_s] = |t - s|, \qquad Var[W_t] = t.$$

Exercise 3.1.4 *Show that a Brownian process B_t is a Wiener process.*

The only property B_t has and W_t seems not to have is that the increments are normally distributed. However, it can be shown that there is no distinction between these two processes, as the famous Lévy theorem states, see section 10.2. From now on, the notations B_t and W_t will be used interchangeably.

Infinitesimal relations In stochastic calculus we often need to use infinitesimal notation and its properties. If dW_t denotes the infinitesimal increment of a Wiener process in the time interval dt, the aforementioned properties become $dW_t \sim N(0, dt)$, $\mathbb{E}[dW_t] = 0$, and $\mathbb{E}[(dW_t)^2] = dt$.

Proposition 3.1.5 *If W_t is a Brownian motion with respect to the information set $\mathcal{F}_t$, then $Y_t = W_t^2 - t$ is a martingale.*

Proof: Y_t is integrable since

$$\mathbb{E}[|Y_t|] \leq \mathbb{E}[W_t^2 + t] = 2t < \infty, \qquad t > 0.$$

Let $s < t$. Using that the increments $W_t - W_s$ and $(W_t - W_s)^2$ are independent of the information set $\mathcal{F}_s$ and applying Proposition 2.12.6 yields

$$\begin{aligned}
\mathbb{E}[W_t^2|\mathcal{F}_s] &= \mathbb{E}[(W_s + W_t - W_s)^2|\mathcal{F}_s] \\
&= \mathbb{E}[W_s^2 + 2W_s(W_t - W_s) + (W_t - W_s)^2|\mathcal{F}_s] \\
&= \mathbb{E}[W_s^2|\mathcal{F}_s] + \mathbb{E}[2W_s(W_t - W_s)|\mathcal{F}_s] + \mathbb{E}[(W_t - W_s)^2|\mathcal{F}_s] \\
&= W_s^2 + 2W_s\mathbb{E}[W_t - W_s|\mathcal{F}_s] + \mathbb{E}[(W_t - W_s)^2|\mathcal{F}_s] \\
&= W_s^2 + 2W_s\mathbb{E}[W_t - W_s] + \mathbb{E}[(W_t - W_s)^2] \\
&= W_s^2 + t - s,
\end{aligned}$$

and hence $\mathbb{E}[W_t^2 - t|\mathcal{F}_s] = W_s^2 - s$, for $s < t$. ∎

The following result states the memoryless property of Brownian motion[1] W_t.

Proposition 3.1.6 *The conditional distribution of W_{t+s}, given the present W_t and the past W_u, $0 \le u < t$, depends only on the present.*

Proof: Using the independent increment assumption, we have

$$\begin{aligned} & P(W_{t+s} \le c|W_t = x, W_u, 0 \le u < t) \\ = \; & P(W_{t+s} - W_t \le c - x|W_t = x, W_u, 0 \le u < t) \\ = \; & P(W_{t+s} - W_t \le c - x) \\ = \; & P(W_{t+s} \le c|W_t = x). \end{aligned}$$

∎

Since W_t is normally distributed with mean 0 and variance t, its density function is

$$\phi_t(x) = \frac{1}{\sqrt{2\pi t}} e^{-\frac{x^2}{2t}}.$$

Then its distribution function is

$$F_t(x) = P(W_t \le x) = \frac{1}{\sqrt{2\pi t}} \int_{-\infty}^{x} e^{-\frac{u^2}{2t}}\, du.$$

The probability that W_t is between the values a and b is given by

$$P(a \le W_t \le b) = \frac{1}{\sqrt{2\pi t}} \int_a^b e^{-\frac{u^2}{2t}}\, du, \qquad a < b.$$

Even if the increments of a Brownian motion are independent, their values are still correlated.

Proposition 3.1.7 *Let $0 \le s \le t$. Then*

1. $Cov(W_s, W_t) = s$;
2. $Corr(W_s, W_t) = \sqrt{\dfrac{s}{t}}$.

[1]These type of processes are called Markov processes.

Proof: 1. Using the properties of covariance

$$\begin{aligned}
Cov(W_s, W_t) &= Cov(W_s, W_s + W_t - W_s) \\
&= Cov(W_s, W_s) + Cov(W_s, W_t - W_s) \\
&= Var(W_s) + \mathbb{E}[W_s(W_t - W_s)] - \mathbb{E}[W_s]\mathbb{E}[W_t - W_s] \\
&= s + \mathbb{E}[W_s]\mathbb{E}[W_t - W_s] \\
&= s,
\end{aligned}$$

since $\mathbb{E}[W_s] = 0$.

We can also arrive at the same result starting from the formula

$$Cov(W_s, W_t) = \mathbb{E}[W_s W_t] - \mathbb{E}[W_s]\mathbb{E}[W_t] = \mathbb{E}[W_s W_t].$$

Using that conditional expectations have the same expectation, factoring out the predictable part, and using that W_t is a martingale, we have

$$\begin{aligned}
\mathbb{E}[W_s W_t] &= \mathbb{E}[\mathbb{E}[W_s W_t | \mathcal{F}_s]] = \mathbb{E}[W_s \mathbb{E}[W_t | \mathcal{F}_s]] \\
&= \mathbb{E}[W_s W_s] = \mathbb{E}[W_s^2] = s,
\end{aligned}$$

so $Cov(W_s, W_t) = s$.

2. The correlation formula yields

$$Corr(W_s, W_t) = \frac{Cov(W_s, W_t)}{\sigma(W_t)\sigma(W_s)} = \frac{s}{\sqrt{s}\sqrt{t}} = \sqrt{\frac{s}{t}}.$$

■

Remark 3.1.8 Removing the order relation between s and t, the previous relations can also be stated as

$$\begin{aligned}
Cov(W_s, W_t) &= \min\{s, t\}; \\
Corr(W_s, W_t) &= \sqrt{\frac{\min\{s, t\}}{\max\{s, t\}}}.
\end{aligned}$$

The following exercises state the translation and the scaling invariance properties of the Brownian motion.

Exercise 3.1.9 *For any $t_0 \geq 0$, show that the process $X_t = W_{t+t_0} - W_{t_0}$ is a Brownian motion. It can also be stated that the Brownian motion is translation invariant.*

Exercise 3.1.10 *For any $\lambda > 0$, show that the process $X_t = \frac{1}{\sqrt{\lambda}} W_{\lambda t}$ is a Brownian motion. This says that the Brownian motion is invariant by scaling.*

Exercise 3.1.11 *Let $0 < s < t < u$. Show the following multiplicative property*

$$Corr(W_s, W_t)Corr(W_t, W_u) = Corr(W_s, W_u).$$

Exercise 3.1.12 *Find the expectations $\mathbb{E}[W_t^3]$ and $\mathbb{E}[W_t^4]$.*

Exercise 3.1.13 *(a) Use the martingale property of $W_t^2 - t$ to find*

$$\mathbb{E}[(W_t^2 - t)(W_s^2 - s)];$$

(b) Evaluate $\mathbb{E}[W_t^2 W_s^2]$;
(c) Compute $Cov(W_t^2, W_s^2)$;
(d) Find $Corr(W_t^2, W_s^2)$.

Exercise 3.1.14 *Consider the process $Y_t = tW_{\frac{1}{t}}$, $t > 0$, and define $Y_0 = 0$.*
(a) Find the distribution of Y_t;
(b) Find the probability density of Y_t;
(c) Find $Cov(Y_s, Y_t)$;
(d) Find $\mathbb{E}[Y_t - Y_s]$ and $Var(Y_t - Y_s)$ for $s < t$.

It is worth noting that the process $Y_t = tW_{\frac{1}{t}}$, $t > 0$ with $Y_0 = 0$ is a Brownian motion, see Exercise 10.2.10 .

Exercise 3.1.15 *The process $X_t = |W_t|$ is called a Brownian motion reflected at the origin. Show that*
(a) $\mathbb{E}[|W_t|] = \sqrt{2t/\pi}$;
(b) $Var(|W_t|) = (1 - \frac{2}{\pi})t$.

Exercise 3.1.16 *Let $0 < s < t$. Find $\mathbb{E}[W_t^2 | \mathcal{F}_s]$.*

Exercise 3.1.17 *Let $0 < s < t$. Show that*
(a) $\mathbb{E}[W_t^3 | \mathcal{F}_s] = 3(t-s)W_s + W_s^3$;
(b) $\mathbb{E}[W_t^4 | \mathcal{F}_s] = 3(t-s)^2 + 6(t-s)W_s^2 + W_s^4$.

Exercise 3.1.18 *Show that $\mathbb{E}\Big[\int_s^t W_u \, du | \mathcal{F}_s\Big] = (t-s)W_s$.*

Exercise 3.1.19 *Show that the process*

$$X_t = W_t^3 - 3\int_0^t W_s \, ds$$

is a martingale with respect to the information set $\mathcal{F}_t = \sigma\{W_s; s \leq t\}$.

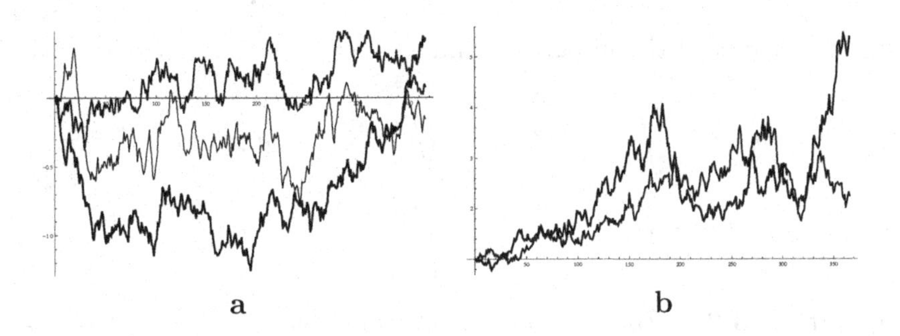

a b

Figure 3.1: (a) *Three simulations of the Brownian motion* W_t. (b) *Two simulations of the exponential Brownian motion* e^{W_t}.

Exercise 3.1.20 *Show that the following processes are Brownian motions*

(*a*) $X_t = W_T - W_{T-t}$, $0 \leq t \leq T$;

(*b*) $Y_t = -W_t$, $t \geq 0$.

Exercise 3.1.21 *Let* W_t *and* $\tilde{W}_t$ *be two independent Brownian motions and* ρ *be a constant with* $|\rho| \leq 1$.

(*a*) *Show that the process* $X_t = \rho W_t + \sqrt{1-\rho^2}\tilde{W}_t$ *is continuous and has the distribution* $N(0,t)$;

(*b*) *Is* X_t *a Brownian motion?*

Exercise 3.1.22 *Let* Y *be a random variable distributed as* $N(0,1)$. *Consider the process* $X_t = \sqrt{t}Y$. *Is* X_t *a Brownian motion?*

3.2 Geometric Brownian Motion

The *geometric Brownian motion* with drift μ and volatility σ is the process

$$X_t = e^{\sigma W_t + (\mu - \frac{\sigma^2}{2})t}, \qquad t \geq 0.$$

In the standard case, when $\mu = 0$ and $\sigma = 1$, the process becomes $X_t = e^{W_t - t/2}$, $t \geq 0$. This driftless process is always a martingale, see Exercise 3.2.4.

The following result will be useful later in the chapter.

Lemma 3.2.1 $\mathbb{E}[e^{\alpha W_t}] = e^{\alpha^2 t/2}$, *for* $\alpha \geq 0$.

Proof: Using the definition of expectation

$$\begin{aligned}
\mathbb{E}[e^{\alpha W_t}] &= \int e^{\alpha x}\phi_t(x)\,dx = \frac{1}{\sqrt{2\pi t}}\int e^{-\frac{x^2}{2t}+\alpha x}\,dx \\
&= e^{\alpha^2 t/2},
\end{aligned}$$

where we have used the integral formula

$$\int e^{-ax^2+bx}\,dx = \sqrt{\frac{\pi}{a}}e^{\frac{b^2}{4a}}, \qquad a > 0$$

with $a = \frac{1}{2t}$ and $b = \alpha$. ■

Proposition 3.2.2 *The exponential Brownian motion $X_t = e^{W_t}$ is log-normally distributed with mean $e^{t/2}$ and variance $e^{2t} - e^t$.*

Proof: Since W_t is normally distributed, then $X_t = e^{W_t}$ will have a log-normal distribution. Using Lemma 3.2.1 we have

$$\begin{aligned} \mathbb{E}[X_t] &= \mathbb{E}[e^{W_t}] = e^{t/2} \\ \mathbb{E}[X_t^2] &= \mathbb{E}[e^{2W_t}] = e^{2t}, \end{aligned}$$

and hence the variance is

$$Var[X_t] = \mathbb{E}[X_t^2] - \mathbb{E}[X_t]^2 = e^{2t} - (e^{t/2})^2 = e^{2t} - e^t.$$

■

A few simulations of the process X_t are contained in Fig. 3.1(b).

The distribution function of $X_t = e^{W_t}$ can be obtained by reducing it to the distribution function of a Brownian motion as in the following.

$$\begin{aligned} F_{X_t}(x) &= P(X_t \le x) = P(e^{W_t} \le x) \\ &= P(W_t \le \ln x) = F_{W_t}(\ln x) \\ &= \frac{1}{\sqrt{2\pi t}} \int_{-\infty}^{\ln x} e^{-\frac{u^2}{2t}}\,du. \end{aligned}$$

The density function of the geometric Brownian motion $X_t = e^{W_t}$ is given by

$$p(x) = \frac{d}{dx}F_{X_t}(x) = \begin{cases} \dfrac{1}{x\sqrt{2\pi t}}e^{-(\ln x)^2/(2t)}, & \text{if } x > 0, \\ 0, & \text{elsewhere.} \end{cases}$$

Exercise 3.2.3 *Show that*

$$\mathbb{E}[e^{W_t - W_s}] = e^{\frac{t-s}{2}}, \qquad s < t.$$

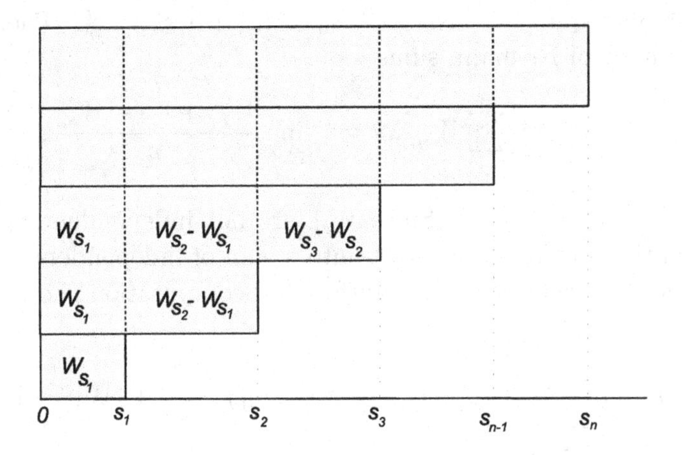

Figure 3.2: *To be used in the proof of formula (3.3.1); the area of the blocks can be counted in two equivalent ways, horizontally and vertically.*

Exercise 3.2.4 *Let* $X_t = e^{W_t}$.

(a) *Show that* X_t *is not a martingale.*

(b) *Show that* $e^{-\frac{t}{2}} X_t$ *is a martingale.*

(c) *Show that for any constant* $c \in \mathbb{R}$, *the process* $Y_t = e^{cW_t - \frac{1}{2}c^2 t}$ *is a martingale.*

Exercise 3.2.5 *If* $X_t = e^{W_t}$, *find* $Cov(X_s, X_t)$

(a) *by direct computation;*

(b) *by using Exercise* 3.2.4 (*b*).

Exercise 3.2.6 *Show that*

$$\mathbb{E}[e^{2W_t^2}] = \begin{cases} (1-4t)^{-1/2}, & 0 \le t < 1/4 \\ \infty, & \textit{otherwise.} \end{cases}$$

3.3 Integrated Brownian Motion

The stochastic process

$$Z_t = \int_0^t W_s \, ds, \qquad t \ge 0$$

is called the *integrated Brownian motion*. Obviously, $Z_0 = 0$.

Let $0 = s_0 < s_1 < \cdots < s_k < \cdots s_n = t$, with $s_k = \frac{kt}{n}$. Then Z_t can be written as a limit of Riemann sums

$$Z_t = \lim_{n\to\infty} \sum_{k=1}^{n} W_{s_k} \Delta s = t \lim_{n\to\infty} \frac{W_{s_1} + \cdots + W_{s_n}}{n},$$

where $\Delta s = s_{k+1} - s_k = \frac{t}{n}$. Since W_{s_k} are not independent, we first need to transform the previous expression into a sum of independent normally distributed random variables. A straightforward computation shows that

$$\begin{aligned} & W_{s_1} + \cdots + W_{s_n} \\ = \; & n(W_{s_1} - W_0) + (n-1)(W_{s_2} - W_{s_1}) + \cdots + (W_{s_n} - W_{s_{n-1}}) \\ = \; & X_1 + X_2 + \cdots + X_n. \end{aligned} \tag{3.3.1}$$

This formula becomes clear if one sums the area of the blocks in Fig. 3.2 horizontally and then vertically. Since the increments of a Brownian motion are independent and normally distributed, we have

$$\begin{aligned} X_1 &\sim N\big(0, n^2 \Delta s\big) \\ X_2 &\sim N\big(0, (n-1)^2 \Delta s\big) \\ X_3 &\sim N\big(0, (n-2)^2 \Delta s\big) \\ &\vdots \\ X_n &\sim N\big(0, \Delta s\big). \end{aligned}$$

Recall now the following well known theorem on the addition formula for Gaussian random variables.

Theorem 3.3.1 *If X_j are independent random variables normally distributed with mean μ_j and variance σ_j^2, then the sum $X_1 + \cdots + X_n$ is also normally distributed with mean $\mu_1 + \cdots + \mu_n$ and variance $\sigma_1^2 + \cdots + \sigma_n^2$.*

Then

$$X_1 + \cdots + X_n \sim N\Big(0, (1 + 2^2 + 3^2 + \cdots + n^2)\Delta s\Big) = N\Big(0, \frac{n(n+1)(2n+1)}{6} \Delta s\Big),$$

with $\Delta s = \dfrac{t}{n}$. Using (3.3.1) yields

$$t \frac{W_{s_1} + \cdots + W_{s_n}}{n} \sim N\Big(0, \frac{(n+1)(2n+1)}{6n^2} t^3\Big).$$

"Taking the limit" with $n \to \infty$, we get

$$Z_t \sim N\Big(0, \frac{t^3}{3}\Big).$$

Proposition 3.3.2 *The integrated Brownian motion Z_t has a normal distribution with mean 0 and variance $t^3/3$.*

Remark 3.3.3 The aforementioned limit was taken heuristically, without specifying the type of the convergence. In order to make this work, the following result is usually used:

If X_n is a sequence of normal random variables that converges in mean square to X, then the limit X is normally distributed, with $\mathbb{E}[X_n] \to \mathbb{E}[X]$ and $Var(X_n) \to Var(X)$, as $n \to \infty$.

The mean and the variance can also be computed in a direct way as follows. By Fubini's theorem we have

$$\begin{aligned}\mathbb{E}[Z_t] &= \mathbb{E}[\int_0^t W_s\,ds] = \int_{\mathbb{R}}\int_0^t W_s\,ds\,dP \\ &= \int_0^t\int_{\mathbb{R}} W_s\,dP\,ds = \int_0^t \mathbb{E}[W_s]\,ds = 0,\end{aligned}$$

since $\mathbb{E}[W_s] = 0$. Then the variance is given by

$$\begin{aligned}Var[Z_t] &= \mathbb{E}[Z_t^2] - \mathbb{E}[Z_t]^2 = \mathbb{E}[Z_t^2] \\ &= \mathbb{E}\Big[\int_0^t W_u\,du \cdot \int_0^t W_v\,dv\Big] = \mathbb{E}\Big[\int_0^t\int_0^t W_u W_v\,dudv\Big] \\ &= \int_0^t\int_0^t \mathbb{E}[W_u W_v]\,dudv = \iint_{[0,t]\times[0,t]} \min\{u,v\}\,dudv \\ &= \iint_{D_1} \min\{u,v\}\,dudv + \iint_{D_2} \min\{u,v\}\,dudv, \end{aligned} \tag{3.3.2}$$

where

$$D_1 = \{(u,v); u > v, 0 \le u \le t\}, \qquad D_2 = \{(u,v); u < v, 0 \le u \le t\}.$$

The first integral can be evaluated using Fubini's theorem

$$\begin{aligned}\iint_{D_1} \min\{u,v\}\,dudv &= \iint_{D_1} v\,dudv \\ &= \int_0^t \Big(\int_0^u v\,dv\Big)\,du = \int_0^t \frac{u^2}{2}\,du = \frac{t^3}{6}.\end{aligned}$$

Similarly, the latter integral is equal to

$$\iint_{D_2} \min\{u,v\}\,dudv = \frac{t^3}{6}.$$

Substituting in (3.3.2) yields

$$Var[Z_t] = \frac{t^3}{6} + \frac{t^3}{6} = \frac{t^3}{3}.$$

For another computation of the variance of Z_t, see Exercise 5.6.2.

Exercise 3.3.4 *(a) Prove that the moment generating function of Z_t is given by*

$$m(u) = e^{u^2 t^3/6}.$$

(b) Use the first part to find the mean and variance of Z_t.

Exercise 3.3.5 *Let $s < t$. Show that the covariance of the integrated Brownian motion is given by*

$$Cov\Big(Z_s, Z_t\Big) = s^2\Big(\frac{t}{2} - \frac{s}{6}\Big), \qquad s < t.$$

Exercise 3.3.6 *Show that*

(a) $Cov(Z_t, Z_t - Z_{t-h}) = \frac{1}{2}t^2 h + o(h)$, where $o(h)$ denotes a quantity such that $\lim_{h\to 0} o(h)/h = 0$;

(b) $Cov(Z_t, W_t) = \dfrac{t^2}{2}$.

Exercise 3.3.7 *Show that*

$$\mathbb{E}[e^{W_s + W_u}] = e^{\frac{u+s}{2}} e^{\min\{s,u\}}.$$

Exercise 3.3.8 *Consider the process $X_t = \displaystyle\int_0^t e^{W_s}\,ds$.*

(a) Find the mean of X_t;
(b) Find the variance of X_t.

In the next exercises $\mathcal{F}_t$ denotes the σ-field generated by the Brownian motion W_t.

Exercise 3.3.9 *Consider the process $Z_t = \displaystyle\int_0^t W_u\,du$, $t > 0$.*

(a) Show that $\mathbb{E}[Z_T|\mathcal{F}_t] = Z_t + W_t(T-t)$, for any $t < T$;
(b) Prove that the process $M_t = Z_t - tW_t$ is an $\mathcal{F}_t$-martingale.

Exercise 3.3.10 *Let $Y_t = \displaystyle\int_0^t W_u^2\,du$, $t > 0$.*

(a) Show that $\mathbb{E}\Big[\displaystyle\int_t^T W_s^2\,ds|\mathcal{F}_t\Big] = W_t^2(T-t) + \frac{1}{2}(T-t)^2$, for any $t < T$;
(b) Prove that the process $M_t = Y_t - tW_t^2 + t^2/2$ is an $\mathcal{F}_t$-martingale.

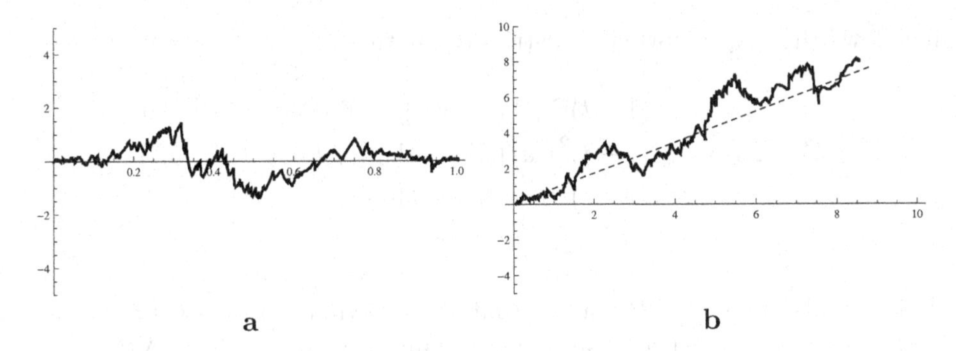

Figure 3.3: (a) *Brownian bridge pinned down at* 0 *and* 1. (b) *Brownian motion with drift* $X_t = \mu t + W_t$, *with positive drift* $\mu > 0$.

3.4 Exponential Integrated Brownian Motion

If $Z_t = \int_0^t W_s\,ds$ denotes the integrated Brownian motion, the process

$$V_t = e^{Z_t}$$

is called *exponential integrated Brownian motion.* The process starts at $V_0 = e^0 = 1$. Since Z_t is normally distributed, then V_t is log-normally distributed. We compute the mean and the variance in a direct way. Using Exercises 3.2.5 and 3.3.4 we have

$$\begin{aligned} \mathbb{E}[V_t] &= \mathbb{E}[e^{Z_t}] = m(1) = e^{\frac{t^3}{6}} \\ \mathbb{E}[V_t^2] &= \mathbb{E}[e^{2Z_t}] = m(2) = e^{\frac{4t^3}{6}} = e^{\frac{2t^3}{3}} \\ Var(V_t) &= \mathbb{E}[V_t^2] - \mathbb{E}[V_t]^2 = e^{\frac{2t^3}{3}} - e^{\frac{t^3}{3}} \\ Cov(V_s, V_t) &= e^{\frac{t+3s}{2}}. \end{aligned}$$

Exercise 3.4.1 *Show that* $\mathbb{E}[V_T|\mathcal{F}_t] = V_t e^{(T-t)W_t + \frac{(T-t)^3}{3}}$ *for* $t < T$.

3.5 Brownian Bridge

The process $X_t = W_t - tW_1$ is called the *Brownian bridge* fixed at both 0 and 1, see Fig. 3.3(a). Since we can also write

$$\begin{aligned} X_t &= W_t - tW_t - tW_1 + tW_t \\ &= (1-t)(W_t - W_0) - t(W_1 - W_t), \end{aligned}$$

using that the increments $W_t - W_0$ and $W_1 - W_t$ are independent and normally distributed, with

$$W_t - W_0 \sim N(0, t), \qquad W_1 - W_t \sim N(0, 1-t),$$

it follows that X_t is normally distributed with

$$\begin{aligned}
\mathbb{E}[X_t] &= (1-t)\mathbb{E}[(W_t - W_0)] - t\mathbb{E}[(W_1 - W_t)] = 0 \\
Var[X_t] &= (1-t)^2 Var[(W_t - W_0)] + t^2 Var[(W_1 - W_t)] \\
&= (1-t)^2(t-0) + t^2(1-t) \\
&= t(1-t).
\end{aligned}$$

This can also be stated by saying that the Brownian bridge tied at 0 and 1 is a Gaussian process with mean 0 and variance $t(1-t)$, so $X_t \sim N\big(0, t(1-t)\big)$.

Exercise 3.5.1 *Let $X_t = W_t - tW_1$, $0 \le t \le 1$ be a Brownian bridge fixed at 0 and 1. Let $Y_t = X_t^2$. Show that $Y_0 = Y_1 = 0$ and find $\mathbb{E}[Y_t]$ and $Var(Y_t)$.*

3.6 Brownian Motion with Drift

The process $Y_t = \mu t + W_t$, $t \ge 0$, is called *Brownian motion with drift*, see Fig. 3.3(b). The process Y_t tends to drift off at a rate μ. It starts at $Y_0 = 0$ and it is a Gaussian process with mean

$$\mathbb{E}[Y_t] = \mu t + \mathbb{E}[W_t] = \mu t$$

and variance

$$Var[Y_t] = Var[\mu t + W_t] = Var[W_t] = t.$$

Exercise 3.6.1 *Find the distribution and the density functions of the process Y_t.*

3.7 Bessel Process

This section deals with the process satisfied by the Euclidean distance from the origin to a particle following a Brownian motion in $\mathbb{R}^n$. More precisely, if $W_1(t), \cdots, W_n(t)$ are independent Brownian motions, consider the n-dimensional Brownian motion $W(t) = (W_1(t), \cdots, W_n(t))$, $n \ge 2$. The process

$$R_t = dist(O, W(t)) = \sqrt{W_1(t)^2 + \cdots + W_n(t)^2}$$

is called the *n-dimensional Bessel process*, see Fig. 3.4.

The probability density of this process is given by the following result.

Proposition 3.7.1 *The probability density function of R_t, $t > 0$ is given by*

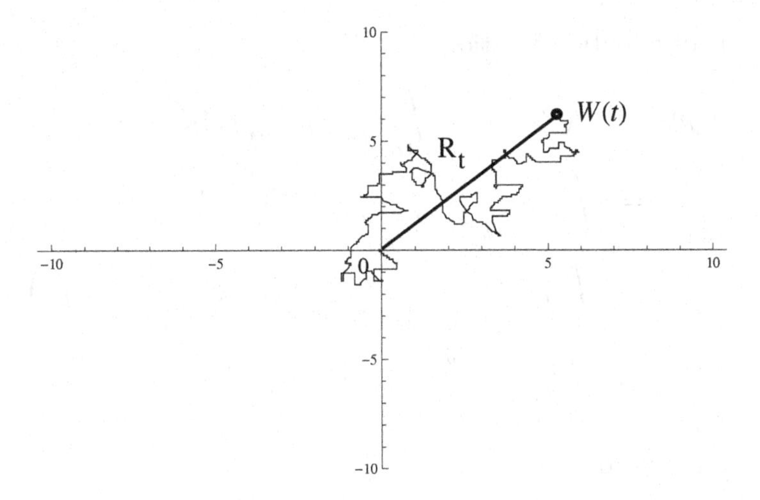

Figure 3.4: *The Bessel process* $R_t = |W(t)|$ *for* $n = 2$.

$$p_t(\rho) = \begin{cases} \dfrac{2}{(2t)^{n/2}\,\Gamma(n/2)}\,\rho^{n-1}e^{-\frac{\rho^2}{2t}}, & \rho \geq 0; \\ 0, & \rho < 0 \end{cases}$$

with

$$\Gamma\Big(\frac{n}{2}\Big) = \begin{cases} (\frac{n}{2}-1)! & \textit{for } n \textit{ even;} \\ (\frac{n}{2}-1)(\frac{n}{2}-2)\cdots\frac{3}{2}\frac{1}{2}\sqrt{\pi}, & \textit{for } n \textit{ odd.} \end{cases}$$

Proof: Since the Brownian motions $W_1(t), \ldots, W_n(t)$ are independent, their joint density function is

$$f_{W_1\cdots W_n}(x) = f_{W_1}(x)\cdots f_{W_n}(x) = \frac{1}{(2\pi t)^{n/2}}\,e^{-(x_1^2+\cdots+x_n^2)/(2t)}, \qquad t > 0.$$

In the next computation we shall use the following formula of integration that follows from the use of polar coordinates

$$\int_{\{|x|\leq\rho\}} f(x)\,dx = \sigma(\mathbb{S}^{n-1})\int_0^\rho r^{n-1}g(r)\,dr, \tag{3.7.3}$$

where $f(x) = g(|x|)$ is a function on $\mathbb{R}^n$ with spherical symmetry, and where $\sigma(\mathbb{S}^{n-1}) = \dfrac{2\pi^{n/2}}{\Gamma(n/2)}$ is the area of the $(n-1)$-dimensional sphere in $\mathbb{R}^n$. Let

$\rho \geq 0$. The distribution function of R_t is

$$
\begin{aligned}
F_R(\rho) &= P(R_t \leq \rho) = \int_{\{R_t \leq \rho\}} f_{W_1 \cdots W_n}(x)\, dx_1 \cdots dx_n \\
&= \int_{x_1^2+\cdots+x_n^2 \leq \rho^2} \frac{1}{(2\pi t)^{n/2}} e^{-(x_1^2+\cdots+x_n^2)/(2t)}\, dx_1 \cdots dx_n \\
&= \int_0^\rho r^{n-1} \left(\int_{\mathbb{S}(0,1)} \frac{1}{(2\pi t)^{n/2}} e^{-(x_1^2+\cdots+x_n^2)/(2t)}\, d\sigma \right) dr \\
&= \frac{\sigma(\mathbb{S}^{n-1})}{(2\pi t)^{n/2}} \int_0^\rho r^{n-1} e^{-r^2/(2t)}\, dr.
\end{aligned}
$$

Differentiating yields

$$
\begin{aligned}
p_t(\rho) &= \frac{d}{d\rho} F_R(\rho) = \frac{\sigma(\mathbb{S}^{n-1})}{(2\pi t)^{n/2}} \rho^{n-1} e^{-\frac{\rho^2}{2t}} \\
&= \frac{2}{(2t)^{n/2}\Gamma(n/2)} \rho^{n-1} e^{-\frac{\rho^2}{2t}}, \qquad \rho > 0, t > 0.
\end{aligned}
$$

■

It is worth noting that in the 2-dimensional case the aforementioned density becomes a particular case of a Weibull distribution with parameters $m = 2$ and $\alpha = 2t$, called *Wald's distribution*

$$
\boxed{p_t(x) = \frac{1}{t} x e^{-\frac{x^2}{2t}}, \qquad x > 0, t > 0.}
$$

Exercise 3.7.2 *Let $P(R_t \leq t)$ be the probability of a 2-dimensional Brownian motion being inside of the disk $D(0, \rho)$ at time $t > 0$. Show that*

$$
\frac{\rho^2}{2t}\left(1 - \frac{\rho^2}{4t}\right) < P(R_t \leq t) < \frac{\rho^2}{2t}.
$$

Exercise 3.7.3 *Let R_t be a 2-dimensional Bessel process. Show that*

(*a*) $\mathbb{E}[R_t] = \sqrt{2\pi t}/2$;

(*b*) $Var(R_t) = 2t\left(1 - \frac{\pi}{4}\right)$.

Exercise 3.7.4 *Let $X_t = \dfrac{R_t}{t}$, $t > 0$, where R_t is a 2-dimensional Bessel process. Show that $X_t \to 0$ as $t \to \infty$ in mean square.*

3.8 The Poisson Process

A *Poisson process* describes the number of occurrences of a certain event before time t, such as: the number of electrons arriving at an anode until time t; the number of cars arriving at a gas station until time t; the number of phone calls received on a certain day until time t; the number of visitors entering a museum on a certain day until time t; the number of earthquakes that occurred in Chile during the time interval $[0, t]$; the number of shocks in the stock market from the beginning of the year until time t; the number of twisters that might hit Alabama from the beginning of the century until time t.

The definition of a Poisson process is stated more precisely in the following. Its graph looks like a stair-type function with unit jumps, see Fig. 3.5.

Definition 3.8.1 *A Poisson process is a stochastic process N_t, $t \geq 0$, which satisfies*

1. The process starts at the origin, $N_0 = 0$;

2. N_t has independent increments;

3. The process N_t is right continuous in t, with left hand limits;

4. The increments $N_t - N_s$, with $0 < s < t$, have a Poisson distribution with parameter $\lambda(t-s)$, i.e.

$$P(N_t - N_s = k) = \frac{\lambda^k (t-s)^k}{k!} e^{-\lambda(t-s)}.$$

It can be shown that condition *4* in the previous definition can be replaced by the following two conditions:

$$P(N_t - N_s = 1) = \lambda(t-s) + o(t-s) \tag{3.8.4}$$

$$P(N_t - N_s \geq 2) = o(t-s), \tag{3.8.5}$$

where $o(h)$ denotes a quantity such that $\lim_{h \to 0} o(h)/h = 0$. Then the probability that a jump of size 1 occurs in the infinitesimal interval dt is equal to λdt, and the probability that at least 2 events occur in the same small interval is zero. This implies that the random variable dN_t may take only two values, 0 and 1, and hence satisfies

$$P(dN_t = 1) = \lambda\, dt \tag{3.8.6}$$

$$P(dN_t = 0) = 1 - \lambda\, dt. \tag{3.8.7}$$

Exercise 3.8.2 *Show that if condition* 4 *is satisfied, then conditions* (3.8.4) *and* (3.8.5) *hold.*

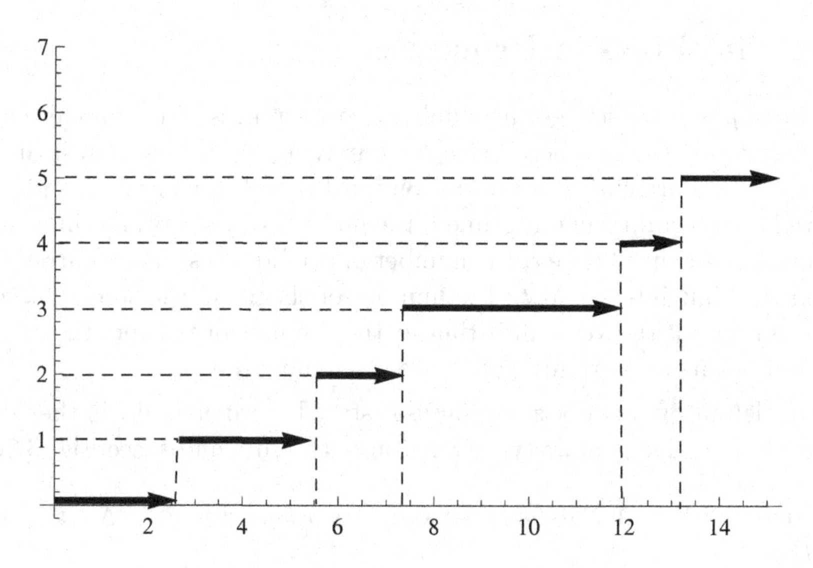

Figure 3.5: *The Poisson process N_t.*

Exercise 3.8.3 *Which of the following expressions are $o(h)$?*

(a) $f(h) = 3h^2 + h$;

(b) $f(h) = \sqrt{h} + 5$;

(c) $f(h) = h \ln |h|$;

(d) $f(h) = he^h$.

Condition *4* also states that N_t has stationary increments. The fact that $N_t - N_s$ is stationary can be stated as

$$P(N_{t+s} - N_s \leq n) = P(N_t - N_0 \leq n) = P(N_t \leq n) = \sum_{k=0}^{n} \frac{(\lambda t)^k}{k!} e^{-\lambda t}.$$

From condition *4* we get the mean and variance of increments

$$\mathbb{E}[N_t - N_s] = \lambda(t - s), \qquad Var[N_t - N_s] = \lambda(t - s).$$

In particular, the random variable N_t is Poisson distributed with $\mathbb{E}[N_t] = \lambda t$ and $Var[N_t] = \lambda t$. The parameter λ is called the *rate* of the process. This means that the events occur at the constant rate λ, with $\lambda > 0$.

Since the increments are independent, we have for $0 < s < t$

$$\begin{aligned} \mathbb{E}[N_s N_t] &= \mathbb{E}[(N_s - N_0)(N_t - N_s) + N_s^2] \\ &= \mathbb{E}[N_s - N_0]\mathbb{E}[N_t - N_s] + \mathbb{E}[N_s^2] \\ &= \lambda s \cdot \lambda(t - s) + (Var[N_s] + \mathbb{E}[N_s]^2) \\ &= \lambda^2 st + \lambda s. \end{aligned} \tag{3.8.8}$$

As a consequence we have the following result:

Proposition 3.8.4 *Let $0 \leq s \leq t$. Then*

1. $Cov(N_s, N_t) = \lambda s$;
2. $Corr(N_s, N_t) = \sqrt{\dfrac{s}{t}}$.

Proof: 1. Using (3.8.8) we have

$$\begin{aligned} Cov(N_s, N_t) &= \mathbb{E}[N_s N_t] - \mathbb{E}[N_s]\mathbb{E}[N_t] \\ &= \lambda^2 st + \lambda s - \lambda s \lambda t \\ &= \lambda s. \end{aligned}$$

2. Using the formula for the correlation yields

$$Corr(N_s, N_t) = \frac{Cov(N_s, N_t)}{(Var[N_s]Var[N_t])^{1/2}} = \frac{\lambda s}{(\lambda s \lambda t)^{1/2}} = \sqrt{\frac{s}{t}}.$$

■

It worth noting the similarity with Proposition 3.1.7.

Proposition 3.8.5 *Let N_t be $\mathcal{F}_t$-adapted. Then the process $M_t = N_t - \lambda t$ is an $\mathcal{F}_t$-martingale.*

Proof: Let $s < t$ and write $N_t = N_s + (N_t - N_s)$. Then

$$\begin{aligned} \mathbb{E}[N_t|\mathcal{F}_s] &= \mathbb{E}[N_s + (N_t - N_s)|\mathcal{F}_s] \\ &= \mathbb{E}[N_s|\mathcal{F}_s] + \mathbb{E}[N_t - N_s|\mathcal{F}_s] \\ &= N_s + \mathbb{E}[N_t - N_s] \\ &= N_s + \lambda(t - s), \end{aligned}$$

where we used that N_s is $\mathcal{F}_s$-measurable (and hence $\mathbb{E}[N_s|\mathcal{F}_s] = N_s$) and that the increment $N_t - N_s$ is independent of previous values of N_s and the information set $\mathcal{F}_s$. Subtracting λt yields

$$\mathbb{E}[N_t - \lambda t|\mathcal{F}_s] = N_s - \lambda s,$$

or $\mathbb{E}[M_t|\mathcal{F}_s] = M_s$. Since it is obvious that M_t is integrable and $\mathcal{F}_t$-adapted, it follows that M_t is a martingale. ■

It is worth noting that the Poisson process N_t is not a martingale. The martingale process $M_t = N_t - \lambda t$ is called the *compensated Poisson process.*

Exercise 3.8.6 *Compute $\mathbb{E}[N_t^2|\mathcal{F}_s]$ for $s < t$. Is the process N_t^2 an $\mathcal{F}_s$-martingale?*

Exercise 3.8.7 *(a) Show that the moment generating function of the random variable N_t is*

$$m_{N_t}(x) = e^{\lambda t(e^x - 1)}.$$

(b) Deduce the expressions for the first few moments

$$\begin{aligned}
\mathbb{E}[N_t] &= \lambda t \\
\mathbb{E}[N_t^2] &= \lambda^2 t^2 + \lambda t \\
\mathbb{E}[N_t^3] &= \lambda^3 t^3 + 3\lambda^2 t^2 + \lambda t \\
\mathbb{E}[N_t^4] &= \lambda^4 t^4 + 6\lambda^3 t^3 + 7\lambda^2 t^2 + \lambda t.
\end{aligned}$$

(c) Show that the first few central moments are given by

$$\begin{aligned}
\mathbb{E}[N_t - \lambda t] &= 0 \\
\mathbb{E}[(N_t - \lambda t)^2] &= \lambda t \\
\mathbb{E}[(N_t - \lambda t)^3] &= \lambda t \\
\mathbb{E}[(N_t - \lambda t)^4] &= 3\lambda^2 t^2 + \lambda t.
\end{aligned}$$

Exercise 3.8.8 *Find the mean and variance of the process $X_t = e^{N_t}$.*

Exercise 3.8.9 *(a) Show that the moment generating function of the random variable M_t is*

$$m_{M_t}(x) = e^{\lambda t(e^x - x - 1)}.$$

(b) Let $s < t$. Verify that

$$\begin{aligned}
\mathbb{E}[M_t - M_s] &= 0, \\
\mathbb{E}[(M_t - M_s)^2] &= \lambda(t - s), \\
\mathbb{E}[(M_t - M_s)^3] &= \lambda(t - s), \\
\mathbb{E}[(M_t - M_s)^4] &= \lambda(t - s) + 3\lambda^2(t - s)^2.
\end{aligned}$$

Exercise 3.8.10 *Let $s < t$. Show that*

$$Var[(M_t - M_s)^2] = \lambda(t - s) + 2\lambda^2(t - s)^2.$$

3.9 Interarrival Times

For each state of the world, ω, the path $t \to N_t(\omega)$ is a step function that exhibits unit jumps. Each jump in the path corresponds to an occurrence of a new event. Let T_1 be the random variable which describes the time of the 1st jump. Let T_2 be the time between the 1st jump and the second one. In general, denote by T_n the time elapsed between the $(n-1)$th and nth jumps. The random variables T_n are called *interarrival times*.

Proposition 3.9.1 *The random variables T_n are independent and exponentially distributed with mean $\mathbb{E}[T_n] = 1/\lambda$.*

Proof: We start by noticing that the events $\{T_1 > t\}$ and $\{N_t = 0\}$ are the same, since both describe the situation that no events occurred until time t. Then

$$P(T_1 > t) = P(N_t = 0) = P(N_t - N_0 = 0) = e^{-\lambda t},$$

and hence the distribution function of T_1 is

$$F_{T_1}(t) = P(T_1 \leq t) = 1 - P(T_1 > t) = 1 - e^{-\lambda t}.$$

Differentiating yields the density function

$$f_{T_1}(t) = \frac{d}{dt} F_{T_1}(t) = \lambda e^{-\lambda t}.$$

It follows that T_1 is has an exponential distribution, with $\mathbb{E}[T_1] = 1/\lambda$.
In order to show that the random variables T_1 and T_2 are independent, it suffices to show that

$$P(T_2 \leq t) = P(T_2 \leq t | T_1 = s),$$

i.e. the distribution function of T_2 is independent of the values of T_1. We note first that from the independent increments property

$$\begin{aligned} &P\big(0 \text{ jumps in } (s, s+t], 1 \text{ jump in } (0, s]\big) = P(N_{s+t} - N_s = 0, N_s - N_0 = 1) \\ &= P(N_{s+t} - N_s = 0) P(N_s - N_0 = 1) \\ &= P\big(0 \text{ jumps in } (s, s+t]\big) P\big(1 \text{ jump in } (0, s]\big). \end{aligned}$$

Then the conditional distribution of T_2 is

$$\begin{aligned} F(t|s) &= P(T_2 \leq t | T_1 = s) = 1 - P(T_2 > t | T_1 = s) \\ &= 1 - \frac{P(T_2 > t, T_1 = s)}{P(T_1 = s)} \\ &= 1 - \frac{P\big(0 \text{ jumps in } (s, s+t], 1 \text{ jump in } (0, s]\big)}{P(T_1 = s)} \\ &= 1 - \frac{P\big(0 \text{ jumps in } (s, s+t]\big) P\big(1 \text{ jump in } (0, s]\big)}{P\big(1 \text{ jump in } (0, s]\big)} \\ &= 1 - P\big(0 \text{ jumps in } (s, s+t]\big) \\ &= 1 - P(N_{s+t} - N_s = 0) = 1 - e^{-\lambda t}, \end{aligned}$$

which is independent of s. Then T_2 is independent of T_1 and exponentially distributed. A similar argument for any T_n leads to the desired result.

■

3.10 Waiting Times

The random variable $S_n = T_1 + T_2 + \cdots + T_n$ is called the *waiting time until the nth jump.* The event $\{S_n \leq t\}$ means that there are n jumps that occurred before or at time t, i.e. there are at least n events that happened up to time t; the event is equal to $\{N_t \geq n\}$. Hence the distribution function of S_n is given by

$$F_{S_n}(t) = P(S_n \leq t) = P(N_t \geq n) = e^{-\lambda t} \sum_{k=n}^{\infty} \frac{(\lambda t)^k}{k!}.$$

Differentiating we obtain the density function of the waiting time S_n

$$f_{S_n}(t) = \frac{d}{dt} F_{S_n}(t) = \frac{\lambda e^{-\lambda t} (\lambda t)^{n-1}}{(n-1)!}.$$

Writing

$$f_{S_n}(t) = \frac{t^{n-1} e^{-\lambda t}}{(1/\lambda)^n \Gamma(n)},$$

it turns out that S_n has a gamma distribution with parameters $\alpha = n$ and $\beta = 1/\lambda$. It follows that

$$\mathbb{E}[S_n] = \frac{n}{\lambda}, \quad Var[S_n] = \frac{n}{\lambda^2}.$$

The relation $\lim_{n \to \infty} \mathbb{E}[S_n] = \infty$ states that the expectation of the waiting time is unbounded as $n \to \infty$.

Exercise 3.10.1 *Prove that* $\frac{d}{dt} F_{S_n}(t) = \frac{\lambda e^{-\lambda t} (\lambda t)^{n-1}}{(n-1)!}$.

Exercise 3.10.2 *Using that the interarrival times $T_1, T_2, \cdots$ are independent and exponentially distributed, compute directly the mean $\mathbb{E}[S_n]$ and variance $Var(S_n)$.*

3.11 The Integrated Poisson Process

The function $u \to N_u$ is continuous with the exception of a set of countable jumps of size 1. It is known that such functions are Riemann integrable, so it makes sense to define the process

$$U_t = \int_0^t N_u \, du,$$

called the *integrated Poisson process.* The next result provides a relation between the process U_t and the partial sum of the waiting times S_k.

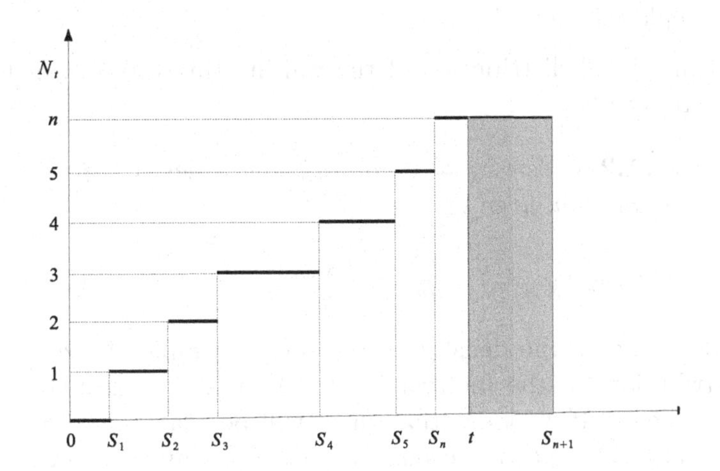

Figure 3.6: *The Poisson process N_t and the waiting times $S_1, S_2, \cdots S_n$. The area of the shaded rectangle is $n(S_{n+1} - t)$.*

Proposition 3.11.1 *The integrated Poisson process can be expressed as*

$$U_t = tN_t - \sum_{k=1}^{N_t} S_k.$$

Let $N_t = n$. Since N_u is equal to k between the waiting times S_k and S_{k+1}, the process U_t, which is equal to the area of the subgraph of N_u between 0 and t, can be expressed as

$$U_t = \int_0^t N_u \, du = 1 \cdot (S_2 - S_1) + 2 \cdot (S_3 - S_2) + \cdots + n(S_{n+1} - S_n) - n(S_{n+1} - t).$$

Since $S_n < t < S_{n+1}$, the difference of the last two terms represents the area of the last rectangle, which has the length $t - S_n$ and the height n. Using associativity, a computation yields

$$1 \cdot (S_2 - S_1) + 2 \cdot (S_3 - S_2) + \cdots + n(S_{n+1} - S_n) = nS_{n+1} - (S_1 + S_2 + \cdots + S_n).$$

Substituting in the aforementioned relation, we get

$$\begin{aligned} U_t &= nS_{n+1} - (S_1 + S_2 + \cdots + S_n) - n(S_{n+1} - t) \\ &= nt - (S_1 + S_2 + \cdots + S_n) \\ &= tN_t - \sum_{k=1}^{N_t} S_k, \end{aligned}$$

where we replaced n by N_t.

The conditional distribution of the waiting times is provided by the following useful result.

Theorem 3.11.2 *Given $N_t = n$, the waiting times $S_1, S_2, \cdots, S_n$ have the joint density function given by*

$$f(s_1, s_2, \cdots, s_n) = \frac{n!}{t^n}, \qquad 0 < s_1 \leq s_2 \leq \cdots \leq s_n < t.$$

This is the same as the density of an ordered sample of size n from a uniform distribution on the interval $(0, t)$. A "naive" explanation of this result is as follows. If we know that there will be exactly n events during the time interval $(0, t)$, since the events can occur at any time, each of them can be considered uniformly distributed, with the density $f(s_k) = 1/t$. Since it makes sense to consider the events independent, taking into consideration all $n!$ possible permutations, the joint density function becomes $f(s_1, \cdots, s_n) = n! f(s_1) \cdots f(s_n) = \dfrac{n!}{t^n}$.

Exercise 3.11.3 *Find the following means*

(a) $\mathbb{E}[U_t]$.

(b) $\mathbb{E}\big[\sum_{k=1}^{N_t} S_k\big]$.

Exercise 3.11.4 *Show that*

(a) $S_1 + S_2 + \cdots + S_n = nT_1 + (n-1)T_2 + \cdots 2T_{n-1} + T_n$;

(b) $\mathbb{E}\Big[\sum_{k=1}^{N_t} S_k | N_t = n\Big] = \dfrac{n(n+1)}{2\lambda}$;

(c) $\mathbb{E}\Big[U_t | N_t = n\Big] = n\Big(t - \dfrac{n+1}{2\lambda}\Big)$.

Exercise 3.11.5 *Show that $Var(U_t) = \dfrac{\lambda t^3}{3}$.*

Exercise 3.11.6 *Can you apply a similar proof as in Proposition 3.3.2 to show that the integrated Poisson process U_t is also a Poisson process?*

Exercise 3.11.7 *Let $Y : \Omega \to \mathbb{N}$ be a discrete random variable. Show that for any random variable X we have*

$$\mathbb{E}[X] = \sum_{y \geq 0} \mathbb{E}[X|Y = y] P(Y = y).$$

Exercise 3.11.8 *Use Exercise 3.11.7 to solve Exercise 3.11.3 (b).*

Exercise 3.11.9 *(a) Let T_k be the kth interarrival time. Show that*

$$\mathbb{E}[e^{-\sigma T_k}] = \frac{\lambda}{\lambda + \sigma}, \qquad \sigma > 0.$$

(b) Let $n = N_t$. Show that

$$U_t = nt - [nT_1 + (n-1)T_2 + \cdots + 2T_{n-1} + T_n].$$

(c) Find the conditional expectation

$$\mathbb{E}\Big[e^{-\sigma U_t}\Big| N_t = n\Big].$$

(Hint: If we know that there are exactly n jumps in the interval $[0, T]$, it makes sense to consider the arrival time of the jumps T_i independent and uniformly distributed on $[0, T]$).
(d) Find the expectation

$$\mathbb{E}\Big[e^{-\sigma U_t}\Big].$$

3.12 Submartingales

A stochastic process X_t on the probability space $(\Omega, \mathcal{F}, P)$ is called a *submartingale* with respect to the filtration $\mathcal{F}_t$ if:

(a) $\int_\Omega |X_t|\, dP < \infty$ (X_t integrable);

(b) X_t is known if $\mathcal{F}_t$ is given (X_t is adaptable to $\mathcal{F}_t$);

(c) $\mathbb{E}[X_{t+s}|\mathcal{F}_t] \geq X_t, \forall t, s \geq 0$ (future predictions exceed the present value).

Example 3.12.1 *We shall prove that the process $X_t = \mu t + \sigma W_t$, with $\mu > 0$ is a submartingale.*

The integrability follows from the inequality $|X_t(\omega)| \leq \mu t + |W_t(\omega)|$ and integrability of W_t. The adaptability of X_t is obvious, and the last property follows from the computation:

$$\begin{aligned}\mathbb{E}[X_{t+s}|\mathcal{F}_t] &= \mathbb{E}[\mu t + \sigma W_{t+s}|\mathcal{F}_t] + \mu s > \mathbb{E}[\mu t + \sigma W_{t+s}|\mathcal{F}_t] \\ &= \mu t + \sigma \mathbb{E}[W_{t+s}|\mathcal{F}_t] = \mu t + \sigma W_t = X_t,\end{aligned}$$

where we used that W_t is a martingale.

Example 3.12.2 *We shall show that the square of the Brownian motion, W_t^2, is a submartingale.*

Using that $W_t^2 - t$ is a martingale, we have

$$\begin{aligned}\mathbb{E}[W_{t+s}^2|\mathcal{F}_t] &= \mathbb{E}[W_{t+s}^2 - (t+s)|\mathcal{F}_t] + t + s = W_t^2 - t + t + s \\ &= W_t^2 + s \geq W_t^2.\end{aligned}$$

The following result supplies examples of submartingales starting from martingales or submartingales.

Proposition 3.12.3 (*a*) *If X_t is a martingale and ϕ a convex function such that $\phi(X_t)$ is integrable, then the process $Y_t = \phi(X_t)$ is a submartingale.*
(*b*) *If X_t is a submartingale and ϕ an increasing convex function such that $\phi(X_t)$ is integrable, then the process $Y_t = \phi(X_t)$ is a submartingale.*
(*c*) *If X_t is a martingale and $f(t)$ is an increasing, integrable function, then $Y_t = X_t + f(t)$ is a submartingale.*

Proof: (*a*) Using Jensen's inequality for conditional probabilities, Exercise 2.13.7, we have

$$\mathbb{E}[Y_{t+s}|\mathcal{F}_t] = \mathbb{E}[\phi(X_{t+s})|\mathcal{F}_t] \geq \phi\Big(\mathbb{E}[X_{t+s}|\mathcal{F}_t]\Big) = \phi(X_t) = Y_t.$$

(*b*) From the submartingale property and monotonicity of ϕ we have

$$\phi\Big(\mathbb{E}[X_{t+s}|\mathcal{F}_t]\Big) \geq \phi(X_t).$$

Then apply a similar computation as in part (*a*).
(*c*) We shall check only the forecast property, since the other properties are obvious.

$$\begin{aligned}\mathbb{E}[Y_{t+s}|\mathcal{F}_t] &= \mathbb{E}[X_{t+s} + f(t+s)|\mathcal{F}_t] = \mathbb{E}[X_{t+s}|\mathcal{F}_t] + f(t+s) \\ &= X_t + f(t+s) \geq X_t + f(t) = Y_t, \qquad \forall s,t>0.\end{aligned}$$

■

Corollary 3.12.4 (*a*) *Let X_t be a right continuous martingale. Then X_t^2, $|X_t|$, e^{X_t} are submartingales.*
(*b*) *Let $\mu > 0$. Then $e^{\mu t + \sigma W_t}$ is a submartingale.*

Proof: (*a*) Results from part (*a*) of Proposition 3.12.3.
(*b*) It follows from Example 3.12.1 and part (*b*) of Proposition 3.12.3. ■

The following result provides important inequalities involving submartingales, see for instance Doob [14].

Proposition 3.12.5 (Doob's Submartingale Inequality) *(a) Let X_t be a non-negative submartingale. Then*

$$P(\sup_{s\leq t} X_s \geq x) \leq \frac{\mathbb{E}[X_t]}{x}, \qquad \forall x > 0.$$

(b) If X_t is a right continuous submartingale, then for any $x > 0$

$$P(\sup_{s\leq t} X_t \geq x) \leq \frac{\mathbb{E}[X_t^+]}{x},$$

where $X_t^+ = \max\{X_t, 0\}$.

Exercise 3.12.6 *Let $x > 0$. Show the inequalities:*
(a) $P(\sup_{s\leq t} W_s^2 \geq x) \leq \frac{t}{x}$.

(b) $P(\sup_{s\leq t} |W_s| \geq x) \leq \frac{\sqrt{2t/\pi}}{x}$.

Exercise 3.12.7 *Show that* $p\text{-}\lim_{t\to\infty} \frac{\sup_{s\leq t}|W_s|}{t} = 0$.

Exercise 3.12.8 *Show that for any martingale X_t we have the inequality*

$$P(\sup_{s\leq t} X_t^2 > x) \leq \frac{\mathbb{E}[X_t^2]}{x}, \qquad \forall x > 0.$$

It is worth noting that Doob's inequality implies Markov's inequality. Since $\sup_{s\leq t} X_s \geq X_t$, then $P(X_t \geq x) \leq P(\sup_{s\leq t} X_s \geq x)$. Then Doob's inequality

$$P(\sup_{s\leq t} X_s \geq x) \leq \frac{\mathbb{E}[X_t]}{x}$$

implies Markov's inequality (see Theorem 2.13.9)

$$P(X_t \geq x) \leq \frac{\mathbb{E}[X_t]}{x}.$$

Exercise 3.12.9 *Let N_t denote the Poisson process and consider the information set $\mathcal{F}_t = \sigma\{N_s; s \leq t\}$.*

(a) Show that N_t is a submartingale;

(b) Is N_t^2 a submartingale?

Exercise 3.12.10 *It can be shown that for any $0 < \sigma < \tau$ we have the inequality*

$$\mathbb{E}\Big[\sum_{\sigma \le t \le \tau} \Big(\frac{N_t}{t} - \lambda\Big)^2\Big] \le \frac{4\tau\lambda}{\sigma^2}.$$

Using this inequality prove that $ms\text{-}\lim_{t\to\infty} \dfrac{N_t}{t} = \lambda$.

The following famous inequality involving expectations was also found by Doob. The proof can be found for instance in Chung and Williams [12].

Theorem 3.12.11 (Doob's inequality) *If X_t is a continuous martingale, then*

$$\mathbb{E}\Big[\sup_{0\le t\le T} X_t^2\Big] \le 4\mathbb{E}[X_t^2].$$

Exercise 3.12.12 *Use Doob's inequality to show*

$$\mathbb{E}\Big[\sup_{0\le t\le T} W_t^2\Big] \le 4T.$$

Exercise 3.12.13 *Find Doob's inequality for the martingale $X_t = W_t^2 - t$.*

Chapter 4

Properties of Stochastic Processes

This chapter presents detailed properties specific to stochastic processes, such as stopping times, hitting times, bounded variation, quadratic variation as well as some results regarding convergence and optimal stopping.

4.1 Stopping Times

Consider the probability space $(\Omega, \mathcal{F}, P)$ and the filtration $(\mathcal{F}_t)_{t\geq 0}$, i.e. an ascending sequence of σ-fields

$$\mathcal{F}_s \subset \mathcal{F}_t \subset \mathcal{F}, \qquad \forall s < t.$$

Assume that the decision to stop playing a game before or at time t is determined by the information $\mathcal{F}_t$ available at time t. Then this decision can be modeled by a random variable $\tau : \Omega \to [0, \infty]$, which satisfies

$$\{\omega; \tau(\omega) \leq t\} \in \mathcal{F}_t, \qquad \forall t \geq 0.$$

This means that given the information set $\mathcal{F}_t$, we know whether the event $\{\omega; \tau(\omega) \leq t\}$ had occurred or not. We note that the possibility $\tau = \infty$ is also included, since the decision to continue the game for ever is a possible event. A random variable τ with the previous properties is called a *stopping time.*

The next example illustrates a few cases when a decision is or is not a stopping time. In order to accomplish this, think of the situation that τ is the time when some random event related to a given stochastic process occurs first.

Example 4.1.1 *Let $\mathcal{F}_t$ be the information available until time t regarding the evolution of a stock. Assume the price of the stock at time $t = 0$ is \$50 per share. The following decisions are stopping times:*

(a) Sell the stock when it reaches for the first time the price of \$100 per share;

(b) Buy the stock when it reaches for the first time the price of \$10 per share;

(c) Sell the stock at the end of the year;

(d) Sell the stock either when it reaches for the first time \$80 or at the end of the year.

(e) Keep the stock either until the initial investment doubles or until the end of the year;

The following decision is not a stopping time:

(f) Sell the stock when it reaches the maximum level it will ever be.

Part (f) is not a stopping time because it requires information about the future that is not contained in $\mathcal{F}_t$. In part (e) there are two conditions; the latter one has the occurring probability equal to 1.

Exercise 4.1.2 *Show that any positive constant, $\tau = c$, is a stopping time with respect to any filtration.*

Exercise 4.1.3 *Let $\tau(\omega) = \inf\{t > 0; |W_t(\omega)| \geq K\}$, with $K > 0$ constant. Show that τ is a stopping time with respect to the filtration $\mathcal{F}_t = \sigma(W_s; s \leq t)$.*

The random variable τ is called the *first exit time* of the Brownian motion W_t from the interval $(-K, K)$. In a similar way one can define the *first exit time* of the process X_t from the interval (a, b):

$$\tau(\omega) = \inf\{t > 0; X_t(\omega) \notin (a, b)\} = \inf\{t > 0; X_t(\omega) \geq b \text{ or } X_t(\omega) \leq a)\}.$$

Let $X_0 < a$. The *first entry time* of X_t in the interval $[a, b]$ is defined as

$$\tau(\omega) = \inf\{t > 0; X_t(\omega) \in [a, b]\}.$$

If we let $b = \infty$, we obtain the *first hitting time* of the level a

$$\tau(\omega) = \inf\{t > 0; X_t(\omega) \geq a)\}.$$

We shall deal with hitting times in more detail in section 4.3.

Exercise 4.1.4 *Let X_t be a continuous stochastic process. Prove that the first exit time of X_t from the interval (a, b) is a stopping time.*

We shall present in the following some properties regarding operations with stopping times. Consider the notations $\tau_1 \vee \tau_2 = \max\{\tau_1, \tau_2\}$, $\tau_1 \wedge \tau_2 = \min\{\tau_1, \tau_2\}$, $\bar{\tau}_n = \sup_{n\geq 1} \tau_n$ and $\underline{\tau}_n = \inf_{n\geq 1} \tau_n$.

Proposition 4.1.5 *Let τ_1 and τ_2 be two stopping times with respect to the filtration $\mathcal{F}_t$. Then*

1. $\tau_1 \vee \tau_2$
2. $\tau_1 \wedge \tau_2$
3. $\tau_1 + \tau_2$

are stopping times.

Proof: 1. We have

$$\{\omega; \tau_1 \vee \tau_2 \le t\} = \{\omega; \tau_1 \le t\} \cap \{\omega; \tau_2 \le t\} \in \mathcal{F}_t,$$

since $\{\omega; \tau_1 \le t\} \in \mathcal{F}_t$ and $\{\omega; \tau_2 \le t\} \in \mathcal{F}_t$. Then $\tau_1 \vee \tau_2$ is a stopping time.

2. The event $\{\omega; \tau_1 \wedge \tau_2 \le t\} \in \mathcal{F}_t$ if and only if $\{\omega; \tau_1 \wedge \tau_2 > t\} \in \mathcal{F}_t$.

$$\{\omega; \tau_1 \wedge \tau_2 > t\} = \{\omega; \tau_1 > t\} \cap \{\omega; \tau_2 > t\} \in \mathcal{F}_t,$$

since $\{\omega; \tau_1 > t\} \in \mathcal{F}_t$ and $\{\omega; \tau_2 > t\} \in \mathcal{F}_t$, as the σ-algebra $\mathcal{F}_t$ is closed to complements.

3. We note that $\tau_1 + \tau_2 \le t$ if there is a $c \in (0, t)$ such that

$$\tau_1 \le c, \qquad \tau_2 \le t - c.$$

Using that the rational numbers are dense in $\mathbb{R}$, we can write

$$\{\omega; \tau_1 + \tau_2 \le t\} = \bigcup_{0<c<t, c \in \mathbb{Q}} \Big(\{\omega; \tau_1 \le c\} \cap \{\omega; \tau_2 \le t - c\}\Big) \in \mathcal{F}_t,$$

since

$$\{\omega; \tau_1 \le c\} \in \mathcal{F}_c \subset \mathcal{F}_t, \qquad \{\omega; \tau_2 \le t - c\} \in \mathcal{F}_{t-c} \subset \mathcal{F}_t.$$

It follows that $\tau_1 + \tau_2$ is a stopping time. ■

A filtration $\mathcal{F}_t$ is called *right-continuous* if $\mathcal{F}_t = \bigcap_{n=1}^{\infty} \mathcal{F}_{t+\frac{1}{n}}$, for $t > 0$. This means that the information available at time t is a good approximation for any future infinitesimal information $\mathcal{F}_{t+\epsilon}$; or, equivalently, nothing more can be learned by peeking infinitesimally far into the future. If denote by $\mathcal{F}_{t+} = \bigcap_{n=1}^{\infty} \mathcal{F}_{t+\frac{1}{n}}$, then the right-continuity can be written conveniently as $\mathcal{F}_t = \mathcal{F}_{t+}$.

Exercise 4.1.6 (*a*) *Let $\mathcal{F}_t = \sigma\{W_s; s \le t\}$ and $\mathcal{G}_t = \sigma\{\int_0^s W_u\, du; s \le t\}$, where W_t is a Brownian motion. Is $\mathcal{F}_t$ right-continuous? What about $\mathcal{G}_t$?*

(*b*) *Let $\mathcal{N}_t = \sigma\{N_s; s \le t\}$, where N_t is a Poisson motion. Is $\mathcal{N}_t$ right-continuous?*

The next result states that in the case of a right-continuous filtration the inequality $\{\tau \leq t\}$ from the definition of the stopping time can be replaced by a strict inequality.

Proposition 4.1.7 *Let $\mathcal{F}_t$ be right-continuous. The following are equivalent:*

(a) τ is a stoping time;

(b) $\{\tau < t\} \in \mathcal{F}_t$, for all $t \geq 0$.

Proof: "(a) $\Longrightarrow$ (b)" Let τ be a stopping time. Then $\{\tau \leq t - \frac{1}{n}\} \subset \mathcal{F}_{t-\frac{1}{n}} \subset \mathcal{F}_t$, and hence $\bigcup_{n\geq 1}\{\tau \leq t - \frac{1}{n}\} \in \mathcal{F}_t$. Then writing

$$\{\tau < t\} = \bigcup_{n\geq 1}\{\tau \leq t - \frac{1}{n}\} \in \mathcal{F}_t$$

it follows that $\{\tau < t\} \in \mathcal{F}_t$, i.e. τ is a stopping time.

"(b) $\Longrightarrow$ (a)" It follows from

$$\{\tau \leq t\} = \bigcap_{n\geq 1}\{\tau < t + \frac{1}{n}\} \in \bigcap_{n\geq 1} \mathcal{F}_{t+\frac{1}{n}} = \mathcal{F}_{t+} = \mathcal{F}_t.$$

■

Proposition 4.1.8 *Let $\mathcal{F}_t$ be right-continuous and $(\tau_n)_{n\geq 1}$ be a sequence of bounded stopping times. Then $\sup_n \tau_n$ and $\inf \tau_n$ are stopping times.*

Proof: The fact that $\bar{\tau}_n = \sup_n \tau_n$ is a stopping time follows from

$$\{\omega; \bar{\tau}_n \leq t\} \subset \bigcap_{n\geq 1}\{\omega; \tau_n \leq t\} \in \mathcal{F}_t.$$

In order to show that $\underline{\tau}_n = \inf \tau_n$ is a stopping time we shall proceed as in the following. Using that $\mathcal{F}_t$ is right-continuous and closed to complements, using proposition 4.1.7, it suffices to show that $\{\omega; \underline{\tau}_n \geq t\} \in \mathcal{F}_t$. This follows from

$$\{\omega; \underline{\tau}_n \geq t\} = \bigcap_{n\geq 1}\{\omega; \tau_n > t\} \in \mathcal{F}_t.$$

■

Exercise 4.1.9 *Let τ be a stopping time.*

(a) Let $c \geq 1$ be a constant. Show that $c\tau$ is a stopping time.

(b) Let $f : [0, \infty) \to \mathbb{R}$ be a continuous, increasing function satisfying $f(t) \geq t$. Prove that $f(\tau)$ is a stopping time.

(c) Show that e^τ is a stopping time.

Exercise 4.1.10 *Let τ be a stopping time and $c > 0$ a constant. Prove that $\tau + c$ is a stopping time.*

Exercise 4.1.11 *Let a be a constant and define $\tau = \inf\{t \geq 0; W_t = a\}$. Is τ a stopping time?*

Exercise 4.1.12 *Let τ be a stopping time. Consider the following sequence $\tau_n = (m+1)2^{-n}$ if $m2^{-n} \leq \tau < (m+1)2^{-n}$ (stop at the first time of the form $k2^{-n}$ after τ). Prove that τ_n is a stopping time.*

4.2 Stopping Theorem for Martingales

The next result states that in a fair game, the expected final fortune of a gambler, who is using a stopping time to quit the game, is the same as the expected initial fortune. From the financial point of view, the theorem says that if you buy an asset at some initial time and adopt a strategy of deciding when to sell it, then the expected price at the selling time is the initial price; so one cannot make money by buying and selling an asset whose price is a martingale. Fortunately, the price of a stock is not a martingale, and people can still expect to make money buying and selling stocks.

If $(M_t)_{t\geq 0}$ is an $\mathcal{F}_t$-martingale, then taking the expectation in

$$\mathbb{E}[M_t|\mathcal{F}_s] = M_s, \qquad \forall s < t$$

and using Example 2.12.4 yields

$$\mathbb{E}[M_t] = \mathbb{E}[M_s], \qquad \forall s < t.$$

In particular, $\mathbb{E}[M_t] = \mathbb{E}[M_0]$, for any $t > 0$. The next result states necessary conditions under which this identity holds if t is replaced by any stopping time τ. The reader can skip the proof at the first reading.

Theorem 4.2.1 (Optional Stopping Theorem) *Let $(M_t)_{t\geq 0}$ be a right continuous $\mathcal{F}_t$-martingale and τ be a stopping time with respect to $\mathcal{F}_t$ such that*

τ is bounded, i.e. $\exists N < \infty$ such that $\tau \leq N$.

Then $\mathbb{E}[M_\tau] = \mathbb{E}[M_0]$. If M_t is an $\mathcal{F}_t$-submartingale, then $\mathbb{E}[M_\tau] \geq \mathbb{E}[M_0]$.

Proof: Consider the following convenient notation for the indicator function of a set

$$\mathbf{1}_{\{\tau>t\}}(\omega) = \begin{cases} 1, & \tau(\omega) > t; \\ 0, & \tau(\omega) \leq t. \end{cases}$$

Taking the expectation in relation

$$M_\tau = M_{\tau\wedge t} + (M_\tau - M_t)\mathbf{1}_{\{\tau>t\}},$$

see Exercise 4.2.3, yields

$$\mathbb{E}[M_\tau] = \mathbb{E}[M_{\tau\wedge t}] + \mathbb{E}[M_\tau \mathbf{1}_{\{\tau>t\}}] - \mathbb{E}[M_t \mathbf{1}_{\{\tau>t\}}].$$

Since $M_{\tau\wedge t}$ is a martingale, see Exercise 4.2.4 (b), then $\mathbb{E}[M_{\tau\wedge t}] = \mathbb{E}[M_0]$. The previous relation becomes

$$\mathbb{E}[M_\tau] = \mathbb{E}[M_0] + \mathbb{E}[M_\tau \mathbf{1}_{\{\tau>t\}}] - \mathbb{E}[M_t \mathbf{1}_{\{\tau>t\}}], \quad \forall t > 0.$$

Taking the limit yields

$$\mathbb{E}[M_\tau] = \mathbb{E}[M_0] + \lim_{t\to\infty} \mathbb{E}[M_\tau \mathbf{1}_{\{\tau>t\}}] - \lim_{t\to\infty} \mathbb{E}[M_t \mathbf{1}_{\{\tau>t\}}]. \tag{4.2.1}$$

We shall show that both limits are equal to zero.

Since $|M_\tau \mathbf{1}_{\{\tau>t\}}| \leq |M_\tau|$, $\forall t > 0$, and M_τ is integrable, see Exercise 4.2.4 (a), by the dominated convergence theorem we have

$$\lim_{t\to\infty} \mathbb{E}[M_\tau \mathbf{1}_{\{\tau>t\}}] = \lim_{t\to\infty} \int_\Omega M_\tau \mathbf{1}_{\{\tau>t\}}\, dP = \int_\Omega \lim_{t\to\infty} M_\tau \mathbf{1}_{\{\tau>t\}}\, dP = 0.$$

For the second limit

$$\lim_{t\to\infty} \mathbb{E}[M_t \mathbf{1}_{\{\tau>t\}}] = \lim_{t\to\infty} \int_\Omega M_t \mathbf{1}_{\{\tau>t\}}\, dP = 0,$$

since for $t > N$ the integrand vanishes. Hence relation (4.2.1) yields $\mathbb{E}[M_\tau] = \mathbb{E}[M_0]$. ■

It is worth noting that the previous theorem is a special case of the more general Optional Stopping Theorem of Doob:

Theorem 4.2.2 *Let M_t be a right continuous martingale and σ, τ be two bounded stopping times, with $\sigma \leq \tau$. Then M_σ, M_τ are integrable and*

$$\mathbb{E}[M_\tau | \mathcal{F}_\sigma] = M_\sigma \quad a.s.$$

In particular, taking expectations, we have

$$\mathbb{E}[M_\tau] = \mathbb{E}[M_\sigma] \quad a.s.$$

In the case when M_t is a submartingale then $\mathbb{E}[M_\tau] \geq \mathbb{E}[M_\sigma]$ a.s.

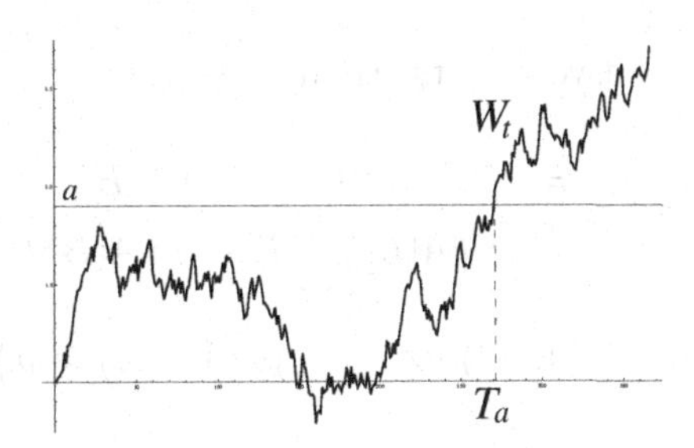

Figure 4.1: *The first hitting time T_a given by $W_{T_a} = a$.*

Exercise 4.2.3 *Show that*

$$M_\tau = M_{\tau \wedge t} + (M_\tau - M_t)\mathbf{1}_{\{\tau > t\}},$$

where $\mathbf{1}_{\{\tau>t\}}$ is the indicator function of the set $\{\tau > t\}$.

Exercise 4.2.4 *Let M_t be a right continuous martingale and τ be a bounded stopping time. Show that*

(*a*) *M_τ is integrable;*

(*b*) *$M_{\tau\wedge t}$ is a martingale.*

Exercise 4.2.5 *Show that letting $\sigma = 0$ in Theorem* 4.2.2 *yields Theorem* 4.2.1.

4.3 The First Passage of Time

The first passage of time is a particular type of hitting time, which is useful in finance when studying barrier options and lookback options. For instance, knock-in options enter into existence when the stock price hits for the first time a certain barrier before option maturity. A lookback option is priced using the maximum value of the stock until the present time. The stock price is not a Brownian motion, but it depends on one. Hence the need for studying the hitting time for the Brownian motion.

The first result deals with the first hitting time for a Brownian motion to reach the barrier $a \in \mathbb{R}$, see Fig. 4.1.

Lemma 4.3.1 *Let T_a be the first time the Brownian motion W_t hits a. Then the distribution function of T_a is given by*

$$\boxed{P(T_a \le t) = \frac{2}{\sqrt{2\pi}} \int_{|a|/\sqrt{t}}^{\infty} e^{-y^2/2}\, dy.}$$

Proof: If A and B are two events, then

$$\begin{aligned} P(A) &= P(A \cap B) + P(A \cap \overline{B}) \\ &= P(A|B)P(B) + P(A|\overline{B})P(\overline{B}). \end{aligned} \tag{4.3.2}$$

Let $a > 0$. Using formula (4.3.2) for $A = \{\omega; W_t(\omega) \geq a\}$ and $B = \{\omega; T_a(\omega) \leq t\}$ yields

$$\begin{aligned} P(W_t \geq a) &= P(W_t \geq a|T_a \leq t)P(T_a \leq t) \\ &\quad + P(W_t \geq a|T_a > t)P(T_a > t) \end{aligned} \tag{4.3.3}$$

If $T_a > t$, the Brownian motion did not reach the barrier a yet, so we must have $W_t < a$. Therefore

$$P(W_t \geq a|T_a > t) = 0.$$

If $T_a \leq t$, then $W_{T_a} = a$. Since the Brownian motion is a Markov process, it starts fresh at T_a. Due to symmetry of the density function of a normal variable, W_t has equal chances to go up or go down after the time interval $t - T_a$. It follows that

$$P(W_t \geq a|T_a \leq t) = \frac{1}{2}.$$

Substituting into (4.3.3) yields

$$\begin{aligned} P(T_a \leq t) &= 2P(W_t \geq a) \\ &= \frac{2}{\sqrt{2\pi t}} \int_a^\infty e^{-x^2/(2t)}\, dx = \frac{2}{\sqrt{2\pi}} \int_{a/\sqrt{t}}^\infty e^{-y^2/2}\, dy. \end{aligned}$$

If $a < 0$, symmetry implies that the distribution of T_a is the same as that of T_{-a}, so we get

$$P(T_a \leq t) = P(T_{-a} \leq t) = \frac{2}{\sqrt{2\pi}} \int_{-a/\sqrt{t}}^\infty e^{-y^2/2}\, dy.$$

■

Remark 4.3.2 The previous proof is based on a more general principle called the *reflection principle:* If τ is a stopping time for the Brownian motion W_t, then the Brownian motion reflected at τ is also a Brownian motion.

Theorem 4.3.3 *Let $a \in \mathbb{R}$ be fixed. Then the Brownian motion hits a (in a finite amount of time) with probability 1.*

Proof: The probability that W_t hits a (in a finite amount of time) is

$$\begin{aligned} P(T_a < \infty) &= \lim_{t\to\infty} P(T_a \le t) = \lim_{t\to\infty} \frac{2}{\sqrt{2\pi}} \int_{|a|/\sqrt{t}}^{\infty} e^{-y^2/2}\, dy \\ &= \frac{2}{\sqrt{2\pi}} \int_0^{\infty} e^{-y^2/2}\, dy = 1, \end{aligned}$$

where we used the well known integral

$$\int_0^{\infty} e^{-y^2/2}\, dy = \frac{1}{2}\int_{-\infty}^{\infty} e^{-y^2/2}\, dy = \frac{1}{2}\sqrt{2\pi}.$$

■

The previous result stated that the Brownian motion hits the barrier a almost surely. The next result shows that the expected time to hit the barrier is infinite.

Proposition 4.3.4 *The random variable T_a has a Pearson 5 distribution given by*

$$\boxed{p(t) = \frac{|a|}{\sqrt{2\pi}} e^{-\frac{a^2}{2t}} t^{-\frac{3}{2}}, \qquad t > 0.}$$

It has the mean $\mathbb{E}[T_a] = \infty$ and the mode $\dfrac{a^2}{3}$.

Proof: Differentiating in the formula of distribution function[1]

$$F_{T_a}(t) = P(T_a \le t) = \frac{2}{\sqrt{2\pi}} \int_{a/\sqrt{t}}^{\infty} e^{-y^2/2}\, dy$$

yields the following probability density function

$$p(t) = \frac{dF_{T_a}(t)}{dt} = \frac{a}{\sqrt{2\pi}} e^{-\frac{a^2}{2t}} t^{-\frac{3}{2}}, \qquad t > 0.$$

This is a Pearson 5 distribution with parameters $\alpha = 1/2$ and $\beta = a^2/2$. The expectation is

$$\mathbb{E}[T_a] = \int_0^{\infty} t p(t)\, dt = \frac{a}{\sqrt{2\pi}} \int_0^{\infty} \frac{1}{\sqrt{t}} e^{-\frac{a^2}{2t}}\, dt.$$

---

[1]One may use Leibniz's formula $\dfrac{d}{dt}\displaystyle\int_{\varphi(t)}^{\psi(t)} f(u)\, du = f(\psi(t))\psi'(t) - f(\varphi(t))\varphi'(t)$.

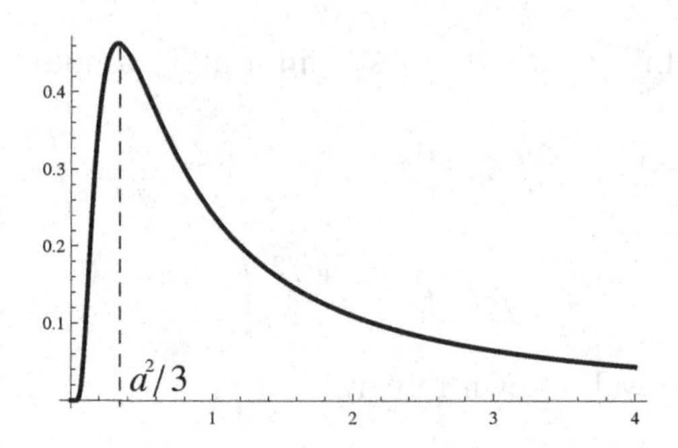

Figure 4.2: *The distribution of the first hitting time T_a.*

Using the inequality $e^{-\frac{a^2}{2t}} > 1 - \dfrac{a^2}{2t}$, $t > 0$, we have the estimation

$$\mathbb{E}[T_a] \;>\; \frac{a}{\sqrt{2\pi}}\int_0^\infty \frac{1}{\sqrt{t}}\,dt - \frac{a^3}{2\sqrt{2\pi}}\int_0^\infty \frac{1}{t^{3/2}}\,dt = \infty, \qquad (4.3.4)$$

since $\int_0^\infty \frac{1}{\sqrt{t}}\,dt$ is divergent and $\int_0^\infty \frac{1}{t^{3/2}}\,dt$ is convergent.

The mode of T_a is given by

$$\frac{\beta}{\alpha+1} = \frac{a^2}{2(\frac{1}{2}+1)} = \frac{a^2}{3}.$$

■

Remark 4.3.5 The distribution has a peak at $a^2/3$. Then if we need to pick a small time interval $[t-dt, t+dt]$ in which the probability that the Brownian motion hits the barrier a is maximum, we need to choose $t = a^2/3$, see Fig. 4.2.

Remark 4.3.6 Formula (4.3.4) states that the expected waiting time for W_t to reach the barrier a is infinite. However, the expected waiting time for the Brownian motion W_t to hit either a or $-a$ is finite, see Exercise 4.3.9.

Corollary 4.3.7 *A Brownian motion process returns to the origin in a finite amount time with probability 1.*

Proof: Choose $a = 0$ and apply Theorem 4.3.3. ■

Exercise 4.3.8 *Try to apply the proof of Lemma 4.3.1 for the following stochastic processes*

(a) $X_t = \mu t + \sigma W_t$, *with* $\mu, \sigma > 0$ *constants;*

(b) $X_t = \displaystyle\int_0^t W_s\,ds.$

Where is the difficulty?

Exercise 4.3.9 *Let $a > 0$ and consider the hitting time*

$$\tau_a = \inf\{t > 0; |W_t| \geq a\}.$$

Prove that $\mathbb{E}[\tau_a] = a^2$.

Exercise 4.3.10 *(a) Show that the distribution function of the process*

$$X_t = \max_{s\in[0,t]} W_s$$

is given by

$$P(X_t \leq a) = \frac{2}{\sqrt{2\pi}} \int_0^{a/\sqrt{t}} e^{-y^2/2}\, dy,$$

and the probability density is

$$p_t(x) = \frac{2}{\sqrt{2\pi t}} e^{-\frac{x^2}{2t}}, \qquad x \geq 0.$$

(b) Show that $\mathbb{E}[X_t] = \sqrt{2t/\pi}$ and $Var(X_t) = t\Big(1 - \dfrac{2}{\pi}\Big)$.

Exercise 4.3.11 *(a) Show that the probability density of the absolute value of a Brownian motion $X_t = |W_t|$, $t \geq 0$, is given by*

$$p_t(x) = \frac{2}{\sqrt{2\pi t}} e^{-\frac{x^2}{2t}}, \qquad x \geq 0.$$

(b) Consider the processes $X_t = |W_t|$ and $Y_t = |B_t|$, with W_t and B_t independent Brownian motions. Use Theorem 2.11.1 to obtain the probability density of the sum process $Z_t = X_t + Y_t$.

The fact that a Brownian motion returns to the origin or hits a barrier almost surely is a property characteristic to the first dimension only. The next result states that in larger dimensions this is no longer possible.

Theorem 4.3.12 *Let $(a, b) \in \mathbb{R}^2$. The 2-dimensional Brownian motion $W(t) = \big(W_1(t), W_2(t)\big)$ (with $W_1(t)$ and $W_2(t)$ independent) hits the point (a, b) with probability zero. The same result is valid for any n-dimensional Brownian motion, with $n \geq 2$.*

However, if the point (a, b) is replaced by the disk

$$D_\epsilon(\mathbf{x}_0) = \{x \in \mathbb{R}^2; |\mathbf{x} - \mathbf{x}_0| \leq \epsilon\},$$

then there is a difference in the behavior of the Brownian motion from $n = 2$ to $n > 2$, as pointed out by the next two results:

Theorem 4.3.13 *The 2-dimensional Brownian motion* $W(t) = \big(W_1(t), W_2(t)\big)$ *hits the disk* $D_\epsilon(\mathbf{x}_0)$ *with probability one.*

Theorem 4.3.14 *Let* $n > 2$*. The* n*-dimensional Brownian motion* $W(t)$ *hits the ball* $D_\epsilon(\mathbf{x}_0)$ *with probability*

$$P = \Big(\frac{|\mathbf{x}_0|}{\epsilon}\Big)^{2-n} < 1.$$

The previous results can be stated by saying that that Brownian motion is transient in $\mathbb{R}^n$, for $n > 2$. If $n = 2$ the previous probability equals 1. We shall come back with proofs to the aforementioned results in a later chapter (see section 9.6).

Remark 4.3.15 If life spreads according to a Brownian motion, the aforementioned results explain why life is more extensive on earth rather than in space. The probability for a form of life to reach a planet of radius R situated at distance d is $\frac{R}{d}$. Since d is large the probability is very small, unlike in the plane, where the probability is always 1.

Exercise 4.3.16 *Is the one-dimensional Brownian motion transient or recurrent in* $\mathbb{R}$*?*

4.4 The Arc-sine Laws

In this section we present a few results which provide certain probabilities related with the behavior of a Brownian motion in terms of the arc-sine of a quotient of two time instances. These results are generally known as the Arc-sine Laws.

The following result will be used in the proof of the first Arc-sine Law.

Proposition 4.4.1 *(a) If* $X : \Omega \to \mathbb{N}$ *is a discrete random variable, then for any subset* $A \subset \Omega$*, we have*

$$P(A) = \sum_{x \in \mathbb{N}} P(A|X = x)P(X = x).$$

(b) If $X : \Omega \to \mathbb{R}$ *is a continuous random variable, then*

$$P(A) = \int P(A|X = x)dP = \int P(A|X = x)f_X(x)\,dx.$$

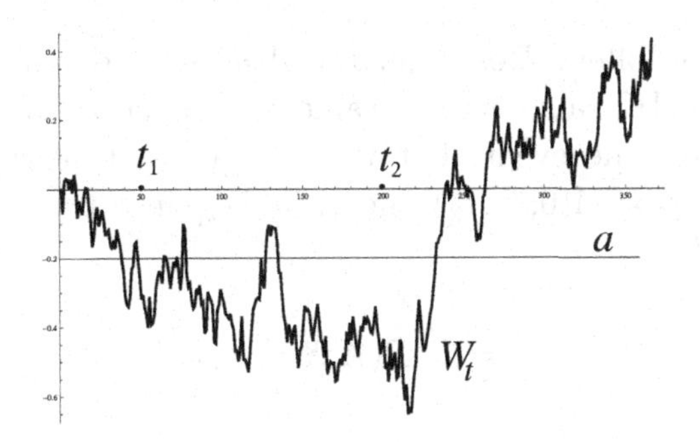

Figure 4.3: *The event $A(a; t_1, t_2)$ in the Arc-sine Law.*

Proof: (a) The sets $X^{-1}(x) = \{X = x\} = \{\omega; X(\omega) = x\}$ form a partition of the sample space Ω, i.e.:

(i) $\Omega = \bigcup_x X^{-1}(x)$;

(ii) $X^{-1}(x) \cap X^{-1}(y) = \emptyset$ for $x \neq y$.

Then $A = \bigcup\limits_x \Big(A \cap X^{-1}(x)\Big) = \bigcup\limits_x \Big(A \cap \{X = x\}\Big)$, and hence

$$\begin{aligned} P(A) &= \sum_x P\Big(A \cap \{X = x\}\Big) \\ &= \sum_x \frac{P(A \cap \{X = x\})}{P(\{X = x\})} P(\{X = x\}) \\ &= \sum_x P(A|X = x) P(X = x). \end{aligned}$$

(b) In the case when X is continuous, the sum is replaced by an integral and the probability $P(\{X = x\})$ by $f_X(x)dx$, where f_X is the density function of X. ■

The *zero set* of a Brownian motion W_t is defined by $\{t \geq 0; W_t = 0\}$. Since W_t is continuous, the zero set is closed with no isolated points almost surely. The next result deals with the probability that the zero set does not intersect the interval (t_1, t_2).

Theorem 4.4.2 (The Arc-sine Law) *The probability that a Brownian motion W_t does not have any zeros in the interval (t_1, t_2) is equal to*

$$\boxed{P(W_t \neq 0, t_1 \leq t \leq t_2) = \frac{2}{\pi} \arcsin \sqrt{\frac{t_1}{t_2}}.}$$

Proof: The proof follows Ross [43]. Let $A(a; t_1, t_2)$ denote the event that the Brownian motion W_t takes on the value a between t_1 and t_2. In particular, $A(0; t_1, t_2)$ denotes the event that W_t has (at least) a zero between t_1 and t_2. Substituting $A = A(0; t_1, t_2)$ and $X = W_{t_1}$ into the formula provided by Proposition 4.4.1

$$P(A) = \int P(A|X = x) f_X(x)\, dx$$

yields

$$\begin{aligned} P\big(A(0; t_1, t_2)\big) &= \int P\big(A(0; t_1, t_2)|W_{t_1} = x\big) f_{W_{t_1}}(x)\, dx \qquad (4.4.5) \\ &= \frac{1}{\sqrt{2\pi t_1}} \int_{-\infty}^{\infty} P\big(A(0; t_1, t_2)|W_{t_1} = x\big) e^{-\frac{x^2}{2t_1}}\, dx. \end{aligned}$$

Using the properties of W_t with respect to time translation and symmetry we have

$$\begin{aligned} P\big(A(0; t_1, t_2)|W_{t_1} = x\big) &= P\big(A(0; 0, t_2 - t_1)|W_0 = x\big) \\ &= P\big(A(-x; 0, t_2 - t_1)|W_0 = 0\big) \\ &= P\big(A(|x|; 0, t_2 - t_1)|W_0 = 0\big) \\ &= P\big(A(|x|; 0, t_2 - t_1)\big) \\ &= P\big(T_{|x|} \le t_2 - t_1\big), \end{aligned}$$

the last identity stating that W_t hits $|x|$ before $t_2 - t_1$. Using Lemma 4.3.1 yields

$$P\big(A(0; t_1, t_2)|W_{t_1} = x\big) = \frac{2}{\sqrt{2\pi(t_2 - t_1)}} \int_{|x|}^{\infty} e^{-\frac{y^2}{2(t_2 - t_1)}}\, dy.$$

Substituting into (4.4.5) we obtain

$$\begin{aligned} P\big(A(0; t_1, t_2)\big) &= \frac{1}{\sqrt{2\pi t_1}} \int_{-\infty}^{\infty} \Big(\frac{2}{\sqrt{2\pi(t_2 - t_1)}} \int_{|x|}^{\infty} e^{-\frac{y^2}{2(t_2 - t_1)}}\, dy \Big) e^{-\frac{x^2}{2t_1}}\, dx \\ &= \frac{1}{\pi\sqrt{t_1(t_2 - t_1)}} \int_0^{\infty} \int_{|x|}^{\infty} e^{-\frac{y^2}{2(t_2 - t_1)} - \frac{x^2}{2t_1}}\, dy dx. \end{aligned}$$

The above integral can be evaluated to get (see Exercise 4.4.3)

$$P\big(A(0; t_1, t_2)\big) = 1 - \frac{2}{\pi} \arcsin\sqrt{\frac{t_1}{t_2}}.$$

Using $P(W_t \neq 0, t_1 \le t \le t_2) = 1 - P\big(A(0; t_1, t_2)\big)$ we obtain the desired result. ∎

Exercise 4.4.3 *Use polar coordinates to show*

$$\frac{1}{\pi\sqrt{t_1(t_2-t_1)}}\int_0^\infty\int_{|x|}^\infty e^{-\frac{y^2}{2(t_2-t_1)}-\frac{x^2}{2t_1}}\,dydx = 1-\frac{2}{\pi}\arcsin\sqrt{\frac{t_1}{t_2}}.$$

Exercise 4.4.4 *Find the probability that a 2-dimensional Brownian motion $W(t) = \big(W_1(t), W_2(t)\big)$ stays in the same quadrant for the time interval $t \in (t_1, t_2)$.*

Exercise 4.4.5 *Find the probability that a Brownian motion W_t does not take the value a in the interval (t_1, t_2).*

Exercise 4.4.6 *Let $a \neq b$. Find the probability that a Brownian motion W_t does not take any of the values $\{a, b\}$ in the interval (t_1, t_2). Formulate and prove a generalization.*

We provide below without proof a few similar results dealing with arc-sine probabilities, whose proofs can be found for instance in Kuo [30]. The first result deals with the amount of time spent by a Brownian motion on the positive half-axis.

Theorem 4.4.7 (Arc-sine Law of Lévy) *Let $L_t^+ = \int_0^t sgn^+ W_s\,ds$ be the amount of time a Brownian motion W_t is positive during the time interval $[0, t]$. Then*

$$P(L_t^+ \le \tau) = \frac{2}{\pi}\arcsin\sqrt{\frac{\tau}{t}}.$$

The next result deals with the Arc-sine Law for the last exit time of a Brownian motion from 0.

Theorem 4.4.8 (Arc-sine Law of exit from 0) *Let $\gamma_t = \sup\{0 \le s \le t; W_s = 0\}$. Then*

$$P(\gamma_t \le \tau) = \frac{2}{\pi}\arcsin\sqrt{\frac{\tau}{t}}, \qquad 0 \le \tau \le t.$$

The Arc-sine Law for the time the Brownian motion attains its maximum on the interval $[0, t]$ is given by the next result.

Theorem 4.4.9 (Arc-sine Law of maximum) *Let $M_t = \max\limits_{0\le s\le t} W_s$ and define*

$$\theta_t = \sup\{0 \le s \le t; W_s = M_t\}.$$

Then

$$P(\theta_t \le s) = \frac{2}{\pi}\arcsin\sqrt{\frac{s}{t}}, \qquad 0 \le s \le t, t > 0.$$

4.5 More on Hitting Times

In this section we shall deal with results regarding hitting times of Brownian motion with drift. These type of results are useful in Mathematical Finance when finding the value of barrier options.

Theorem 4.5.1 *Let $X_t = \mu t + W_t$ denote a Brownian motion with nonzero drift rate μ, and consider $\alpha, \beta > 0$. Then*

$$\boxed{P(X_t \text{ goes up to } \alpha \text{ before down to } -\beta) = \frac{e^{2\mu\beta} - 1}{e^{2\mu\beta} - e^{-2\mu\alpha}}.}$$

Proof: Let $T = \inf\{t > 0; X_t \geq \alpha \text{ or } X_t \leq -\beta\}$ be the first exit time of X_t from the interval $(-\beta, \alpha)$, which is a stopping time, see Exercise 4.1.4. The exponential process

$$M_t = e^{cW_t - \frac{c^2}{2}t}, \qquad t \geq 0$$

is a martingale, see Exercise 3.2.4(*c*). Then $\mathbb{E}[M_t] = \mathbb{E}[M_0] = 1$. By the Optional Stopping Theorem (see Theorem 4.2.1), we get $\mathbb{E}[M_T] = 1$. This can be written as

$$1 = \mathbb{E}[e^{cW_T - \frac{1}{2}c^2 T}] = \mathbb{E}[e^{cX_T - (c\mu + \frac{1}{2}c^2)T}]. \tag{4.5.6}$$

Choosing $c = -2\mu$ yields $\mathbb{E}[e^{-2\mu X_T}] = 1$. Since the random variable X_T takes only the values α and $-\beta$, if we let $p_\alpha = P(X_T = \alpha)$, the previous relation becomes

$$e^{-2\mu\alpha}p_\alpha + e^{2\mu\beta}(1 - p_\alpha) = 1.$$

Solving for p_α yields

$$p_\alpha = \frac{e^{2\mu\beta} - 1}{e^{2\mu\beta} - e^{-2\mu\alpha}}. \tag{4.5.7}$$

Noting that

$$p_\alpha = P(X_t \text{ goes up to } \alpha \text{ before down to } -\beta)$$

leads to the desired answer. ■

It is worth noting how the previous formula changes in the case when the drift rate is zero, i.e. when $\mu = 0$, and $X_t = W_t$. The previous probability is computed by taking the limit $\mu \to 0$ and using L'Hospital's rule

$$\lim_{\mu \to 0} \frac{e^{2\mu\beta} - 1}{e^{2\mu\beta} - e^{-2\mu\alpha}} = \lim_{\mu \to 0} \frac{2\beta e^{2\mu\beta}}{2\beta e^{2\mu\beta} + 2\alpha e^{-2\mu\alpha}} = \frac{\beta}{\alpha + \beta}.$$

Hence

$$\boxed{P(W_t \text{ goes up to } \alpha \text{ before down to } -\beta) = \frac{\beta}{\alpha + \beta}.}$$

Taking the limit $\beta \to \infty$ we recover the following result

$$\boxed{P(W_t \text{ hits } \alpha) = 1.}$$

If $\alpha = \beta$ we obtain

$$\boxed{P(W_t \text{ goes up to } \alpha \text{ before down to } -\alpha) = \frac{1}{2},}$$

which shows that the Brownian motion is equally likely to go up or down an amount α in a given time interval.

If T_α and T_β denote the times when the process X_t reaches α and β, respectively, then the aforementioned probabilities can be written using inequalities. For instance, the first identity becomes

$$P(T_\alpha \le T_{-\beta}) = \frac{e^{2\mu\beta} - 1}{e^{2\mu\beta} - e^{-2\mu\alpha}}.$$

Exercise 4.5.2 *Let $X_t = \mu t + W_t$ denote a Brownian motion with nonzero drift rate μ, and consider $\alpha > 0$.*
(a) If $\mu > 0$ show that

$$P(X_t \text{ goes up to } \alpha) = 1.$$

(b) If $\mu < 0$ show that

$$P(X_t \text{ goes up to } \alpha) = e^{2\mu\alpha} < 1.$$

Formula (a) can be written equivalently as

$$P(\sup_{t\ge 0}(W_t + \mu t) \ge \alpha) = 1, \qquad \mu \ge 0,$$

while formula (b) becomes

$$P(\sup_{t\ge 0}(W_t + \mu t) \ge \alpha) = e^{2\mu\alpha}, \qquad \mu < 0,$$

or

$$P(\sup_{t\ge 0}(W_t - \gamma t) \ge \alpha) = e^{-2\gamma\alpha}, \qquad \gamma > 0,$$

which is known as one of the Doob's inequalities. This can also be described in terms of stopping times as follows. Define the stopping time

$$\tau_\alpha = \inf\{t > 0; W_t - \gamma t \ge \alpha\}.$$

Using

$$P(\tau_\alpha < \infty) = P\big(\sup_{t\geq 0}(W_t - \gamma t) \geq \alpha\big)$$

yields the identities

$$\begin{aligned} P(\tau_\alpha < \infty) &= e^{-2\alpha\gamma}, \qquad \gamma > 0, \\ P(\tau_\alpha < \infty) &= 1, \qquad \gamma \leq 0. \end{aligned}$$

Exercise 4.5.3 *Let $X_t = \mu t + W_t$ denote a Brownian motion with nonzero drift rate μ, and consider $\beta > 0$. Show that the probability that X_t never hits $-\beta$ is given by*

$$\begin{cases} 1 - e^{-2\mu\beta}, & \text{if } \mu > 0 \\ 0, & \text{if } \mu < 0. \end{cases}$$

Recall that T is the first time when the process X_t hits α or $-\beta$.

Exercise 4.5.4 (*a*) *Show that*

$$\mathbb{E}[X_T] = \frac{\alpha e^{2\mu\beta} + \beta e^{-2\mu\alpha} - \alpha - \beta}{e^{2\mu\beta} - e^{-2\mu\alpha}}.$$

(*b*) *Find* $\mathbb{E}[X_T^2]$;

(*c*) *Compute* $Var(X_T)$.

The next result deals with the time one has to wait (in expectation) for the process $X_t = \mu t + W_t$ to reach either α or $-\beta$.

Proposition 4.5.5 *The expected value of T is*

$$\mathbb{E}[T] = \frac{\alpha e^{2\mu\beta} + \beta e^{-2\mu\alpha} - \alpha - \beta}{\mu(e^{2\mu\beta} - e^{-2\mu\alpha})}.$$

Proof: Using that W_t is a martingale, with $\mathbb{E}[W_t] = \mathbb{E}[W_0] = 0$, applying the Optional Stopping Theorem, Theorem 4.2.1, yields

$$0 = \mathbb{E}[W_T] = \mathbb{E}[X_T - \mu T] = \mathbb{E}[X_T] - \mu\mathbb{E}[T].$$

Then by Exercise 4.5.4(*a*) we get

$$\mathbb{E}[T] = \frac{\mathbb{E}[X_T]}{\mu} = \frac{\alpha e^{2\mu\beta} + be^{-2\mu\alpha} - \alpha - \beta}{\mu(e^{2\mu\beta} - e^{-2\mu\alpha})}.$$

■

Exercise 4.5.6 *Take the limit $\mu \to 0$ in the formula provided by Proposition 4.5.5 to find the expected time for a Brownian motion to hit either α or $-\beta$.*

Exercise 4.5.7 *Find $\mathbb{E}[T^2]$ and $Var(T)$.*

Exercise 4.5.8 (Wald's identities) *Let T be a finite and bounded stopping time for the Brownian motion W_t. Show that:*
(a) $\mathbb{E}[W_T] = 0$;
(b) $\mathbb{E}[W_T^2] = \mathbb{E}[T]$.

The previous techniques can also be applied to right continuous martingales. Let $a > 0$ and consider the hitting time of the Poisson process for the barrier a

$$\tau = \inf\{t > 0; N_t \geq a\}.$$

Proposition 4.5.9 *The expected waiting time for N_t to reach the barrier a is $\mathbb{E}[\tau] = \frac{a}{\lambda}$.*

Proof: Since $M_t = N_t - \lambda t$ is a right continuous martingale, by the Optional Stopping Theorem $\mathbb{E}[M_\tau] = \mathbb{E}[M_0] = 0$. Then $\mathbb{E}[N_\tau - \lambda\tau] = 0$ and hence $\mathbb{E}[\tau] = \frac{1}{\lambda}\mathbb{E}[N_\tau] = \frac{a}{\lambda}$. ■

4.6 The Inverse Laplace Transform Method

In this section we shall use the Optional Stopping Theorem in conjunction with the inverse Laplace transform to obtain the probability density functions for hitting times.

The case of standard Brownian motion Let $x > 0$. The first hitting time $\tau = T_x = \inf\{t > 0; W_t \geq x\}$ is a stopping time. Since $M_t = e^{cW_t - \frac{1}{2}c^2 t}$, $t \geq 0$, is a martingale, with $\mathbb{E}[M_t] = \mathbb{E}[M_0] = 1$, by the Optional Stopping Theorem, see Theorem 4.2.1, we have

$$\mathbb{E}[M_\tau] = 1.$$

This can be written equivalently as $\mathbb{E}[e^{cW_\tau}e^{-\frac{1}{2}c^2\tau}] = 1$. Using $W_\tau = x$, we get

$$\mathbb{E}[e^{-\frac{1}{2}c^2\tau}] = e^{-cx}.$$

It is worth noting that $c > 0$. This is implied from the fact that $e^{-\frac{1}{2}c^2\tau} < 1$ and $\tau, x > 0$.
Substituting $s = \frac{1}{2}c^2$, the previous relation becomes

$$\mathbb{E}[e^{-s\tau}] = e^{-\sqrt{2s}x}. \tag{4.6.8}$$

This relation has a couple of useful applications.

Proposition 4.6.1 *The moments of the first hitting time are all infinite*

$$\mathbb{E}[\tau^n] = \infty, \qquad n \geq 1.$$

Proof: The nth moment of τ can be obtained by differentiating and taking $s = 0$

$$\frac{d^n}{ds^n}\mathbb{E}[e^{-s\tau}]\Big|_{s=0} = \mathbb{E}[(-\tau)^n e^{-s\tau}]\Big|_{s=0} = (-1)^n\mathbb{E}[\tau^n].$$

Using (4.6.8) yields

$$\mathbb{E}[\tau^n] = (-1)^n \frac{d^n}{ds^n} e^{-\sqrt{2s}x}\Big|_{s=0}.$$

Since by induction we have

$$\frac{d^n}{ds^n} e^{-\sqrt{2s}x} = (-1)^n e^{-\sqrt{2s}x} \sum_{k=0}^{n-1} \frac{M_k}{2^{r_k/2}} \frac{x^{n-k}}{s^{(n+k)/2}},$$

with M_k, r_k positive integers, it easily follows that $\mathbb{E}[\tau^n] = \infty$.

For instance, in the case $n = 1$, we have

$$\mathbb{E}[\tau] = -\frac{d}{ds} e^{-\sqrt{2s}x}\Big|_{s=0} = \lim_{s \to 0+} e^{-\sqrt{2s}x} \frac{x}{2\sqrt{2sx}} = +\infty.$$

■

Another application involves the inverse Laplace transform to get the probability density. This way we can retrieve the result of Proposition 4.3.4.

Proposition 4.6.2 *The probability density of the hitting time τ is given by*

$$p(t) = \frac{|x|}{\sqrt{2\pi t^3}} e^{-\frac{x^2}{2t}}, \qquad t > 0. \tag{4.6.9}$$

Proof: Let $x > 0$. The expectation

$$\mathbb{E}[e^{-s\tau}] = \int_0^\infty e^{-s\tau} p(\tau)\, d\tau = \mathcal{L}\{p(\tau)\}(s)$$

is the Laplace transform of $p(\tau)$. Applying the inverse Laplace transform yields

$$\begin{aligned} p(\tau) &= \mathcal{L}^{-1}\{\mathbb{E}[e^{-s\tau}]\}(\tau) = \mathcal{L}^{-1}\{e^{-\sqrt{2s}x}\}(\tau) \\ &= \frac{x}{\sqrt{2\pi\tau^3}} e^{-\frac{x^2}{2\tau}}, \qquad \tau > 0. \end{aligned}$$

In the case $x < 0$ we obtain

$$p(\tau) = \frac{-x}{\sqrt{2\pi\tau^3}} e^{-\frac{x^2}{2\tau}}, \qquad \tau > 0,$$

which leads to (4.6.9). ■

The computation on the inverse Laplace transform $\mathcal{L}^{-1}\{e^{-\sqrt{2s}x}\}(\tau)$ is beyond the goal of this book. The reader can obtain the value of this inverse Laplace transform using the Mathematica software. However, the more mathematically interested reader is referred to consult the method of complex integration in a book on inverse Laplace transforms.

Another application of formula (4.6.8) is the following inequality.

Proposition 4.6.3 (Chernoff bound) *Let τ be the first hitting time when the Brownian motion W_t hits the barrier x, $x > 0$. Then*

$$P(\tau \le \lambda) \le e^{-\frac{x^2}{2\lambda}}, \qquad \forall \lambda > 0.$$

Proof: Let $s = -t$ in part 2 of Theorem 2.13.11 and use (4.6.8) to get

$$P(\tau \le \lambda) \le \frac{\mathbb{E}[e^{tX}]}{e^{\lambda t}} = \frac{\mathbb{E}[e^{-sX}]}{e^{-\lambda s}} = e^{\lambda s - x\sqrt{2s}}, \qquad \forall s > 0.$$

Then $P(\tau \le \lambda) \le e^{\min_{s>0} f(s)}$, where $f(s) = \lambda s - x\sqrt{2s}$. Since $f'(s) = \lambda - \frac{x}{\sqrt{2s}}$, then $f(s)$ reaches its minimum at the critical point $s_0 = \frac{x^2}{2\lambda^2}$. The minimum value is

$$\min_{s>0} f(s) = f(s_0) = -\frac{x^2}{2\lambda}.$$

Substituting in the previous inequality leads to the required result. ■

The case of Brownian motion with drift Consider the Brownian motion with drift $X_t = \mu t + \sigma W_t$, with $\mu, \sigma > 0$. Let

$$\tau = \inf\{t > 0; X_t \ge x\}$$

denote the first hitting time of the barrier x, with $x > 0$. We shall compute the distribution of the random variable τ and its first two moments.

Applying the Optional Stopping Theorem (Theorem 4.2.1) to the martingale $M_t = e^{cW_t - \frac{1}{2}c^2 t}$ yields

$$\mathbb{E}[M_\tau] = \mathbb{E}[M_0] = 1.$$

Using that $W_\tau = \frac{1}{\sigma}(X_\tau - \mu\tau)$ and $X_\tau = x$, the previous relation becomes

$$\mathbb{E}[e^{-(\frac{c\mu}{\sigma} + \frac{1}{2}c^2)\tau}] = e^{-\frac{c}{\sigma}x}. \tag{4.6.10}$$

Substituting $s = \dfrac{c\mu}{\sigma} + \dfrac{1}{2}c^2$ and completing to a square yields

$$2s + \frac{\mu^2}{\sigma^2} = \Big(c + \frac{\mu}{\sigma}\Big)^2.$$

Solving for c we get the solutions

$$c = -\frac{\mu}{\sigma} + \sqrt{2s + \frac{\mu^2}{\sigma^2}}, \qquad c = -\frac{\mu}{\sigma} - \sqrt{2s + \frac{\mu^2}{\sigma^2}}.$$

Assume $c < 0$. Then substituting the second solution into (4.6.10) yields

$$\mathbb{E}[e^{-s\tau}] = e^{\frac{1}{\sigma^2}(\mu + \sqrt{2s\sigma^2 + \mu^2})x}.$$

This relation is contradictory since $e^{-s\tau} < 1$ while $e^{\frac{1}{\sigma^2}(\mu + \sqrt{2s\sigma^2 + \mu^2})x} > 1$, where we used that $s, x, \tau > 0$. Hence it follows that $c > 0$. Substituting the first solution into (4.6.10) leads to

$$\mathbb{E}[e^{-s\tau}] = e^{\frac{1}{\sigma^2}(\mu - \sqrt{2s\sigma^2 + \mu^2})x}.$$

We arrive at the following result:

Proposition 4.6.4 *Assume $\mu, x > 0$. Let τ be the time the process $X_t = \mu t + \sigma W_t$ hits x for the first time. Then we have*

$$\mathbb{E}[e^{-s\tau}] = e^{\frac{1}{\sigma^2}(\mu - \sqrt{2s\sigma^2 + \mu^2})x}, \qquad s > 0. \tag{4.6.11}$$

Proposition 4.6.5 *Let τ be the time the process $X_t = \mu t + \sigma W_t$ hits x, with $x > 0$ and $\mu > 0$.*

(a) Then the density function of τ is given by

$$p(\tau) = \frac{x}{\sigma\sqrt{2\pi}\tau^{3/2}} e^{-\frac{(x-\mu\tau)^2}{2\tau\sigma^2}}, \qquad \tau > 0. \tag{4.6.12}$$

(b) The mean and variance of τ are

$$\mathbb{E}[\tau] = \frac{x}{\mu}, \qquad Var(\tau) = \frac{x\sigma^2}{\mu^3}.$$

Proof: (*a*) Let $p(\tau)$ be the density function of τ. Since

$$\mathbb{E}[e^{-s\tau}] = \int_0^\infty e^{-s\tau} p(\tau)\,d\tau = \mathcal{L}\{p(\tau)\}(s)$$

is the Laplace transform of $p(\tau)$, applying the inverse Laplace transform yields

$$\begin{aligned} p(\tau) &= \mathcal{L}^{-1}\{\mathbb{E}[e^{-s\tau}]\} = \mathcal{L}^{-1}\{e^{\frac{1}{\sigma^2}(\mu - \sqrt{2s\sigma^2+\mu^2})x}\} \\ &= \frac{x}{\sigma\sqrt{2\pi}\tau^{3/2}} e^{-\frac{(x-\mu\tau)^2}{2\tau\sigma^2}}, \qquad \tau > 0. \end{aligned}$$

It is worth noting that the computation of the previous inverse Laplace transform is non-elementary; however, it can be easily computed using the Mathematica software.

(b) The moments are obtained by differentiating the moment generating function and taking the value at $s = 0$

$$\begin{aligned} \mathbb{E}[\tau] &= -\frac{d}{ds}\mathbb{E}[e^{-s\tau}]\Big|_{s=0} = -\frac{d}{ds} e^{\frac{1}{\sigma^2}(\mu - \sqrt{2s\sigma^2+\mu^2})x}\Big|_{s=0} \\ &= \frac{x}{\sqrt{\mu^2 + 2s\mu}} e^{\frac{1}{\sigma^2}(\mu - \sqrt{2s\sigma^2+\mu^2})x}\Big|_{s=0} \\ &= \frac{x}{\mu}. \end{aligned}$$

$$\begin{aligned} \mathbb{E}[\tau^2] &= (-1)^2\frac{d^2}{ds^2}\mathbb{E}[e^{-s\tau}]\Big|_{s=0} = \frac{d^2}{ds^2} e^{\frac{1}{\sigma^2}(\mu - \sqrt{2s\sigma^2+\mu^2})x}\Big|_{s=0} \\ &= \frac{x(\sigma^2 + x\sqrt{\mu^2 + 2s\sigma^2})}{(\mu^2 + 2s\sigma^2)^{3/2}} e^{\frac{1}{\sigma^2}(\mu - \sqrt{2s\sigma^2+\mu^2})x}\Big|_{s=0} \\ &= \frac{x\sigma^2}{\mu^3} + \frac{x^2}{\mu^2}. \end{aligned}$$

Hence

$$Var(\tau) = \mathbb{E}[\tau^2] - \mathbb{E}[\tau]^2 = \frac{x\sigma^2}{\mu^3}.$$

It is worth noting that we can arrive at the formula $\mathbb{E}[\tau] = \frac{x}{\mu}$ in the following heuristic way. Taking the expectation in the equation $\mu\tau + \sigma W_\tau = x$ yields $\mu\mathbb{E}[\tau] = x$, where we used that $\mathbb{E}[W_\tau] = 0$ for any finite stopping time τ (see Exercise 4.5.8 (a)). Solving for $\mathbb{E}[\tau]$ yields the aforementioned formula. ■

Even if the computations are more or less similar to the previous result, we shall treat next the case of the negative barrier in its full length. This is because of its particular importance in being useful in practical problems, such as pricing perpetual American puts.

Proposition 4.6.6 *Assume $\mu, x > 0$. Let τ be the time the process $X_t = \mu t + \sigma W_t$ hits $-x$ for the first time.*

(a) We have

$$\mathbb{E}[e^{-s\tau}] = e^{\frac{-1}{\sigma^2}(\mu+\sqrt{2s\sigma^2+\mu^2})x}, \qquad s > 0. \tag{4.6.13}$$

(b) Then the density function of τ is given by

$$p(\tau) = \frac{x}{\sigma\sqrt{2\pi}\tau^{3/2}} e^{-\frac{(x+\mu\tau)^2}{2\tau\sigma^2}}, \qquad \tau > 0. \tag{4.6.14}$$

(c) The mean of τ is

$$\mathbb{E}[\tau] = \frac{x}{\mu} e^{-\frac{2\mu x}{\sigma^2}}.$$

Proof: (*a*) Consider the stopping time $\tau = \inf\{t > 0; X_t = -x\}$. By the Optional Stopping Theorem (Theorem 4.2.1) applied to the martingale $M_t = e^{cW_t - \frac{c^2}{2}t}$ yields

$$\begin{aligned} 1 &= M_0 = \mathbb{E}[M_\tau] = E\big[e^{cW_\tau - \frac{c^2}{2}\tau}\big] = E\big[e^{\frac{c}{\sigma}(X_\tau - \mu\tau) - \frac{c^2}{2}\tau}\big] \\ &= E\big[e^{-\frac{c}{\sigma}x - \frac{c\mu}{\sigma}\tau - \frac{c^2}{2}\tau}\big] = e^{-\frac{c}{\sigma}x} E\big[e^{-(\frac{c\mu}{\sigma} + \frac{c^2}{2})\tau}\big]. \end{aligned}$$

Therefore

$$E\big[e^{-(\frac{c\mu}{\sigma} + \frac{c^2}{2})\tau}\big] = e^{\frac{c}{\sigma}x}. \tag{4.6.15}$$

If let $s = \frac{c\mu}{\sigma} + \frac{c^2}{2}$, then solving for c yields $c = -\frac{\mu}{\sigma} \pm \sqrt{2s + \frac{\mu^2}{\sigma^2}}$, but only the negative solution works out; this comes from the fact that both terms of the equation (4.6.15) have to be less than 1. Hence (4.6.15) becomes

$$\mathbb{E}[e^{-s\tau}] = e^{\frac{-1}{\sigma^2}(\mu+\sqrt{2s\sigma^2+\mu^2})x}, \qquad s > 0.$$

(*b*) Relation (4.6.13) can be written equivalently as

$$\mathcal{L}\big(p(\tau)\big) = e^{\frac{-1}{\sigma^2}(\mu+\sqrt{2s\sigma^2+\mu^2})x}.$$

Taking the inverse Laplace transform, and using Mathematica software to compute it, we obtain

$$p(\tau) = \mathcal{L}^{-1}\Big(e^{\frac{-1}{\sigma^2}(\mu+\sqrt{2s\sigma^2+\mu^2})x}\Big)(\tau) = \frac{x}{\sigma\sqrt{2\pi}\tau^{3/2}} e^{-\frac{(x+\mu\tau)^2}{2\tau\sigma^2}}, \qquad \tau > 0.$$

(*c*) Differentiating and evaluating at $s = 0$ we obtain

$$\mathbb{E}[\tau] = -\frac{d}{ds}\mathbb{E}[e^{-s\tau}]\Big|_{s=0} = e^{-\frac{2\mu}{\sigma^2}x}\frac{x}{\sigma^2}\frac{\sigma^2}{\sqrt{\mu^2}} = \frac{x}{\mu}e^{-\frac{2\mu x}{\sigma^2}}.$$

∎

Exercise 4.6.7 *Assume the hypothesis of Proposition 4.6.6 are satisfied. Find* $Var(\tau)$.

Exercise 4.6.8 *Find the modes of distributions* (4.6.12) *and* (4.6.14). *What do you notice?*

Exercise 4.6.9 *Let* $X_t = 2t + 3W_t$ *and* $Y_t = 2t + W_t$.

(*a*) *Show that the expected times for* X_t *and* Y_t *to reach any barrier* $x > 0$ *are the same.*

(*b*) *If* X_t *and* Y_t *model the prices of two stocks, which one would you like to own?*

Exercise 4.6.10 *Does* $4t + 2W_t$ *hit* 9 *faster (in expectation) than* $5t + 3W_t$ *hits* 14*?*

Exercise 4.6.11 *Let* τ *be the first time the Brownian motion with drift* $X_t = \mu t + W_t$ *hits* x, *where* $\mu, x > 0$. *Prove the inequality*

$$P(\tau \leq \lambda) \leq e^{-\frac{x^2+\lambda^2\mu^2}{2\lambda}+\mu x}, \quad \forall \lambda > 0.$$

The double barrier case In the following we shall consider the case of double barrier. Consider the Brownian motion with drift $X_t = \mu t + W_t$, $\mu > 0$. Let $\alpha, \beta > 0$ and define the stopping time

$$T = \inf\{t > 0; X_t \geq \alpha \text{ or } X_t \leq -\beta\}.$$

Relation (4.5.6) states

$$\mathbb{E}[e^{cX_T} e^{-(c\mu+\frac{1}{2}c^2)T}] = 1.$$

Since the random variables T and X_T are independent (why?), we have

$$\mathbb{E}[e^{cX_T}]\mathbb{E}[e^{-(c\mu+\frac{1}{2}c^2)T}] = 1.$$

Using $\mathbb{E}[e^{cX_T}] = e^{c\alpha}p_\alpha + e^{-c\beta}(1 - p_\alpha)$, with p_α given by (4.5.7), then

$$\mathbb{E}[e^{-(c\mu+\frac{1}{2}c^2)T}] = \frac{1}{e^{c\alpha}p_\alpha + e^{-c\beta}(1 - p_\alpha)}.$$

If we substitute $s = c\mu + \frac{1}{2}c^2$, then

$$\mathbb{E}[e^{-sT}] = \frac{1}{e^{(-\mu+\sqrt{2s+\mu^2})\alpha}p_\alpha + e^{-(-\mu+\sqrt{2s+\mu^2})\beta}(1 - p_\alpha)}. \tag{4.6.16}$$

The probability density of the stopping time T is obtained by taking the inverse Laplace transform of the right side expression

$$p(T) = \mathcal{L}^{-1}\left(\frac{1}{e^{(-\mu+\sqrt{2s+\mu^2})\alpha}p_\alpha + e^{-(-\mu+\sqrt{2s+\mu^2})\beta}(1-p_\alpha)}\right)(\tau),$$

an expression which is not feasible for having a closed form solution. However, expression (4.6.16) would be useful for computing the price for double barrier derivatives.

Exercise 4.6.12 *Use formula* (4.6.16) *to find the expectation* $\mathbb{E}[T]$.

Exercise 4.6.13 *Let* $T_x = \inf\{t > 0; |W_t| \geq x\}$, *for* $x \geq 0$.
(*a*) *Show that* $\mathbb{E}[e^{cW_{T_x}}] = \cosh(cx)$, *for any* $c > 0$.
(*b*) *Prove that*

$$\mathbb{E}[e^{-\lambda T_x}] = \text{sech}(\sqrt{2\lambda}x), \qquad \forall \lambda \geq 0.$$

Exercise 4.6.14 *Denote by* $M_t = N_t - \lambda t$ *the compensated Poisson process and let* $c > 0$ *be a constant.*
(*a*) *Show that*

$$X_t = e^{cM_t - \lambda t(e^c - c - 1)}$$

is an $\mathcal{F}_t$*-martingale, with* $\mathcal{F}_t = \sigma(N_u; u \leq t)$.
(*b*) *Let* $a > 0$ *and* $T = \inf\{t > 0; M_t > a\}$ *be the first hitting time of the level* a. *Use the Optional Stopping Theorem to show that*

$$\mathbb{E}[e^{-\lambda s T}] = e^{-\varphi(s)a}, \qquad s > 0,$$

where $\varphi : [0, \infty) \to [0, \infty)$ *is the inverse function of* $f(x) = e^x - x - 1$.
(*c*) *Show that* $\mathbb{E}[T] = \infty$.
(*d*) *Can you use the inverse Laplace transform to find the probability density function of* T?

4.7 The Theorems of Lévy and Pitman

This section presents two of the most famous results on Brownian motions, which are Lévy's and Pitman's theorems. They deal with surprising equivalences in law involving a Brownian motion and its maximum and minimum.

Let W_t be a standard Brownian motion and denote its *running maximum* by

$$M_t = \max_{0 \leq s \leq t} W_s.$$

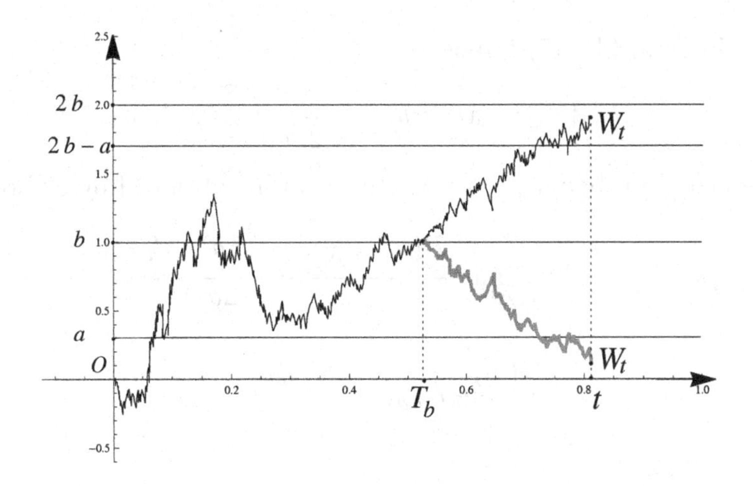

Figure 4.4: *The reflection principle for a Brownian motion.*

From Exercise 4.3.10 the probability density of M_t is given by

$$p_t(x) = \frac{2}{\sqrt{2\pi t}} e^{-\frac{x^2}{2t}}, \quad x \geq 0.$$

Lemma 4.7.1 *The joint density function of* (W_t, M_t) *is given by*

$$f(a, b) = \frac{2(2b-a)}{\sqrt{2\pi} t^{3/2}} e^{-\frac{(2b-a)^2}{2t}}, \qquad a \leq b,\ b \geq 0.$$

Proof: Let $a \leq b$. Assume $M_t \geq b$, i.e. W_s takes values larger than or equal to b in the interval $[0, t]$. By the reflection principle, see Fig. 4.4, the Brownian motion reflected at the stopping time T_b is also a Brownian motion. The probabilities to increase or decrease by an amount of at least $b - a$ are then equal

$$P(W_t \geq 2b - a \,|\, M_t \geq b) = P(W_t \leq a \,|\, M_t \geq b).$$

Multiplying by $P(M_t \geq b)$ and using the conditional probability formula yields

$$P(W_t \geq 2b - a,\ M_t \geq b) = P(W_t \leq a,\ M_t \geq b). \tag{4.7.17}$$

Conditions $W_t \geq 2b - a$ and $2b - a \geq b$ imply $M_t \geq b$, so

$$P(M_t \geq b \,|\, W_t \geq 2b - a) = 1.$$

Then the left side of (4.7.17) can be computed as

$$\begin{aligned} P(W_t \geq 2b - a,\ M_t \geq b) &= P(W_t \geq 2b - a) P(M_t \geq b \,|\, W_t \geq 2b - a) \\ &= P(W_t \geq 2b - a) \\ &= \frac{1}{\sqrt{2\pi t}} \int_{2b-a}^{\infty} e^{-\frac{x^2}{2t}}\, dx. \end{aligned}$$

Substituting into (4.7.17) implies

$$P(W_t \leq a,\, M_t \geq b) = \frac{1}{\sqrt{2\pi t}} \int_{2b-a}^{\infty} e^{-\frac{x^2}{2t}}\, dx.$$

The associated probability density $f(a,b)$ can be obtained by differentiation

$$\begin{aligned} f(a,b) &= -\lim_{\Delta a \to 0} \lim_{\Delta b \to 0} \frac{P\Big(W_t \in (a, a-\Delta a),\, M_t \in (b, b+\Delta b)\Big)}{\Delta a\, \Delta b} \\ &= -\frac{\partial^2}{\partial a \partial b}\Big(\frac{1}{\sqrt{2\pi t}} \int_{2b-a}^{\infty} e^{-\frac{x^2}{2t}}\, dx\Big) \\ &= \frac{2(2b-a)}{\sqrt{2\pi t^3}} e^{-\frac{(2b-a)^2}{2t}}. \end{aligned}$$

■

The following equivalence in law was found by Lévy [33] in 1948.

Theorem 4.7.2 *The processes $X_t = M_t - W_t$ and $|W_t|$, $t \geq 0$ have the same probability law.*

Proof: The probability density of $|W_t|$ is given by Exercise 4.3.11 (a)

$$p_t(x) = \frac{2}{\sqrt{2\pi t}} e^{-\frac{x^2}{2t}}, \quad x \geq 0. \tag{4.7.18}$$

Using a direct computation and Lemma 4.7.1 we shall show that $X_t = M_t - W_t$ also has the density (4.7.18). For $u \geq 0$ we have

$$\begin{aligned} P(X_t \leq u) &= P(M_t - W_t \leq u) = \iint_{\{0 \leq y-x \leq u, y \geq 0\}} f(x,y)\, dx\, dy \\ &= I_1 - I_2, \end{aligned}$$

with

$$I_1 = \iint_{\{y-x \leq u, y \geq 0\}} f(x,y)\, dx\, dy, \quad I_2 = \iint_{\{0 \leq y \leq x\}} f(x,y)\, dx\, dy.$$

This writes the integral over a strip as a difference of integrals over the interior of two angles, see Fig. 4.5(a). A great simplification of computation is done by observing that the second integral vanishes

$$\begin{aligned} I_2 &= \int_0^{\infty} \int_0^{x} f(x,y)\, dy\, dx \\ &= \frac{2}{\sqrt{2\pi} t^{3/2}} \int_0^{\infty} \int_0^{x} (2y-x) e^{-\frac{(2y-x)^2}{2t}}\, dy\, dx \\ &= \frac{1}{\sqrt{2\pi} t^{3/2}} \int_0^{\infty} \int_{-x}^{x} z e^{-\frac{z^2}{2t}}\, dz\, dx = 0, \end{aligned}$$

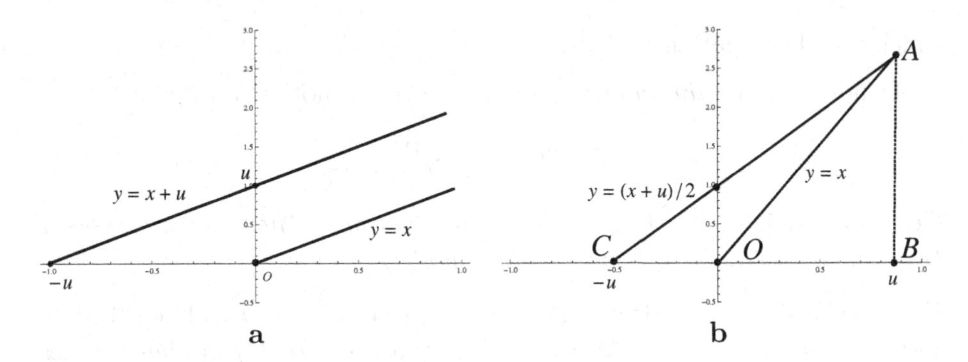

Figure 4.5: (a) *The integration strip for the proof of Lévy's theorem.* (b) *The domains D_{AOC}, D_{AOB}, and D_{ABC} for the proof of Pitman's theorem.*

as an integral of an odd function over a symmetric interval. Next we shall compute the first integral

$$\begin{aligned}
I_1 &= \int_{-u}^{\infty} \int_0^{u+x} f(x,y)\,dy\,dx \\
&= \frac{2}{\sqrt{2\pi}t^{3/2}} \int_{-u}^{\infty} \int_0^{u+x} (2y-x)e^{-\frac{(2y-x)^2}{2t}}\,dy\,dx \\
&= \frac{2}{\sqrt{2\pi t}} \int_{-u}^{\infty} \int_{-x/\sqrt{2t}}^{(2u+x)/\sqrt{2t}} ze^{-z^2}\,dz\,dx,
\end{aligned}$$

where we substituted $z = 2y - x$. Changing the variable in $r = z^2$, we get

$$\begin{aligned}
I_1 &= \frac{1}{\sqrt{2\pi t}} \int_{-u}^{\infty} \int_{x^2/(2t)}^{(2u+x)^2/(2t)} e^{-r}\,dr\,dx \\
&= \frac{1}{\sqrt{2\pi t}} \int_{-u}^{\infty} \left(e^{-\frac{x^2}{2t}} - e^{-\frac{(2u+x)^2}{2t}}\right) dx \\
&= \frac{1}{\sqrt{2\pi t}} \int_{-u}^{\infty} e^{-\frac{x^2}{2t}}\,dx - \frac{1}{\sqrt{2\pi t}} \int_{u}^{\infty} e^{-\frac{v^2}{2t}}\,dv \\
&= \frac{1}{\sqrt{2\pi t}} \int_{-u}^{u} e^{-\frac{x^2}{2t}}\,dx = \frac{2}{\sqrt{2\pi t}} \int_0^{u} e^{-\frac{x^2}{2t}}\,dx.
\end{aligned}$$

Therefore $P(X_t \leq u) = I_1 = \frac{2}{\sqrt{2\pi t}} \int_0^u e^{-\frac{x^2}{2t}}\,dx$. Differentiate with respect to u we obtain the probability density of X_t

$$\begin{aligned}
p_{X_t}(u) &= \frac{d}{du}P(X_t \leq u) = \frac{d}{du}\frac{2}{\sqrt{2\pi t}} \int_0^u e^{-\frac{x^2}{2t}}\,dx \\
&= \frac{2}{\sqrt{2\pi t}} e^{-\frac{u^2}{2t}}, \quad u \geq 0,
\end{aligned}$$

which matches relation (4.7.18). ■

Denote the *running minimum* of a Brownian motion W_t by

$$m_t = \min_{0 \le s \le t} W_s.$$

Corollary 4.7.3 *Let B_t be a Brownian motion. Then the processes $Y_t = B_t - m_t$ and $|B_t|$, $t \geq 0$ have the same probability law.*

Proof: Consider the reflected Brownian motion $W_t = -B_t$. Then $|W_t| = |B_t|$ and $B_t - m_t = M_t - W_t$. Applying Theorem 4.7.2 for W_t implies the desired result. ■

It is worth concluding that the processes $|W_t|$, M_t and $M_t - W_t$ have the same law.

Before getting to the second result of this section, we shall recall a few results. The 3-dimensional Bessel process R_t is the process satisfied by the Euclidean distance from the origin to a 3-dimensional Brownian motion $W(t) = (W_1(t), W_2(t), W_2(t))$ in $\mathbb{R}^3$

$$R_t = \sqrt{W_1(t)^2 + W_2(t)^2 + W_2(t)^3}.$$

Its probability density is given by Proposition 3.7.1

$$p_t(\rho) = \frac{2}{\sqrt{2\pi} t^{3/2}} \rho^2 e^{-\frac{\rho^2}{2t}}, \qquad \rho > 0. \tag{4.7.19}$$

In the following we present another striking identity in law, similar to Lévy's, which was found by Pitman [38] in 1975.

Theorem 4.7.4 *The process $Z_t = 2M_t - W_t$, $t \geq 0$ is distributed as a 3-dimensional Bessel process R_t.*

Proof: We shall follow the same idea as in the proof of Theorem 4.7.2. Let $u \geq 0$ and consider the domains, see Fig. 4.5(b).

$$\begin{aligned} D_{AOC} &= \{2y - x \le u, y \ge 0, x \le y\} \\ D_{AOB} &= \{0 \le x, x \ge y\} \\ D_{ABC} &= \{2y - x \le u, y \ge 0\}. \end{aligned}$$

The probability function of Z_t can be evaluated as

$$\begin{aligned} P(Z_t \le u) &= P(2M_t - W_t \le u) = \iint_{D_{AOC}} f(x, y)\, dx\, dy \qquad (4.7.20) \\ &= \iint_{D_{ABC}} f(x, y)\, dx\, dy - \iint_{D_{AOB}} f(x, y)\, dx\, dy \\ &= I_1(u) - I_2(u). \end{aligned}$$

We note first that the second integral vanishes, as an integral of an odd function over a symmetric interval

$$\begin{aligned} I_2(u) &= \int_0^u \int_0^x \frac{2}{\sqrt{2\pi}t^{3/2}}(2y-x)e^{-\frac{(2y-x)^2}{2t}}\,dy\,dx \\ &= \frac{1}{\sqrt{2\pi}t^{3/2}}\int_0^u \int_{-x}^x ze^{\frac{z^2}{2t}}\,dz\,dx = 0. \end{aligned}$$

The first integral is computed using the substitutions $z = 2y - x$ and $r = z^2$

$$\begin{aligned} I_1(u) &= \int_{-u}^u \int_0^{(x+u)/2} \frac{2}{\sqrt{2\pi}t^{3/2}}(2y-x)e^{-\frac{(2y-x)^2}{2t}}\,dy\,dx \\ &= \frac{1}{\sqrt{2\pi}t^{3/2}}\int_{-u}^u \int_{-x}^u ze^{\frac{z^2}{2t}}\,dz\,dx = \frac{1}{2\sqrt{2\pi}t^{3/2}}\int_{-u}^u \int_{x^2}^{u^2} e^{-\frac{r}{2t}}\,dr\,dx \\ &= \frac{t}{\sqrt{2\pi}t^{3/2}}\int_{-u}^u \left(e^{-\frac{x^2}{2t}} - e^{-\frac{u^2}{2t}}\right)dx \\ &= \frac{1}{\sqrt{2\pi t}}\int_{-u}^u e^{-\frac{x^2}{2t}}\,dx - \frac{2}{\sqrt{2\pi t}}ue^{-\frac{u^2}{2t}} \\ &= \frac{2}{\sqrt{2\pi t}}\left\{\int_0^u e^{-\frac{x^2}{2t}}\,dx - ue^{-\frac{u^2}{2t}}\right\}. \end{aligned}$$

Differentiate using the Fundamental Theorem of Calculus and the product rule to obtain

$$\frac{d}{du}I_1(u) = \frac{2}{\sqrt{2\pi}\,t^{3/2}}u^2e^{-\frac{u^2}{2t}}.$$

Using (4.7.21) yields

$$\frac{d}{du}P(Z_t \le u) = \frac{2}{\sqrt{2\pi}\,t^{3/2}}u^2e^{-\frac{u^2}{2t}},$$

which retrieves the density function (4.7.19) of a 3-dimensional Bessel process, and hence the theorem is proved.

■

Corollary 4.7.5 *The process $W_t - 2m_t$, $t \ge 0$ is distributed as a 3-dimensional Bessel process R_t.*

Exercise 4.7.6 *Prove Corollary 4.7.5.*

4.8 Limits of Stochastic Processes

Let $(X_t)_{t\geq 0}$ be a stochastic process. We can make sense of the limit expression $X = \lim\limits_{t\to\infty} X_t$, in a similar way as we did in section 2.14 for sequences of random variables. We shall rewrite the definitions for the continuous case.

Almost Certain Limit The process X_t converges *almost certainly* to X, if for all states of the world ω, except a set of probability zero, we have

$$\lim_{t\to\infty} X_t(\omega) = X(\omega).$$

We shall write $\text{ac-}\lim\limits_{t\to\infty} X_t = X$. It is also sometimes called *strong convergence.*

Mean Square Limit We say that the process X_t converges to X in the *mean square* if

$$\lim_{t\to\infty} \mathbb{E}[(X_t - X)^2] = 0.$$

In this case we write $\text{ms-}\lim\limits_{t\to\infty} X_t = X$.

Limit in Probability The stochastic process X_t converges in *probability* to X if

$$\lim_{t\to\infty} P\big(\omega; |X_t(\omega) - X(\omega)| > \epsilon\big) = 0.$$

This limit is abbreviated by $\text{p-}\lim\limits_{t\to\infty} X_t = X$.

It is worth noting that, like in the case of sequences of random variables, both almost certain convergence and convergence in mean square imply the convergence in probability, which implies the limit in distribution.

Limit in Distribution We say that X_t converges *in distribution* to X if for any continuous bounded function $\varphi(x)$ we have

$$\lim_{t\to\infty} \mathbb{E}[\varphi(X_t)] = \mathbb{E}[\varphi(X)].$$

It is worth noting that the stochastic convergence implies the convergence in distribution.

4.9 Mean Square Convergence

The following property is a reformulation of Proposition 2.14.1 in the continuous setup, the main lines of the proof remaining the same.

Proposition 4.9.1 *Consider a stochastic process X_t such that $\mathbb{E}[X_t] \to k$, a constant, and $Var(X_t) \to 0$ as $t \to \infty$. Then $\text{ms-}\lim\limits_{t\to\infty} X_t = k$.*

It is worthy to note that the previous statement holds true if the limit to infinity is replaced by a limit to any other number.

Next we shall provide a few applications.

Application 4.9.2 *If $\alpha > 1/2$, then*

$$ms\text{-}\lim_{t\to\infty} \frac{W_t}{t^\alpha} = 0.$$

Proof: Let $X_t = \dfrac{W_t}{t^\alpha}$. Then $\mathbb{E}[X_t] = \dfrac{\mathbb{E}[W_t]}{t^\alpha} = 0$, and

$$Var[X_t] = \frac{1}{t^{2\alpha}} Var[W_t] = \frac{t}{t^{2\alpha}} = \frac{1}{t^{2\alpha-1}},$$

for any $t > 0$. Since $\dfrac{1}{t^{2\alpha-1}} \to 0$ as $t \to \infty$, applying Proposition 4.9.1 yields $\text{ms-}\lim\limits_{t\to\infty} \dfrac{W_t}{t^\alpha} = 0$. ■

Corollary 4.9.3 *We have* $ms\text{-}\lim\limits_{t\to\infty} \dfrac{W_t}{t} = 0$.

Application 4.9.4 *Let $Z_t = \int_0^t W_s\, ds$. If $\beta > 3/2$, then*

$$ms\text{-}\lim_{t\to\infty} \frac{Z_t}{t^\beta} = 0.$$

Proof: Let $X_t = \dfrac{Z_t}{t^\beta}$. Then $\mathbb{E}[X_t] = \dfrac{\mathbb{E}[Z_t]}{t^\beta} = 0$, and

$$Var[X_t] = \frac{1}{t^{2\beta}} Var[Z_t] = \frac{t^3}{3t^{2\beta}} = \frac{1}{3t^{2\beta-3}},$$

for any $t > 0$. Since $\dfrac{1}{3t^{2\beta-3}} \to 0$ as $t \to \infty$, applying Proposition 4.9.1 leads to the desired result. ■

Application 4.9.5 *For any $p > 0$, $c \geq 1$ we have*

$$ms\text{-}\lim_{t\to\infty} \frac{e^{W_t - ct}}{t^p} = 0.$$

Proof: Consider the process $X_t = \dfrac{e^{W_t - ct}}{t^p} = \dfrac{e^{W_t}}{t^p e^{ct}}$. Since

$$\begin{aligned}
\mathbb{E}[X_t] &= \frac{\mathbb{E}[e^{W_t}]}{t^p e^{ct}} = \frac{e^{t/2}}{t^p e^{ct}} = \frac{1}{e^{(c-\frac{1}{2})t}} \frac{1}{t^p} \to 0, \quad \text{as } t \to \infty \\
Var[X_t] &= \frac{Var[e^{W_t}]}{t^{2p} e^{2ct}} = \frac{e^{2t} - e^t}{t^{2p} e^{2ct}} = \frac{1}{t^{2p}} \left(\frac{1}{e^{2t(c-1)}} - \frac{1}{e^{t(2c-1)}} \right) \to 0,
\end{aligned}$$

as $t \to \infty$, Proposition 4.9.1 leads to the desired result. ■

Application 4.9.6 *Show that*

$$ms\text{-}\lim_{t\to\infty} \frac{\max\limits_{0\le s\le t} W_s}{t} = 0.$$

Proof: Let $X_t = \dfrac{\max\limits_{0\le s\le t} W_s}{t}$. Since by Exercise 4.3.10

$$\begin{aligned} E[\max_{0\le s\le t} W_s] &= 0 \\ Var(\max_{0\le s\le t} W_s) &= 2\sqrt{t}, \end{aligned}$$

then

$$\begin{aligned} \mathbb{E}[X_t] &= 0 \\ Var[X_t] &= \frac{2\sqrt{t}}{t^2} \to 0, \qquad t\to\infty. \end{aligned}$$

Apply Proposition 4.9.1 to get the desired result. ■

Remark 4.9.7 One of the strongest results regarding limits of Brownian motions is called *the law of iterated logarithms* and was first proved by Lamperti:

$$\limsup_{t\to\infty} \frac{W_t}{\sqrt{2t\ln(\ln t)}} = 1,$$

almost certainly.

Exercise 4.9.8 *Find a stochastic process X_t such that the following both conditions are satisfied:*

(*i*) $ms\text{-}\lim\limits_{t\to\infty} X_t = 0$

(*ii*) $ms\text{-}\lim\limits_{t\to\infty} X_t^2 \neq 0.$

Exercise 4.9.9 *Let X_t be a stochastic process. Show that*

$$ms\text{-}\lim_{t\to\infty} X_t = 0 \Longleftrightarrow ms\text{-}\lim_{t\to\infty} |X_t| = 0.$$

4.10 The Martingale Convergence Theorem

We state now, without proof, a result which is a powerful way of proving the almost sure convergence. We start with the discrete version:

Theorem 4.10.1 *Let X_n be a martingale with bounded means*

$$\exists M > 0 \text{ such that } \mathbb{E}[|X_n|] \leq M, \qquad \forall n \geq 1. \tag{4.10.21}$$

Then there is a random variable L such that as-$\lim_{n\to\infty} X_n = L$, i.e.

$$P\big(\omega; \lim_{n\to\infty} X_n(\omega) = L(\omega)\big) = 1.$$

Since $\mathbb{E}[|X_n|]^2 \leq \mathbb{E}[X_n^2]$, the condition (4.10.21) can be replaced by its stronger version

$$\exists M > 0 \text{ such that } \mathbb{E}[X_n^2] \leq M, \qquad \forall n \geq 1.$$

The following result deals with the continuous version of the Martingale Convergence Theorem. Denote the infinite knowledge by $\mathcal{F}_\infty = \sigma\Big(\cup_t \mathcal{F}_t\Big)$.

Theorem 4.10.2 *Let X_t be an $\mathcal{F}_t$-martingale such that*

$$\exists M > 0 \text{ such that } \mathbb{E}[|X_t|] < M, \qquad \forall t > 0.$$

Then there is an $\mathcal{F}_\infty$-measurable random variable X_∞ such that $X_t \to X_\infty$ a.s. as $t \to \infty$.

The next exercise involves a process that is as-convergent but is not ms-convergent.

Exercise 4.10.3 *It is known that $X_t = e^{W_t - t/2}$ is a martingale. Since*

$$\mathbb{E}[|X_t|] = \mathbb{E}[e^{W_t - t/2}] = e^{-t/2}\mathbb{E}[e^{W_t}] = e^{-t/2}e^{t/2} = 1,$$

by the Martingale Convergence Theorem there is a random variable L such that $X_t \to L$ a.s. as $t \to \infty$.

(a) *What is the limit L? How did you make your guess?*

(b) *Show that*

$$\mathbb{E}[|X_t - 1|^2] = Var(X_t) + \Big(E(X_t) - 1\Big)^2.$$

(c) *Show that X_t does not converge in the mean square to 1.*

(d) *Prove that the sequence X_t is as-convergent but it is not ms-convergent.*

Exercise 4.10.4 *Let $X_t = W_t + 1$, where W_t is a Brownian motion and consider*

$$T = \inf\{t > 0; X_t \leq 0\}.$$

(a) *Is T a stopping time?*

(b) *Is $Y_t = X_{T\wedge t}$ a continuous martingale?*

(c) *Show that $\mathbb{E}[Y_t] = 1$, $\forall t > 0$.*

(d) *Verify the limit as-$\lim_{t\to\infty} Y_t = 0$.*

(e) *Is this contradicting the Optional Stopping Theorem?*

4.11 Quadratic Variation

In order to gauge the roughness of stochastic processes we shall consider the sum of k-powers of consecutive increments in the process, as the norm of the partition decreases to zero.

Definition 4.11.1 *Let X_t be a continuous stochastic process on Ω, and $k > 0$ be a constant. The kth variation process of X_t, denoted by $\langle X, X\rangle_t^{(k)}$ is defined by the following limit in probability*

$$\langle X, X\rangle_t^{(k)}(\omega) = p\text{-}\lim_{\max_i |t_{i+1}-t_i|} \sum_{i=0}^{n-1} |X_{t_{i+1}}(\omega) - X_{t_i}(\omega)|^k,$$

where $0 = t_1 < t_2 < \cdots < t_n = t$.

If $k = 1$, the process $\langle X, X\rangle_t^{(1)}$ is called the total variation process of X_t.

For $k = 2$, $\langle X, X\rangle_t = \langle X, X\rangle_t^{(2)}$ is called the quadratic variation process of X_t.

It can be shown that the quadratic variation exists and is unique (up to indistinguishability) for continuous square integrable martingales X_t, i.e. martingales satisfying $X_0 = 0$ a.s. and $\mathbb{E}[X_t^2] < \infty$, for all $t \geq 0$. Furthermore, the quadratic variation, $\langle X, X\rangle_t$, of a square integrable martingale X_t is an increasing continuous process satisfying

(*i*) $\langle X, X\rangle_0 = 0$;

(*ii*) $X_t^2 - \langle X, X\rangle_t$ is a martingale.

Next we introduce a symmetric and bilinear operation.

Definition 4.11.2 *The quadratic covariation of two continuous square integrable martingales X_t and Y_t is defined as*

$$\langle X, Y\rangle_t = \frac{1}{4}\Big(\langle X+Y, X+Y\rangle_t - \langle X-Y, X-Y\rangle_t\Big).$$

Exercise 4.11.3 *Prove that:*

(*a*) $\langle X, Y\rangle_t = \langle Y, X\rangle_t$;

(*b*) $\langle aX + bY, Z\rangle_t = a\langle X, Z\rangle_t + b\langle Y, Z\rangle_t$.

Exercise 4.11.4 *Let M_t and N_t be square integrable martingales. Prove that the process $M_tN_t - \langle M, N\rangle_t$ is a martingale.*

Exercise 4.11.5 *Prove that the total variation on the interval $[0, t]$ of a Brownian motion is infinite a.s.*

We shall encounter in the following a few important examples that will be useful when dealing with stochastic integrals.

4.11.1 The quadratic variation of W_t

The next result states that the *quadratic variation* of the Brownian motion W_t on the interval $[0, T]$ is T. More precisely, we have:

Proposition 4.11.6 *Let $T > 0$ and consider the equidistant partition $0 = t_0 < t_1 < \cdots t_{n-1} < t_n = T$. Then*

$$\boxed{ms\text{-}\lim_{n\to\infty} \sum_{i=0}^{n-1} (W_{t_{i+1}} - W_{t_i})^2 = T.} \qquad (4.11.22)$$

Proof: Consider the random variable

$$X_n = \sum_{i=0}^{n-1} (W_{t_{i+1}} - W_{t_i})^2.$$

Since the increments of a Brownian motion are independent, Proposition 5.2.1 yields

$$\begin{aligned} \mathbb{E}[X_n] &= \sum_{i=0}^{n-1} \mathbb{E}[(W_{t_{i+1}} - W_{t_i})^2] = \sum_{i=0}^{n-1} (t_{i+1} - t_i) \\ &= t_n - t_0 = T; \end{aligned}$$

$$\begin{aligned} Var(X_n) &= \sum_{i=0}^{n-1} Var[(W_{t_{i+1}} - W_{t_i})^2] = \sum_{i=0}^{n-1} 2(t_{i+1} - t_i)^2 \\ &= n \cdot 2\Big(\frac{T}{n}\Big)^2 = \frac{2T^2}{n}, \end{aligned}$$

where we used that the partition is equidistant. Since X_n satisfies the conditions

$$\begin{aligned} \mathbb{E}[X_n] &= T, \quad \forall n \geq 1; \\ Var[X_n] &\to 0, \quad n \to \infty, \end{aligned}$$

by Proposition 4.9.1 we obtain $\text{ms-}\lim_{n\to\infty} X_n = T$, or

$$\text{ms-}\lim_{n\to\infty} \sum_{i=0}^{n-1} (W_{t_{i+1}} - W_{t_i})^2 = T. \qquad (4.11.23)$$

■

Since the mean square convergence implies convergence in probability, it follows that

$$\text{p-}\lim_{n\to\infty}\sum_{i=0}^{n-1}(W_{t_{i+1}} - W_{t_i})^2 = T,$$

i.e. the quadratic variation of W_t on $[0, T]$ is T.

Exercise 4.11.7 *Prove that the quadratic variation of the Brownian motion W_t on $[a, b]$ is equal to $b - a$.*

The Fundamental Relation $dW_t^2 = dt$ The relation discussed in this section can be regarded as the fundamental relation of Stochastic Calculus. We shall start by recalling relation (4.11.23)

$$\text{ms-}\lim_{n\to\infty}\sum_{i=0}^{n-1}(W_{t_{i+1}} - W_{t_i})^2 = T. \tag{4.11.24}$$

The right side can be regarded as a regular Riemann integral

$$T = \int_0^T dt,$$

while the left side can be regarded as a stochastic integral with respect to dW_t^2

$$\int_0^T (dW_t)^2 = \text{ms-}\lim_{n\to\infty}\sum_{i=0}^{n-1}(W_{t_{i+1}} - W_{t_i})^2.$$

Substituting into (4.11.24) yields

$$\int_0^T (dW_t)^2 = \int_0^T dt, \quad \forall T > 0.$$

The differential form of this integral equation is

$$\boxed{dW_t^2 = dt.}$$

In fact, this expression also holds in the mean square sense, as it can be inferred from the next exercise.

Exercise 4.11.8 *Show that*

(a) $\mathbb{E}[dW_t^2 - dt] = 0$;

(b) $Var(dW_t^2 - dt) = o(dt)$;

(c) $ms\text{-}\lim_{dt\to 0}(dW_t^2 - dt) = 0$.

Roughly speaking, the process dW_t^2, which is the square of infinitesimal increments of a Brownian motion, is deterministic. This relation plays a central role in Stochastic Calculus and will be useful when dealing with Ito's lemma.

The following exercise states that $dt\, dW_t = 0$, which is another important stochastic relation useful in Ito's lemma.

Exercise 4.11.9 *Consider the equidistant partition* $0 = t_0 < t_1 < \cdots t_{n-1} < t_n = T$. *Show that*

$$ms\text{-}\lim_{n\to\infty} \sum_{i=0}^{n-1} (W_{t_{i+1}} - W_{t_i})(t_{i+1} - t_i) = 0. \tag{4.11.25}$$

4.12 The Total Variation of Brownian Motion

The total variation of a Brownian motion W_t is defined as

$$V(W_t) = \sup_{t_k} \sum_{k=0}^{n-1} |W_{t_{k+1}} - W_{t_k}|,$$

for all partitions $0 = t_0 < t_1 < \cdots < t_{n-1} < t_n = T$. Without losing generality, we may assume the partition is equidistant, i.e. $t_{k+1} - t_k = \frac{T}{n}$. Equivalently,

$$V(W_t) = \lim_{n\to\infty} Y_n,$$

where

$$Y_n = \sum_{k=0}^{n-1} |W_{t_{k+1}} - W_{t_k}|.$$

Using Exercise 3.1.15 and the independent increments of the Brownian motion provides the mean and variance of the random variable Y_n

$$\begin{aligned}
\mu &= \mathbb{E}[Y_n] = \sum_{k=0}^{n-1} \mathbb{E}[|W_{t_{k+1}} - W_{t_k}|] = \sqrt{\frac{2}{\pi}} \sum_{k=0}^{n-1} \sqrt{(t_{k+1} - t_k)} = \sqrt{\frac{2nT}{\pi}} \\
\sigma^2 &= Var(Y_n) = \sum_{k=0}^{n-1} Var[|W_{t_{k+1}} - W_{t_k}|] = \Big(1 - \frac{2}{\pi}\Big) \sum_{k=0}^{n-1} (t_{k+1} - t_k) \\
&= \Big(1 - \frac{2}{\pi}\Big) T.
\end{aligned}$$

Since

$$\{\omega; Y_n(\omega) < \mu - k\sigma\} \subset \{\omega; |Y_n(\omega) - \mu| > k\sigma\}$$

Tchebychev's inequality, Theorem 2.13.10, provides

$$P(Y_n < \mu - k\sigma) \leq P(|Y_n - \mu| > k\sigma) < \frac{1}{k^2}, \qquad \forall k \geq 1.$$

This states that the probability of the left tail is smaller than the probability of both tails, the second being bounded by $\frac{1}{k^2}$. Using the probability of the complement event, the foregoing relation implies

$$P(Y_n \geq \mu - k\sigma) \geq 1 - \frac{1}{k^2}, \qquad \forall k \geq 1.$$

Substituting for μ and σ yields

$$P\Big(Y_n \geq \sqrt{\frac{2nT}{\pi}} - k\sqrt{\Big(1 - \frac{2}{\pi}\Big)T}\Big) \geq 1 - \frac{1}{k^2}, \qquad \forall k \geq 1.$$

Considering $k = \sqrt{n}$, we get

$$P\Big(Y_n \geq C\sqrt{n}\Big) \geq 1 - \frac{1}{n}, \qquad \forall n \geq 1,$$

where

$$C = \sqrt{\frac{2T}{\pi}} - \sqrt{\Big(1 - \frac{2}{\pi}\Big)T} > 0.$$

Then for any constant $M > 0$, there is an integer n such that $C\sqrt{n} \leq M$ and hence

$$P(Y_n \geq M) \geq P\Big(Y_n \geq C\sqrt{n}\Big) \geq 1 - \frac{1}{n}, \qquad \forall n \geq n_0.$$

Taking the limit over n yields

$$\lim_{n\to\infty} P(Y_n \geq M) = 1.$$

Hence, the total variation of a Brownian motion is infinite, almost surely

$$P(V(W_t) = \infty) = 1.$$

The rest of this chapter deals with similar properties regarding quadratic variation of compensated Poisson process and can be skipped at a first reading.

4.12.1 The quadratic variation of $N_t - \lambda t$

The following result deals with the quadratic variation of the compensated Poisson process $M_t = N_t - \lambda t$.

Proposition 4.12.1 *Let $a < b$ and consider the partition $a = t_0 < t_1 < \cdots < t_{n-1} < t_n = b$. Then*

$$\boxed{ms\text{-}\lim_{\|\Delta_n\|\to 0}\sum_{k=0}^{n-1}(M_{t_{k+1}} - M_{t_k})^2 = N_b - N_a,} \tag{4.12.26}$$

where $\|\Delta_n\| := \sup_{0\le k\le n-1}(t_{k+1} - t_k)$.

Proof: For the sake of simplicity we shall use the following notations

$$\Delta t_k = t_{k+1} - t_k, \quad \Delta M_k = M_{t_{k+1}} - M_{t_k}, \quad \Delta N_k = N_{t_{k+1}} - N_{t_k}.$$

The relation we need to prove can also be written as

$$\text{ms-}\lim_{n\to\infty}\sum_{k=0}^{n-1}\left[(\Delta M_k)^2 - \Delta N_k\right] = 0.$$

Let

$$Y_k = (\Delta M_k)^2 - \Delta N_k = (\Delta M_k)^2 - \Delta M_k - \lambda\Delta t_k.$$

It suffices to show that

$$E\Big[\sum_{k=0}^{n-1} Y_k\Big] = 0, \tag{4.12.27}$$

$$\lim_{n\to\infty} Var\Big[\sum_{k=0}^{n-1} Y_k\Big] = 0. \tag{4.12.28}$$

The first identity follows from the properties of Poisson processes (see Exercise 3.8.9)

$$\begin{aligned} E\Big[\sum_{k=0}^{n-1} Y_k\Big] &= \sum_{k=0}^{n-1}\mathbb{E}[Y_k] = \sum_{k=0}^{n-1}\mathbb{E}[(\Delta M_k)^2] - \mathbb{E}[\Delta N_k] \\ &= \sum_{k=0}^{n-1}(\lambda\Delta t_k - \lambda\Delta t_k) = 0. \end{aligned}$$

For the proof of the identity (4.12.28) we need to first find the variance of Y_k.

$$\begin{aligned} Var[Y_k] &= Var[(\Delta M_k)^2 - (\Delta M_k + \lambda\Delta t_k)] = Var[(\Delta M_k)^2 - \Delta M_k] \\ &= Var[(\Delta M_k)^2] + Var[\Delta M_k] - 2Cov[\Delta M_k^2, \Delta M_k] \\ &= \lambda\Delta t_k + 2\lambda^2\Delta t_k^2 + \lambda\Delta t_k \\ &\quad -2\Big[\mathbb{E}[(\Delta M_k)^3] - \mathbb{E}[(\Delta M_k)^2]\mathbb{E}[\Delta M_k]\Big] \\ &= 2\lambda^2(\Delta t_k)^2, \end{aligned}$$

where we used Exercise 3.8.9 and the fact that $\mathbb{E}[\Delta M_k] = 0$. Since M_t is a process with independent increments, then $Cov[Y_k, Y_j] = 0$ for $i \neq j$. Then

$$\begin{aligned} Var\Big[\sum_{k=0}^{n-1} Y_k\Big] &= \sum_{k=0}^{n-1} Var[Y_k] + 2\sum_{k\neq j} Cov[Y_k, Y_j] = \sum_{k=0}^{n-1} Var[Y_k] \\ &= 2\lambda^2 \sum_{k=0}^{n-1} (\Delta t_k)^2 \leq 2\lambda^2 \|\Delta_n\| \sum_{k=0}^{n-1} \Delta t_k = 2\lambda^2 (b-a)\|\Delta_n\|, \end{aligned}$$

and hence $Var\Big[\sum_{k=0}^{n-1} Y_n\Big] \to 0$ as $\|\Delta_n\| \to 0$. According to the Proposition 2.14.1, we obtain the desired limit in mean square. ■

The previous result states that the quadratic variation of the martingale M_t between a and b is equal to the jump of the Poisson process between a and b.

The Relation $dM_t^2 = dN_t$ Recall relation (4.12.26)

$$\text{ms-}\lim_{n\to\infty} \sum_{k=0}^{n-1} (M_{t_{k+1}} - M_{t_k})^2 = N_b - N_a. \tag{4.12.29}$$

The right side can be regarded as a Riemann-Stieltjes integral

$$N_b - N_a = \int_a^b dN_t,$$

while the left side can be regarded as a stochastic integral with respect to $(dM_t)^2$

$$\int_a^b (dM_t)^2 := \text{ms-}\lim_{n\to\infty} \sum_{k=0}^{n-1} (M_{t_{k+1}} - M_{t_k})^2.$$

Substituting in (4.12.29) yields

$$\int_a^b (dM_t)^2 = \int_a^b dN_t,$$

for any $a < b$. The equivalent differential form is

$$\boxed{(dM_t)^2 = dN_t.} \tag{4.12.30}$$

The Relations $dt\, dM_t = 0$, $dW_t\, dM_t = 0$ In order to show that $dt\, dM_t = 0$ in the mean square sense, we need to prove the limit

$$\text{ms-}\lim_{n\to\infty} \sum_{k=0}^{n-1} (t_{k+1} - t_k)(M_{t_{k+1}} - M_{t_k}) = 0. \tag{4.12.31}$$

This can be thought of as a vanishing integral of the increment process dM_t with respect to dt

$$\int_a^b dM_t\, dt = 0, \qquad \forall a, b \in \mathbb{R}.$$

Denote

$$X_n = \sum_{k=0}^{n-1} (t_{k+1} - t_k)(M_{t_{k+1}} - M_{t_k}) = \sum_{k=0}^{n-1} \Delta t_k \Delta M_k.$$

In order to show (4.12.31) it suffices to prove that

1. $\mathbb{E}[X_n] = 0$;
2. $\lim\limits_{n\to\infty} Var[X_n] = 0$.

Using the additivity of the expectation and Exercise 3.8.9 (b)

$$\mathbb{E}[X_n] = E\Big[\sum_{k=0}^{n-1} \Delta t_k \Delta M_k\Big] = \sum_{k=0}^{n-1} \Delta t_k \mathbb{E}[\Delta M_k] = 0.$$

Since the Poisson process N_t has independent increments, the same property holds for the compensated Poisson process M_t. Then $\Delta t_k \Delta M_k$ and $\Delta t_j \Delta M_j$ are independent for $k \neq j$, and using the properties of variance we have

$$Var[X_n] = Var\Big[\sum_{k=0}^{n-1} \Delta t_k \Delta M_k\Big] = \sum_{k=0}^{n-1} (\Delta t_k)^2 Var[\Delta M_k] = \lambda \sum_{k=0}^{n-1} (\Delta t_k)^3,$$

where we used

$$Var[\Delta M_k] = \mathbb{E}[(\Delta M_k)^2] - (\mathbb{E}[\Delta M_k])^2 = \lambda \Delta t_k,$$

see Exercise 3.8.9 (b). If we let $\|\Delta_n\| = \max\limits_k \Delta t_k$, then

$$Var[X_n] = \lambda \sum_{k=0}^{n-1} (\Delta t_k)^3 \le \lambda \|\Delta_n\|^2 \sum_{k=0}^{n-1} \Delta t_k = \lambda(b-a)\|\Delta_n\|^2 \to 0$$

as $n \to \infty$. Hence we proved the stochastic differential relation

$$\boxed{dt\, dM_t = 0.} \tag{4.12.32}$$

For showing the relation $dW_t\, dM_t = 0$, we need to prove

$$\text{ms-}\lim_{n\to\infty} Y_n = 0, \tag{4.12.33}$$

where we have denoted

$$Y_n = \sum_{k=0}^{n-1}(W_{k+1} - W_k)(M_{t_{k+1}} - M_{t_k}) = \sum_{k=0}^{n-1}\Delta W_k \Delta M_k.$$

Since the Brownian motion W_t and the process M_t have independent increments and ΔW_k is independent of ΔM_k, we have

$$\mathbb{E}[Y_n] = \sum_{k=0}^{n-1}\mathbb{E}[\Delta W_k \Delta M_k] = \sum_{k=0}^{n-1}\mathbb{E}[\Delta W_k]\mathbb{E}[\Delta M_k] = 0,$$

where we used $\mathbb{E}[\Delta W_k] = \mathbb{E}[\Delta M_k] = 0$. Using also $\mathbb{E}[(\Delta W_k)^2] = \Delta t_k$, $\mathbb{E}[(\Delta M_k)^2] = \lambda \Delta t_k$, and invoking the independence of ΔW_k and ΔM_k, we get

$$\begin{aligned} Var[\Delta W_k \Delta M_k] &= \mathbb{E}[(\Delta W_k)^2(\Delta M_k)^2] - (\mathbb{E}[\Delta W_k \Delta M_k])^2 \\ &= \mathbb{E}[(\Delta W_k)^2]\mathbb{E}[(\Delta M_k)^2] - \mathbb{E}[\Delta W_k]^2\mathbb{E}[\Delta M_k]^2 \\ &= \lambda(\Delta t_k)^2. \end{aligned}$$

Then using the independence of the terms in the sum, we get

$$\begin{aligned} Var[Y_n] &= \sum_{k=0}^{n-1} Var[\Delta W_k \Delta M_k] = \lambda \sum_{k=0}^{n-1}(\Delta t_k)^2 \\ &\le \lambda \|\Delta_n\| \sum_{k=0}^{n-1}\Delta t_k = \lambda(b-a)\|\Delta_n\| \to 0, \end{aligned}$$

as $n \to \infty$. Since Y_n is a random variable with mean zero and variance decreasing to zero, it follows that $Y_n \to 0$ in the mean square sense. Hence we proved that

$$\boxed{dW_t\, dM_t = 0.} \qquad (4.12.34)$$

Exercise 4.12.2 *Show the following stochastic differential relations:*

(*a*) $dt\, dN_t = 0$; (*b*) $dW_t\, dN_t = 0$; (*c*) $dt\, dW_t = 0$;

(*d*) $(dN_t)^2 = dN_t$; (*e*) $(dM_t)^2 = dN_t$; (*f*) $(dM_t)^4 = dN_t$.

The relations proved in this section are useful when developing stochastic models for a stock price that exhibits jumps modeled by a Poisson process. We can represent all these rules in the following multiplication table:

$\times$	dt	dW_t	dN_t	dM_t
dt	0	0	0	0
dW_t	0	dt	0	0
dN_t	0	0	dN_t	dN_t
dM_t	0	0	dN_t	dN_t

Chapter 5

Stochastic Integration

This chapter deals with one of the most useful stochastic integrals, called the *Ito integral.* This type of integral was introduced in 1944 by the Japanese mathematician Ito [24], [25], and was originally motivated by a construction of diffusion processes. We shall keep the presentation to a maximum simplicity, integrating with respect to a Brownian motion or Poisson process only. The reader interested in details regarding a larger class of integrators may consult Protter [40] or Kuo [30]. For a more formal introduction into stochastic integration see Revuz and Yor [41].

Here is a motivation for studying an integral of stochastic type. The Riemann integral $\int_a^b F(x)\,dx$ represents the work done by the force F between positions $x = a$ and $x = b$. The element $F(x)\,dx$ represents the work done by the force for the infinitesimal displacement dx. Similarly, $F(t)\,dW_t$ represents the work done by F during an infinitesimal Brownian jump dW_t. The cummulative effect is described by the object $\int_a^b F(t)\,dW_t$, which will be studied in this chapter. This represents the work effect of the force F done along the trajectory of a particle modeled by a Brownian motion during the time interval $[a, b]$.

5.1 Nonanticipating Processes

Consider the Brownian motion W_t. A process F_t is called a *nonanticipating process* if F_t is independent of any future increment $W_{t'} - W_t$, for any t and t' with $t < t'$. Consequently, the process F_t is independent of the behavior of the Brownian motion in the future, i.e. it cannot anticipate the future. For instance, W_t, e^{W_t}, $W_t^2 - W_t + t$ are examples of nonanticipating processes, while W_{t+1} or $\frac{1}{2}(W_{t+1} - W_t)^2$ are not.

Nonanticipating processes are important because the Ito integral concept can be easily applied to them.

If $\mathcal{F}_t$ denotes the information known until time t, where this information is generated by the Brownian motion $\{W_s; s \leq t\}$, then any $\mathcal{F}_t$-adapted process F_t is nonanticipating.

5.2 Increments of Brownian Motions

In this section we shall discuss a few basic properties of the increments of a Brownian motion, which will be useful when computing stochastic integrals.

Proposition 5.2.1 *Let W_t be a Brownian motion. If $s < t$, we have*

1. $\mathbb{E}[(W_t - W_s)^2] = t - s$.
2. $Var[(W_t - W_s)^2] = 2(t-s)^2$.

Proof: 1. Using that $W_t - W_s \sim N(0, t-s)$, we have

$$\mathbb{E}[(W_t - W_s)^2] = \mathbb{E}[(W_t - W_s)^2] - (\mathbb{E}[W_t - W_s])^2 = Var(W_t - W_s) = t - s.$$

2. Dividing by the standard deviation yields the standard normal random variable $\frac{W_t - W_s}{\sqrt{t-s}} \sim N(0,1)$. Its square, $\frac{(W_t - W_s)^2}{t-s}$ is χ^2-distributed with 1 degree of freedom.[1] Its mean is 1 and its variance is 2. This implies

$$\begin{aligned} \mathbb{E}\Big[\frac{(W_t - W_s)^2}{t-s}\Big] &= 1 \Longrightarrow \mathbb{E}[(W_t - W_s)^2] = t - s; \\ Var\Big[\frac{(W_t - W_s)^2}{t-s}\Big] &= 2 \Longrightarrow Var[(W_t - W_s)^2] = 2(t-s)^2. \end{aligned}$$

■

Remark 5.2.2 The infinitesimal version of the previous result is obtained by replacing $t - s$ with dt

1. $\mathbb{E}[dW_t^2] = dt$;
2. $Var[dW_t^2] = 2dt^2 = 0$.

Exercise 5.2.3 *Show that*

(a) $\mathbb{E}[(W_t - W_s)^4] = 3(t-s)^2$;

(b) $\mathbb{E}[(W_t - W_s)^6] = 15(t-s)^3$.

[1] A χ^2-distributed random variable with n degrees of freedom has mean n and variance $2n$.

5.3 The Ito Integral

The Ito integral will be defined in a way that is similar to the Riemann integral. The Ito integral is taken with respect to infinitesimal increments of a Brownian motion, dW_t, which are random variables, while the Riemann integral considers integration with respect to the deterministic infinitesimal changes dt. It is worth noting that the Ito integral is a random variable, while the Riemann integral is just a real number. Despite this fact, we shall see that there are several common properties and relations between these two types of integrals.

Consider $0 \leq a < b$ and let $F_t = f(W_t, t)$ be a nonanticipating process satisfying the "non-explosive" condition

$$\mathbb{E}\Big[\int_a^b F_t^2\,dt\Big] < \infty. \tag{5.3.1}$$

The role of the previous condition will be made more clear when we discuss the martingale property of the Ito integral, see Proposition 5.5.7. Divide the interval $[a, b]$ into n subintervals using the partition points

$$a = t_0 < t_1 < \cdots < t_{n-1} < t_n = b,$$

and consider the partial sums

$$S_n = \sum_{i=0}^{n-1} F_{t_i}(W_{t_{i+1}} - W_{t_i}).$$

We emphasize that the intermediate points are the left endpoints of each interval, and this is the way they should always be chosen. Since the process F_t is nonanticipative, the random variables F_{t_i} and $W_{t_{i+1}} - W_{t_i}$ are always independent; this is an important feature in the definition of the Ito integral.

The *Ito integral* is the limit of the partial sums S_n

$$\text{ms-}\lim_{n\to\infty} S_n = \int_a^b F_t\,dW_t,$$

provided the limit exists. It can be shown that the choice of partition does not influence the value of the Ito integral. This is the reason why, for practical purposes, it suffices to assume the intervals equidistant, i.e.

$$t_{i+1} - t_i = \frac{(b-a)}{n}, \qquad i = 0, 1, \cdots, n-1.$$

The previous convergence is taken in the mean square sense, i.e.

$$\lim_{n\to\infty} \mathbb{E}\Big[\Big(S_n - \int_a^b F_t\,dW_t\Big)^2\Big] = 0.$$

Existence of the Ito integral

It is known that the Ito stochastic integral $\int_a^b F_t\,dW_t$ exists if the process $F_t = f(W_t, t)$ satisfies the following properties:

1. The paths $t \to F_t(\omega)$ are continuous on $[a, b]$ for any state of the world $\omega \in \Omega$;
2. The process F_t is nonanticipating for $t \in [a, b]$;
3. $\mathbb{E}\Big[\int_a^b F_t^2\,dt\Big] < \infty$.

For instance, the following stochastic integrals make sense:

$$\int_0^T W_t^2\,dW_t, \quad \int_0^T \sin(W_t)\,dW_t, \quad \int_a^b \frac{\cos(W_t)}{t}\,dW_t.$$

5.4 Examples of Ito Integrals

As in the case of the Riemann integral, using the definition is not an efficient way of computing integrals. The same philosophy applies to Ito integrals. We shall compute in the following two simple Ito integrals. In later sections we shall introduce more efficient methods for computing these types of stochastic integrals.

5.4.1 The case $F_t = c$, constant

In this case the partial sums can be computed explicitly

$$\begin{aligned} S_n &= \sum_{i=0}^{n-1} F_{t_i}(W_{t_{i+1}} - W_{t_i}) = \sum_{i=0}^{n-1} c(W_{t_{i+1}} - W_{t_i}) \\ &= c(W_b - W_a), \end{aligned}$$

and since the answer does not depend on n, we have

$$\int_a^b c\,dW_t = c(W_b - W_a).$$

In particular, taking $c = 1$, $a = 0$, and $b = T$, since the Brownian motion starts at 0, we have the following formula:

$$\boxed{\int_0^T dW_t = W_T.}$$

5.4.2 The case $F_t = W_t$

We shall integrate the process W_t between 0 and T. Considering an equidistant partition, we take $t_k = \dfrac{kT}{n}$, $k = 0, 1, \cdots, n-1$. The partial sums are given by

$$S_n = \sum_{i=0}^{n-1} W_{t_i}(W_{t_{i+1}} - W_{t_i}).$$

Since

$$xy = \frac{1}{2}[(x+y)^2 - x^2 - y^2],$$

letting $x = W_{t_i}$ and $y = W_{t_{i+1}} - W_{t_i}$ yields

$$W_{t_i}(W_{t_{i+1}} - W_{t_i}) = \frac{1}{2}W_{t_{i+1}}^2 - \frac{1}{2}W_{t_i}^2 - \frac{1}{2}(W_{t_{i+1}} - W_{t_i})^2.$$

Then after pair cancelations the sum becomes

$$\begin{aligned} S_n &= \frac{1}{2}\sum_{i=0}^{n-1} W_{t_{i+1}}^2 - \frac{1}{2}\sum_{i=0}^{n-1} W_{t_i}^2 - \frac{1}{2}\sum_{i=0}^{n-1}(W_{t_{i+1}} - W_{t_i})^2 \\ &= \frac{1}{2}W_{t_n}^2 - \frac{1}{2}\sum_{i=0}^{n-1}(W_{t_{i+1}} - W_{t_i})^2. \end{aligned}$$

Using $t_n = T$, we get

$$S_n = \frac{1}{2}W_T^2 - \frac{1}{2}\sum_{i=0}^{n-1}(W_{t_{i+1}} - W_{t_i})^2.$$

Since the first term on the right side is independent of n, using Proposition 4.11.6, we have

$$\begin{aligned} \text{ms-}\lim_{n\to\infty} S_n &= \frac{1}{2}W_T^2 - \text{ms-}\lim_{n\to\infty} \frac{1}{2}\sum_{i=0}^{n-1}(W_{t_{i+1}} - W_{t_i})^2 &\quad (5.4.2) \\ &= \frac{1}{2}W_T^2 - \frac{1}{2}T. &\quad (5.4.3) \end{aligned}$$

We have now obtained the following explicit formula of a stochastic integral:

$$\boxed{\int_0^T W_t\, dW_t = \frac{1}{2}W_T^2 - \frac{1}{2}T.}$$

In a similar way one can obtain

$$\boxed{\int_a^b W_t\, dW_t = \frac{1}{2}(W_b^2 - W_a^2) - \frac{1}{2}(b-a).}$$

It is worth noting that the right side contains random variables depending on the limits of integration a and b.

Exercise 5.4.1 *Show the following identities:*

(a) $\mathbb{E}[\int_0^T dW_t] = 0;$

(b) $\mathbb{E}[\int_0^T W_t dW_t] = 0;$

(c) $Var[\int_0^T W_t dW_t] = \frac{T^2}{2}.$

5.5 Properties of the Ito Integral

We shall start with some properties which are similar to those of the Riemannian integral.

Proposition 5.5.1 *Let $f(W_t,t)$, $g(W_t,t)$ be nonanticipating processes and $c \in \mathbb{R}$. Then we have*

1. Additivity:

$$\int_0^T [f(W_t,t) + g(W_t,t)]\, dW_t = \int_0^T f(W_t,t)\, dW_t + \int_0^T g(W_t,t)\, dW_t.$$

2. Homogeneity:

$$\int_0^T cf(W_t,t)\, dW_t = c\int_0^T f(W_t,t)\, dW_t.$$

3. Partition property:

$$\int_0^T f(W_t,t)\, dW_t = \int_0^u f(W_t,t)\, dW_t + \int_u^T f(W_t,t)\, dW_t, \qquad \forall 0 < u < T.$$

Proof: 1. Consider the partial sum sequences

$$\begin{aligned} X_n &= \sum_{i=0}^{n-1} f(W_{t_i},t_i)(W_{t_{i+1}} - W_{t_i}) \\ Y_n &= \sum_{i=0}^{n-1} g(W_{t_i},t_i)(W_{t_{i+1}} - W_{t_i}). \end{aligned}$$

Since $\text{ms-}\lim_{n\to\infty} X_n = \int_0^T f(W_t,t)\,dW_t$ and $\text{ms-}\lim_{n\to\infty} Y_n = \int_0^T g(W_t,t)\,dW_t$, using Proposition 2.15.2 yields

$$\begin{aligned}
&\int_0^T \Big(f(W_t,t)+g(W_t,t)\Big)\,dW_t \\
&= \text{ms-}\lim_{n\to\infty} \sum_{i=0}^{n-1} \Big(f(W_{t_i},t_i)+g(W_{t_i},t_i)\Big)(W_{t_{i+1}}-W_{t_i}) \\
&= \text{ms-}\lim_{n\to\infty} \Big[\sum_{i=0}^{n-1} \Big(f(W_{t_i},t_i)(W_{t_{i+1}}-W_{t_i}) + \sum_{i=0}^{n-1} g(W_{t_i},t_i)(W_{t_{i+1}}-W_{t_i})\Big] \\
&= \text{ms-}\lim_{n\to\infty}(X_n+Y_n) = \text{ms-}\lim_{n\to\infty} X_n + \text{ms-}\lim_{n\to\infty} Y_n \\
&= \int_0^T f(W_t,t)\,dW_t + \int_0^T g(W_t,t)\,dW_t.
\end{aligned}$$

The proofs of parts 2 and 3 are left as an exercise for the reader. ∎

Some other properties, such as monotonicity, do not hold in general. It is possible to have a non-negative random variable F_t for which the random variable $\int_0^T F_t\,dW_t$ has negative values. More precisely, let $F_t = 1$. Then $F_t > 0$ but $\int_0^T 1\,dW_t = W_T$ is not always positive. The probability to be negative is $P(W_T < 0) = 1/2$.

Some of the random variable properties of the Ito integral are given by the following result:

Proposition 5.5.2 *We have*
1. Zero mean:

$$\mathbb{E}\Big[\int_a^b f(W_t,t)\,dW_t\Big] = 0.$$

2. Isometry:

$$\mathbb{E}\Big[\Big(\int_a^b f(W_t,t)\,dW_t\Big)^2\Big] = \mathbb{E}\Big[\int_a^b f(W_t,t)^2\,dt\Big].$$

3. Covariance:

$$\mathbb{E}\Big[\Big(\int_a^b f(W_t,t)\,dW_t\Big)\Big(\int_a^b g(W_t,t)\,dW_t\Big)\Big] = \mathbb{E}\Big[\int_a^b f(W_t,t)g(W_t,t)\,dt\Big].$$

We shall discuss the previous properties giving rough reasons of proof. The detailed proofs are beyond the goal of this book.

1. The Ito integral $I = \int_a^b f(W_t,t)\,dW_t$ is the mean square limit of the partial sums $S_n = \sum_{i=0}^{n-1} f_{t_i}(W_{t_{i+1}} - W_{t_i})$, where we denoted $f_{t_i} = f(W_{t_i}, t_i)$. Since

$f(W_t, t)$ is a nonanticipative process, then f_{t_i} is independent of the increments $W_{t_{i+1}} - W_{t_i}$, and hence we have

$$\begin{aligned} \mathbb{E}[S_n] &= \mathbb{E}\Big[\sum_{i=0}^{n-1} f_{t_i}(W_{t_{i+1}} - W_{t_i})\Big] = \sum_{i=0}^{n-1} \mathbb{E}[f_{t_i}(W_{t_{i+1}} - W_{t_i})] \\ &= \sum_{i=0}^{n-1} \mathbb{E}[f_{t_i}]\mathbb{E}[(W_{t_{i+1}} - W_{t_i})] = 0, \end{aligned}$$

because the increments have mean zero. Applying the Squeeze Theorem in the double inequality

$$0 \le \Big(\mathbb{E}[S_n - I]\Big)^2 \le \mathbb{E}[(S_n - I)^2] \to 0, \qquad n \to \infty$$

yields $\mathbb{E}[S_n] - \mathbb{E}[I] \to 0$. Since $\mathbb{E}[S_n] = 0$ it follows that $\mathbb{E}[I] = 0$, i.e. the Ito integral has zero mean.

2. Since the square of the sum of partial sums can be written as

$$\begin{aligned} S_n^2 &= \Big(\sum_{i=0}^{n-1} f_{t_i}(W_{t_{i+1}} - W_{t_i})\Big)^2 \\ &= \sum_{i=0}^{n-1} f_{t_i}^2 (W_{t_{i+1}} - W_{t_i})^2 + 2\sum_{i \ne j} f_{t_i}(W_{t_{i+1}} - W_{t_i}) f_{t_j}(W_{t_{j+1}} - W_{t_j}), \end{aligned}$$

using the independence yields

$$\begin{aligned} \mathbb{E}[S_n^2] &= \sum_{i=0}^{n-1} \mathbb{E}[f_{t_i}^2]\mathbb{E}[(W_{t_{i+1}} - W_{t_i})^2] \\ &\quad +2\sum_{i \ne j} \mathbb{E}[f_{t_i}]\mathbb{E}[(W_{t_{i+1}} - W_{t_i})]\mathbb{E}[f_{t_j}]\mathbb{E}[(W_{t_{j+1}} - W_{t_j})] \\ &= \sum_{i=0}^{n-1} \mathbb{E}[f_{t_i}^2](t_{i+1} - t_i), \end{aligned}$$

which are the Riemann sums of the integral $\displaystyle\int_a^b \mathbb{E}[f_t^2]\,dt = \mathbb{E}\Big[\int_a^b f_t^2\,dt\Big]$, where the last identity follows from Fubini's theorem. Hence $\mathbb{E}[S_n^2]$ converges to the aforementioned integral.

3. Consider the partial sums

$$S_n = \sum_{i=0}^{n-1} f_{t_i}(W_{t_{i+1}} - W_{t_i}), \qquad V_n = \sum_{j=0}^{n-1} g_{t_j}(W_{t_{j+1}} - W_{t_j}).$$

Their product is

$$\begin{aligned} S_n V_n &= \Big(\sum_{i=0}^{n-1} f_{t_i}(W_{t_{i+1}} - W_{t_i})\Big)\Big(\sum_{j=0}^{n-1} g_{t_j}(W_{t_{j+1}} - W_{t_j})\Big) \\ &= \sum_{i=0}^{n-1} f_{t_i} g_{t_i}(W_{t_{i+1}} - W_{t_i})^2 + \sum_{i \neq j}^{n-1} f_{t_i} g_{t_j}(W_{t_{i+1}} - W_{t_i})(W_{t_{j+1}} - W_{t_j}). \end{aligned}$$

Using that f_t and g_t are nonanticipative and that

$$\begin{aligned} \mathbb{E}[(W_{t_{i+1}} - W_{t_i})(W_{t_{j+1}} - W_{t_j})] &= \mathbb{E}[W_{t_{i+1}} - W_{t_i}]\mathbb{E}[W_{t_{j+1}} - W_{t_j}] = 0, \ i \neq j \\ \mathbb{E}[(W_{t_{i+1}} - W_{t_i})^2] &= t_{i+1} - t_i, \end{aligned}$$

it follows that

$$\begin{aligned} \mathbb{E}[S_n V_n] &= \sum_{i=0}^{n-1} \mathbb{E}[f_{t_i} g_{t_i}]\mathbb{E}[(W_{t_{i+1}} - W_{t_i})^2] \\ &= \sum_{i=0}^{n-1} \mathbb{E}[f_{t_i} g_{t_i}](t_{i+1} - t_i), \end{aligned}$$

which is the Riemann sum for the integral $\int_a^b \mathbb{E}[f_t g_t]\, dt$.

From 1 and 2 it follows that the random variable $\int_a^b f(W_t, t)\, dW_t$ has mean zero and variance

$$Var\Big[\int_a^b f(W_t, t)\, dW_t\Big] = \mathbb{E}\Big[\int_a^b f(W_t, t)^2\, dt\Big].$$

From 1 and 3 it follows that

$$Cov\Big[\int_a^b f(W_t, t)\, dW_t, \int_a^b g(W_t, t)\, dW_t\Big] = \int_a^b \mathbb{E}[f(W_t, t) g(W_t, t)]\, dt.$$

Corollary 5.5.3 (Cauchy's integral inequality) *Let $f_t = f(W_t, t)$ and $g_t = g(W_t, t)$. Then*

$$\Big(\int_a^b \mathbb{E}[f_t g_t]\, dt\Big)^2 \leq \Big(\int_a^b \mathbb{E}[f_t^2]\, dt\Big)\Big(\int_a^b \mathbb{E}[g_t^2]\, dt\Big).$$

Proof: It follows from the previous theorem and from the correlation formula $|Corr(X, Y)| = \dfrac{|Cov(X, Y)|}{[Var(X)Var(Y)]^{1/2}} \leq 1$. ∎

Let $\mathcal{F}_t$ be the information set at time t. This implies that f_{t_i} and $W_{t_{i+1}} - W_{t_i}$ are known at time t, for any $t_{i+1} \leq t$. It follows that the partial sum

$S_n = \sum_{i=0}^{n-1} f_{t_i}(W_{t_{i+1}} - W_{t_i})$ is $\mathcal{F}_t$-measurable. The following result, whose proof is omitted for technical reasons, states that this is also valid after taking the limit in the mean square:

Proposition 5.5.4 *The Ito integral $\int_0^t f_s \, dW_s$ is $\mathcal{F}_t$-measurable.*

The following two results state that if the upper limit of an Ito integral is replaced by the parameter t we obtain a continuous martingale.

Proposition 5.5.5 *For any $s < t$ we have*

$$\mathbb{E}\Big[\int_0^t f(W_u, u)\, dW_u | \mathcal{F}_s\Big] = \int_0^s f(W_u, u)\, dW_u.$$

Proof: Using part 3 of Proposition 5.5.2 we get

$$\begin{aligned} & \mathbb{E}\Big[\int_0^t f(W_u, u)\, dW_u | \mathcal{F}_s\Big] \\ = \ & \mathbb{E}\Big[\int_0^s f(W_u, u)\, dW_u + \int_s^t f(W_u, u)\, dW_u | \mathcal{F}_s\Big] \\ = \ & \mathbb{E}\Big[\int_0^s f(W_u, u)\, dW_u | \mathcal{F}_s\Big] + \mathbb{E}\Big[\int_s^t f(W_u, u)\, dW_u | \mathcal{F}_s\Big]. \qquad (5.5.4) \end{aligned}$$

Since $\int_0^s f(W_u, u)\, dW_u$ is $\mathcal{F}_s$-measurable (see Proposition 5.5.4), by part 2 of Proposition 2.12.6

$$\mathbb{E}\Big[\int_0^s f(W_u, u)\, dW_u | \mathcal{F}_s\Big] = \int_0^s f(W_u, u)\, dW_u.$$

Since $\int_s^t f(W_u, u)\, dW_u$ contains only information between s and t, it is independent of the information set $\mathcal{F}_s$, so we can drop the condition in the expectation; using that Ito integrals have zero mean we obtain

$$\mathbb{E}\Big[\int_s^t f(W_u, u)\, dW_u | \mathcal{F}_s\Big] = \mathbb{E}\Big[\int_s^t f(W_u, u)\, dW_u\Big] = 0.$$

Substituting into (5.5.4) yields the desired result. ■

Proposition 5.5.6 *Consider the process $X_t = \int_0^t f(W_s, s)\, dW_s$. Then X_t is continuous, i.e. for almost any state of the world $\omega \in \Omega$, the path $t \to X_t(\omega)$ is continuous.*

Proof: A rigorous proof is beyond the purpose of this book. We shall provide just a rough sketch. Assume the process $f(W_t, t)$ satisfies $\mathbb{E}[f(W_t, t)^2] < M$, for some $M > 0$. Let t_0 be fixed and consider $h > 0$. Consider the increment $Y_h = X_{t_0+h} - X_{t_0}$. Using the aforementioned properties of the Ito integral we have

$$\begin{aligned}
\mathbb{E}[Y_h] &= \mathbb{E}[X_{t_0+h} - X_{t_0}] = \mathbb{E}\Big[\int_{t_0}^{t_0+h} f(W_t, t)\, dW_t\Big] = 0 \\
\mathbb{E}[Y_h^2] &= E\Big[\Big(\int_{t_0}^{t_0+h} f(W_t, t)\, dW_t\Big)^2\Big] = \int_{t_0}^{t_0+h} \mathbb{E}[f(W_t, t)^2]\, dt \\
&< M \int_{t_0}^{t_0+h} dt = Mh.
\end{aligned}$$

The process Y_h has zero mean for any $h > 0$ and its variance tends to 0 as $h \to 0$. Using a convergence theorem yields that Y_h tends to 0 in mean square, as $h \to 0$. This is equivalent to the continuity of X_t at t_0. ■

Proposition 5.5.7 *Let* $X_t = \int_0^t f(W_s, s)\, dW_s$, *with* $E\big[\int_0^\infty f^2(W_s, s)\, ds\big] < \infty$. *Then* X_t *is a continuous* $\mathcal{F}_t$*-martingale.*

Proof: We shall check in the following the properties of a martingale.
Integrability: Using properties of Ito integrals

$$\begin{aligned}
\mathbb{E}[X_t^2] &= \mathbb{E}\Big[\Big(\int_0^t f(W_s, s)\, dW_s\Big)^2\Big] = \mathbb{E}\Big[\int_0^t f^2(W_s, s)\, ds\Big] \\
&< \mathbb{E}\Big[\int_0^\infty f^2(W_s, s)\, ds\Big] < \infty,
\end{aligned}$$

and then from the inequality $\mathbb{E}[|X_t|]^2 \le \mathbb{E}[X_t^2]$ we obtain $\mathbb{E}[|X_t|] < \infty$, for all $t \ge 0$.
Measurability: X_t is $\mathcal{F}_t$-measurable from Proposition 5.5.4.
Forecast: $\mathbb{E}[X_t|\mathcal{F}_s] = X_s$ for $s < t$ by Proposition 5.5.5.
Continuity: See Proposition 5.5.6. ■

5.6 The Wiener Integral

The *Wiener integral* is a particular case of the Ito stochastic integral. It is obtained by replacing the nonanticipating stochastic process $f(W_t, t)$ by the deterministic function $f(t)$. The Wiener integral $\int_a^b f(t)\, dW_t$ is the mean square limit of the partial sums

$$S_n = \sum_{i=0}^{n-1} f(t_i)(W_{t_{i+1}} - W_{t_i}).$$

All properties of Ito integrals also hold for Wiener integrals. The Wiener integral is a random variable with zero mean

$$\mathbb{E}\Big[\int_a^b f(t)\,dW_t\Big] = 0$$

and variance

$$\mathbb{E}\Big[\Big(\int_a^b f(t)\,dW_t\Big)^2\Big] = \int_a^b f(t)^2\,dt.$$

However, in the case of Wiener integrals we can say something more about their distribution.

Proposition 5.6.1 *The Wiener integral $I(f) = \int_a^b f(t)\,dW_t$ is a normal random variable with mean 0 and variance*

$$Var[I(f)] = \int_a^b f(t)^2\,dt := \|f\|_{L^2}^2.$$

Proof: Since increments $W_{t_{i+1}} - W_{t_i}$ are normally distributed with mean 0 and variance $t_{i+1} - t_i$, then

$$f(t_i)(W_{t_{i+1}} - W_{t_i}) \sim N(0, f(t_i)^2(t_{i+1} - t_i)).$$

Since these random variables are independent, by Theorem 3.3.1, their sum is also normally distributed, with

$$S_n = \sum_{i=0}^{n-1} f(t_i)(W_{t_{i+1}} - W_{t_i}) \sim N\Big(0, \sum_{i=0}^{n-1} f(t_i)^2(t_{i+1} - t_i)\Big).$$

Taking $n \to \infty$ and $\max_i \|t_{i+1} - t_i\| \to 0$, the normal distribution tends to

$$N\Big(0, \int_a^b f(t)^2\,dt\Big).$$

The previous convergence holds in distribution, and it still needs to be shown in the mean square. However, we shall omit this essential proof detail. ■

Exercise 5.6.2 *Let $Z_t = \int_0^t W_s\,ds$.*
(a) Use integration by parts to show that

$$Z_t = \int_0^t (t-s)\,dW_s.$$

(b) Use the properties of Wiener integrals to show that

$$Var(Z_t) = \frac{t^3}{3}.$$

Exercise 5.6.3 *Show that the random variable* $X = \int_1^T \frac{1}{\sqrt{t}}\, dW_t$ *is normally distributed with mean* 0 *and variance* $\ln T$.

Exercise 5.6.4 *Let* $Y = \int_1^T \sqrt{t}\, dW_t$. *Show that* Y *is normally distributed with mean* 0 *and variance* $(T^2-1)/2$.

Exercise 5.6.5 *Find the distribution of the integral* $\int_0^t e^{t-s}\, dW_s$.

Exercise 5.6.6 *Show that* $X_t = \int_0^t (2t-u)\, dW_u$ *and* $Y_t = \int_0^t (3t-4u)\, dW_u$ *are Gaussian processes with mean* 0 *and variance* $\frac{7}{3}t^3$.

Exercise 5.6.7 *Show that* $ms\text{-}\lim_{t\to 0} \frac{1}{t}\int_0^t u\, dW_u = 0.$

Exercise 5.6.8 *Find all constants* a, b *such that* $X_t = \int_0^t \left(a + \frac{bu}{t}\right) dW_u$ *is normally distributed with variance* t.

Exercise 5.6.9 *Let* n *be a positive integer. Prove that*

$$Cov\Big(W_t, \int_0^t u^n\, dW_u\Big) = \frac{t^{n+1}}{n+1}.$$

Formulate and prove a more general result.

5.7 Poisson Integration

In this section we deal with the integration with respect to the compensated Poisson process $M_t = N_t - \lambda t$, which is a martingale. Consider $0 \le a < b$ and let $F_t = F(t, M_t)$ be a nonanticipating process with

$$\mathbb{E}\Big[\int_a^b F_t^2\, dt\Big] < \infty.$$

Consider the partition

$$a = t_0 < t_1 < \cdots < t_{n-1} < t_n = b$$

of the interval $[a, b]$, and associate the partial sums

$$S_n = \sum_{i=0}^{n-1} F_{t_i-}(M_{t_{i+1}} - M_{t_i}),$$

where F_{t_i-} is the left-hand limit at t_{i-}. We note that the intermediate points are the left-handed limit to the endpoints of each interval. Since the process F_t is nonanticipative, the random variables F_{t_i-} and $M_{t_{i+1}} - M_{t_i}$ are independent.

The integral of F_{t-} with respect to M_t is the mean square limit of the partial sum S_n

$$\text{ms-}\lim_{n\to\infty} S_n = \int_0^T F_{t-}\, dM_t,$$

provided the limit exists. More precisely, this convergence means that

$$\lim_{n\to\infty} \mathbb{E}\Big[\Big(S_n - \int_a^b F_{t-}\, dM_t\Big)^2\Big] = 0.$$

Exercise 5.7.1 *Let c be a constant. Show that* $\displaystyle\int_a^b c\, dM_t = c(M_b - M_a)$.

5.8 The case $F_t = M_t$

We shall integrate the process M_{t-} between 0 and T with respect to M_t. Consider the equidistant partition points $t_k = \dfrac{kT}{n}$, $k = 0, 1, \cdots, n-1$. Then the partial sums are given by

$$S_n = \sum_{i=0}^{n-1} M_{t_i-}(M_{t_{i+1}} - M_{t_i}).$$

Using $xy = \frac{1}{2}[(x+y)^2 - x^2 - y^2]$, by letting $x = M_{t_i-}$ and $y = M_{t_{i+1}} - M_{t_i}$, we get

$$M_{t_i-}(M_{t_{i+1}} - M_{t_i}) = \frac{1}{2}(M_{t_{i+1}} - M_{t_i} + M_{t_i-})^2 - \frac{1}{2}M_{t_i-}^2 - \frac{1}{2}(M_{t_{i+1}} - M_{t_i})^2.$$

Let J be the set of jump instances between 0 and T. Using that $M_{t_i} = M_{t_i-}$ for $t_i \notin J$, and $M_{t_i} = 1 + M_{t_i-}$ for $t_i \in J$ yields

$$M_{t_{i+1}} - M_{t_i} + M_{t_i-} = \begin{cases} M_{t_{i+1}}, & \text{if } t_i \notin J \\ M_{t_{i+1}} - 1, & \text{if } t_i \in J. \end{cases}$$

Splitting the sum, canceling in pairs, and applying the difference of squares formula we have

$$
\begin{aligned}
S_n &= \frac{1}{2}\sum_{i=0}^{n-1}(M_{t_{i+1}} - M_{t_i} + M_{t_{i-}})^2 - \frac{1}{2}\sum_{i=0}^{n-1} M_{t_{i-}}^2 - \frac{1}{2}\sum_{i=0}^{n-1}(M_{t_{i+1}} - M_{t_i})^2 \\
&= \frac{1}{2}\sum_{t_i\in J}(M_{t_{i+1}} - 1)^2 + \frac{1}{2}\sum_{t_i\notin J} M_{t_{i+1}}^2 - \frac{1}{2}\sum_{t_i\notin J} M_{t_i}^2 - \frac{1}{2}\sum_{t_i\in J} M_{t_{i-}}^2 \\
&\quad -\frac{1}{2}\sum_{i=0}^{n-1}(M_{t_{i+1}} - M_{t_i})^2 \\
&= \frac{1}{2}\sum_{t_i\in J}\Big((M_{t_{i+1}} - 1)^2 - M_{t_{i-}}^2\Big) + \frac{1}{2}M_{t_n}^2 - \frac{1}{2}\sum_{i=0}^{n-1}(M_{t_{i+1}} - M_{t_i})^2 \\
&= \frac{1}{2}\sum_{t_i\in J}\underbrace{(M_{t_{i+1}} - 1 - M_{t_{i-}})}_{=0}(M_{t_{i+1}} - 1 + M_{t_{i-}}) \\
&\quad +\frac{1}{2}M_{t_n}^2 - \frac{1}{2}\sum_{i=0}^{n-1}(M_{t_{i+1}} - M_{t_i})^2 \\
&= \frac{1}{2}M_{t_n}^2 - \frac{1}{2}\sum_{i=0}^{n-1}(M_{t_{i+1}} - M_{t_i})^2.
\end{aligned}
$$

Hence we have arrived at the following formula

$$
\boxed{\int_0^T M_{t-}\,dM_t = \frac{1}{2}M_T^2 - \frac{1}{2}N_T.}
$$

Similarly, one can obtain

$$
\boxed{\int_a^b M_{t-}\,dM_t = \frac{1}{2}(M_b^2 - M_a^2) - \frac{1}{2}(N_b - N_a).}
$$

Exercise 5.8.1 (*a*) *Show that* $\mathbb{E}\Big[\int_a^b M_{t-}\,dM_t\Big] = 0.$

(*b*) *Find* $Var\Big[\int_a^b M_t\,dM_t\Big]$.

Remark 5.8.2 (*a*) Let ω be a fixed state of the world and assume the sample path $t \to N_t(\omega)$ has a jump in the interval (a, b). Even if beyond the scope of this book, it can be shown that the integral

$$
\int_a^b N_t(\omega)\,dN_t
$$

does not exist in the Riemann-Stieltjes sense.

(*b*) Let N_{t-} denote the left-hand limit of N_t. It can be shown that N_{t-} is measurable, while N_t is not.

The previous remarks provide the reason why in the following we shall work with M_{t-} instead of M_t: the integral $\int_a^b M_t\, dN_t$ might not exist, while $\int_a^b M_{t-}\, dN_t$ does exist.

Exercise 5.8.3 *Show that*

$$\int_0^T N_{t-}\, dM_t = \frac{1}{2}(N_t^2 - N_t) - \lambda \int_0^t N_t\, dt.$$

Exercise 5.8.4 *Find the variance of*

$$\int_0^T N_{t-}\, dM_t.$$

The following integrals with respect to a Poisson process N_t are considered in the Riemann-Stieltjes sense. The following result can be found in Bertoin [6].

Proposition 5.8.5 *For any continuous function f we have*

(*a*) $\mathbb{E}\Big[\int_0^t f(s)\, dN_s\Big] = \lambda \int_0^t f(s)\, ds;$

(*b*) $\mathbb{E}\Big[\Big(\int_0^t f(s)\, dN_s\Big)^2\Big] = \lambda \int_0^t f(s)^2\, ds + \lambda^2 \Big(\int_0^t f(s)\, ds\Big)^2;$

(*c*) $\mathbb{E}\Big[e^{\int_0^t f(s)\, dN_s}\Big] = e^{\lambda \int_0^t (e^{f(s)} - 1)\, ds}.$

Proof: (*a*) Consider the equidistant partition $0 = s_0 < s_1 < \cdots < s_n = t$, with $s_{k+1} - s_k = \Delta s$. Then

$$\begin{aligned}
\mathbb{E}\Big[\int_0^t f(s)\, dN_s\Big] &= \lim_{n\to\infty} \mathbb{E}\Big[\sum_{i=0}^{n-1} f(s_i)(N_{s_{i+1}} - N_{s_i})\Big] \\
&= \lim_{n\to\infty} \sum_{i=0}^{n-1} f(s_i)\mathbb{E}\Big[N_{s_{i+1}} - N_{s_i}\Big] \\
&= \lambda \lim_{n\to\infty} \sum_{i=0}^{n-1} f(s_i)(s_{i+1} - s_i) = \lambda \int_0^t f(s)\, ds.
\end{aligned}$$

(b) Using that N_t is stationary and has independent increments, we have respectively

$$\begin{aligned}\mathbb{E}[(N_{s_{i+1}} - N_{s_i})^2] &= \mathbb{E}[N^2_{s_{i+1}-s_i}] = \lambda(s_{i+1} - s_i) + \lambda^2(s_{i+1} - s_i)^2 \\ &= \lambda\,\Delta s + \lambda^2(\Delta s)^2, \\ \mathbb{E}[(N_{s_{i+1}} - N_{s_i})(N_{s_{j+1}} - N_{s_j})] &= \mathbb{E}[(N_{s_{i+1}} - N_{s_i})]\mathbb{E}[(N_{s_{j+1}} - N_{s_j})] \\ &= \lambda(s_{i+1} - s_i)\lambda(s_{j+1} - s_j) = \lambda^2(\Delta s)^2.\end{aligned}$$

Applying the expectation to the formula

$$\begin{aligned}\Big(\sum_{i=0}^{n-1} f(s_i)(N_{s_{i+1}} - N_{s_i})\Big)^2 &= \sum_{i=0}^{n-1} f(s_i)^2(N_{s_{i+1}} - N_{s_i})^2 \\ &\quad +2\sum_{i\neq j} f(s_i)f(s_j)(N_{s_{i+1}} - N_{s_i})(N_{s_{j+1}} - N_{s_j})\end{aligned}$$

yields

$$\begin{aligned}&\mathbb{E}\Big[\Big(\sum_{i=0}^{n-1} f(s_i)(N_{s_{i+1}} - N_{s_i})\Big)^2\Big] \\ &= \sum_{i=0}^{n-1} f(s_i)^2(\lambda\Delta s + \lambda^2(\Delta s)^2) + 2\sum_{i\neq j} f(s_i)f(s_j)\lambda^2(\Delta s)^2 \\ &= \lambda\sum_{i=0}^{n-1} f(s_i)^2\Delta s + \lambda^2\Big[\sum_{i=0}^{n-1} f(s_i)^2(\Delta s)^2 + 2\sum_{i\neq j} f(s_i)f(s_j)(\Delta s)^2\Big] \\ &= \lambda\sum_{i=0}^{n-1} f(s_i)^2\Delta s + \lambda^2\Big(\sum_{i=0}^{n-1} f(s_i)\,\Delta s\Big)^2 \\ &\to \lambda\int_0^t f(s)^2\,ds + \lambda^2\Big(\int_0^t f(s)\,ds\Big)^2, \text{ as } n\to\infty.\end{aligned}$$

(c) Using that N_t is stationary with independent increments and has the moment generating function $\mathbb{E}[e^{kN_t}] = e^{\lambda(e^k-1)t}$, we have

$$\begin{aligned}
&\mathbb{E}\Big[e^{\int_0^t f(s)\,dN_s}\Big] \\
&= \lim_{n\to\infty} \mathbb{E}\Big[e^{\sum_{i=0}^{n-1} f(s_i)(N_{s_{i+1}}-N_{s_i})}\Big] = \lim_{n\to\infty} \mathbb{E}\Big[\prod_{i=0}^{n-1} e^{f(s_i)(N_{s_{i+1}}-N_{s_i})}\Big] \\
&= \lim_{n\to\infty} \prod_{i=0}^{n-1} \mathbb{E}\Big[e^{f(s_i)(N_{s_{i+1}}-N_{s_i})}\Big] = \lim_{n\to\infty} \prod_{i=0}^{n-1} \mathbb{E}\Big[e^{f(s_i)(N_{s_{i+1}-s_i})}\Big] \\
&= \lim_{n\to\infty} \prod_{i=0}^{n-1} e^{\lambda(e^{f(s_i)}-1)(s_{i+1}-s_i)} = \lim_{n\to\infty} e^{\lambda\sum_{i=0}^{n-1}(e^{f(s_i)}-1)(s_{i+1}-s_i)} \\
&= e^{\lambda\int_0^t (e^{f(s)}-1)\,ds}.
\end{aligned}$$

■

Since f is continuous, the Poisson integral $\int_0^t f(s)\,dN_s$ can be computed in terms of the waiting times S_k

$$\int_0^t f(s)\,dN_s = \sum_{k=1}^{N_t} f(S_k).$$

This formula can be used to give a proof for the previous result. For instance, taking the expectation and using conditions over $N_t = n$, yields

$$\begin{aligned}
\mathbb{E}\Big[\int_0^t f(s)\,dN_s\Big] &= \mathbb{E}\Big[\sum_{k=1}^{N_t} f(S_k)\Big] = \sum_{n\geq 0}\mathbb{E}\Big[\sum_{k=1}^{n} f(S_k)|N_t = n\Big]P(N_t = n) \\
&= \sum_{n\geq 0} \frac{n}{t}\int_0^t f(x)\,dx\,\frac{(\lambda t)^n}{n!}e^{-\lambda t} \\
&= e^{-\lambda t}\int_0^t f(x)\,dx\,\frac{1}{t}\sum_{n\geq 0}\frac{(\lambda t)^n}{(n-1)!} \\
&= e^{-\lambda t}\int_0^t f(x)\,dx\,\lambda e^{\lambda t} = \lambda\int_0^t f(x)\,dx.
\end{aligned}$$

Exercise 5.8.6 *Solve parts (b) and (c) of Proposition 5.8.5 using a similar idea with the one presented above.*

Exercise 5.8.7 *Show that*

$$\mathbb{E}\Big[\Big(\int_0^t f(s)\,dM_s\Big)^2\Big] = \lambda\int_0^t f(s)^2\,ds,$$

where $M_t = N_t - \lambda t$ is the compensated Poisson process.

Exercise 5.8.8 *Prove that*

$$Var\Big(\int_0^t f(s)\,dN_s\Big) = \lambda \int_0^t f(s)^2\,dN_s.$$

Exercise 5.8.9 *Find*

$$\mathbb{E}\Big[e^{\int_0^t f(s)\,dM_s}\Big].$$

Proposition 5.8.10 *Let $\mathcal{F}_t = \sigma(N_s; 0 \le s \le t)$. Then for any constant c, the process*

$$M_t = e^{cN_t + \lambda(1-e^c)t}, \qquad t \ge 0$$

is an $\mathcal{F}_t$-martingale.

Proof: Let $s < t$. Since $N_t - N_s$ is independent of $\mathcal{F}_s$ and N_t is stationary, we have

$$\begin{aligned}\mathbb{E}[e^{c(N_t-N_s)}|\mathcal{F}_s] &= \mathbb{E}[e^{c(N_t-N_s)}] = \mathbb{E}[e^{cN_{t-s}}] \\ &= e^{\lambda(e^c-1)(t-s)}.\end{aligned}$$

On the other side, taking out the deterministic part yields

$$\mathbb{E}[e^{c(N_t-N_s)}|\mathcal{F}_s] = e^{-cN_s}\mathbb{E}[e^{cN_t}|\mathcal{F}_s].$$

Equating the last two relations we arrive at

$$\mathbb{E}[e^{cN_t+(1-e^c)t}|\mathcal{F}_s] = e^{cN_s+\lambda(1-e^c)s},$$

which is equivalent to the martingale condition $\mathbb{E}[M_t|\mathcal{F}_s] = M_s$. ∎

We shall present an application of the previous result. Consider the waiting time until the nth jump, $S_n = \inf\{t > 0; N_t = n\}$, which is a stopping time, and the filtration $\mathcal{F}_t = \sigma(N_s; 0 \le s \le t)$. Since

$$M_t = e^{cN_t+\lambda(1-e^c)t}$$

is an $\mathcal{F}_t$-martingale, by the Optional Stopping Theorem (Theorem 4.2.1) we have $\mathbb{E}[M_{S_n}] = \mathbb{E}[M_0] = 1$, which is equivalent to $\mathbb{E}[e^{\lambda(1-e^c)S_n}] = e^{-cn}$. Substituting $s = -\lambda(1-e^c)$, then $c = \ln(1+\frac{s}{\lambda})$. Since $s, \lambda > 0$, then $c > 0$. The previous expression becomes

$$\mathbb{E}[e^{-sS_n}] = e^{-n\ln(1+\frac{s}{\lambda})} = \Big(\frac{\lambda}{\lambda+s}\Big)^x.$$

Since the expectation on the left side is the Laplace transform of the probability density of S_n, then

$$\begin{aligned}p(S_n) &= \mathcal{L}^{-1}\{\mathbb{E}[e^{-sS_n}]\} = \mathcal{L}^{-1}\Big\{\Big(\frac{\lambda}{\lambda+s}\Big)^x\Big\} \\ &= \frac{e^{-t\lambda}t^{n-1}\lambda^n}{\Gamma(n)},\end{aligned}$$

which shows that S_n has a gamma distribution.

5.9 The Distribution Function of $X_T = \int_0^T g(t)\, dN_t$

In this section we consider the function $g(t)$ continuous. Let $S_1 < S_2 < \cdots < S_{N_t}$ denote the waiting times until time t. Since the increments dN_t are equal to 1 at S_k and 0 otherwise, the integral can be written as

$$X_T = \int_0^T g(t)\, dN_t = g(S_1) + \cdots + g(S_{N_t}).$$

The distribution function of the random variable $X_T = \int_0^T g(t)\, dN_t$ can be obtained conditioning over the N_t

$$\begin{aligned} P(X_T \le u) &= \sum_{k\ge 0} P(X_T \le u | N_T = k)\, P(N_T = k) \\ &= \sum_{k\ge 0} P(g(S_1) + \cdots + g(S_{N_t}) \le u | N_T = k)\, P(N_T = k) \\ &= \sum_{k\ge 0} P(g(S_1) + \cdots + g(S_k) \le u)\, P(N_T = k). \end{aligned} \qquad (5.9.5)$$

Considering $S_1, S_2, \cdots, S_k$ independent and uniformly distributed over the interval $[0, T]$, we have

$$P\big(g(S_1) + \cdots + g(S_k) \le u\big) = \int_{D_k} \frac{1}{T^k}\, dx_1 \cdots dx_k = \frac{vol(D_k)}{T^k},$$

where

$$D_k = \{g(x_1) + g(x_2) + \cdots + g(x_k) \le u\} \cap \{0 \le x_1, \cdots, x_k \le T\}.$$

Substituting back in (5.9.5) yields

$$\begin{aligned} P(X_T \le u) &= \sum_{k\ge 0} P(g(S_1) + \cdots + g(S_k) \le u)\, P(N_T = k) \\ &= \sum_{k\ge 0} \frac{vol(D_k)}{T^k} \frac{\lambda^k T^k}{k!} e^{-\lambda T} = e^{-\lambda T} \sum_{k\ge 0} \frac{\lambda^k vol(D_k)}{k!}. \end{aligned} \qquad (5.9.6)$$

In general, the volume of the k-dimensional solid D_k is not easy to obtain. However, there are simple cases when this can be computed explicitly.

A Particular Case We shall do an explicit computation of the partition function of $X_T = \int_0^T s^2\, dN_s$. In this case the solid D_k is the intersection between the k-dimensional ball of radius $\sqrt{u}$ centered at the origin and the k-dimensional cube $[0, T]^k$. There are three possible shapes for D_k, which depend on the size of $\sqrt{u}$:

(a) if $0 \le \sqrt{u} < T$, then D_k is a $\frac{1}{2^k}$-part of a k-dimensional sphere;

(b) if $T \le \sqrt{u} < T\sqrt{k}$, then D_k has a complicated shape;

(c) if $T\sqrt{k} \le \sqrt{u}$, then D_k is the entire k-dimensional cube, and then $vol(D_k) = T^k$.

Since the volume of the k-dimensional ball of radius R is given by $\dfrac{\pi^{k/2}R^k}{\Gamma(\frac{k}{2}+1)}$, then the volume of D_k in case (**a**) becomes

$$vol(D_k) = \frac{\pi^{k/2}u^{k/2}}{2^k\Gamma(\frac{k}{2}+1)}.$$

Substituting in (5.9.6) yields

$$P(X_T \le u) = e^{-\lambda T}\sum_{k\ge 0}\frac{(\lambda^2\pi u)^{k/2}}{k!\Gamma(\frac{k}{2}+1)}, \qquad 0 \le \sqrt{u} < T.$$

It is worth noting that for $u \to \infty$, the inequality $T\sqrt{k} \le \sqrt{u}$ is satisfied for all $k \ge 0$; hence relation (5.9.6) yields

$$\lim_{u\to\infty} P(X_T \le u) = e^{-\lambda T}\sum_{k\ge 0}\frac{\lambda^k T^k}{k!} = e^{-kT}e^{kT} = 1.$$

The computation in case (**b**) is more complicated and will be omitted.

Exercise 5.9.1 *Calculate the expectation* $\mathbb{E}\Big[\int_0^T e^{ks}\,dN_s\Big]$ *and the variance* $Var\Big(\int_0^T e^{ks}\,dN_s\Big)$.

Exercise 5.9.2 *Compute the distribution function of* $X_t = \int_0^T s\,dN_s$.

Exercise 5.9.3 *The following stochastic differential equation has been used in [2] to model the depreciation value of a car with stochastic repair payments*

$$dV_t = -kV_t\,dt - \rho\,dN_t,$$

where $k > 0$ is the depreciation rate, $\rho > 0$ is the average repair payment, and N_t is a Poisson process with rate λ.

(a) Show that the solution is given by

$$V_t = V_0e^{-kt} - \rho e^{-kt}\int_0^t e^{ks}\,dN_s;$$

(b) *Consider the stopping time*

$$\tau = \inf\{t > 0; V_t \leq K\}.$$

Show that

$$\mathbb{E}[e^{k\tau}] = \frac{kV_0 + \lambda\rho}{\lambda\rho + kK}.$$

Chapter 6

Stochastic Differentiation

Most stochastic processes are not differentiable. For instance, the Brownian motion process W_t is a continuous process which is nowhere differentiable. Hence, derivatives like $\frac{dW_t}{dt}$ do not make sense in stochastic calculus. The only quantities allowed to be used are the infinitesimal changes of the process, in our case, dW_t.

The infinitesimal change of a process The change in the process X_t between instances t and $t + \Delta t$ is given by $\Delta X_t = X_{t+\Delta t} - X_t$. When Δt is infinitesimally small, we obtain the infinitesimal change of a process X_t

$$\boxed{dX_t = X_{t+dt} - X_t.}$$

Sometimes it is useful to use the equivalent formula $X_{t+dt} = X_t + dX_t$.

6.1 Basic Rules

The following rules are the analog of some familiar differentiation rules from the elementary Calculus.

The constant multiple rule If X_t is a stochastic process and c is a constant, then

$$\boxed{d(c\,X_t) = c\,dX_t.}$$

The verification follows from a straightforward application of the infinitesimal change formula

$$d(c\,X_t) = c\,X_{t+dt} - c\,X_t = c(X_{t+dt} - X_t) = c\,dX_t.$$

The sum rule If X_t and Y_t are two stochastic processes, then

$$\boxed{d(X_t + Y_t) = dX_t + dY_t.}$$

The verification is as in the following:

$$\begin{aligned} d(X_t + Y_t) &= (X_{t+dt} + Y_{t+dt}) - (X_t + Y_t) \\ &= (X_{t+dt} - X_t) + (Y_{t+dt} - Y_t) \\ &= dX_t + dY_t. \end{aligned}$$

The difference rule If X_t and Y_t are two stochastic processes, then

$$\boxed{d(X_t - Y_t) = dX_t - dY_t.}$$

The proof is similar to the one for the sum rule.

The product rule If X_t and Y_t are two stochastic processes, then

$$\boxed{d(X_t Y_t) = X_t\, dY_t + Y_t\, dX_t + dX_t\, dY_t.}$$

The proof is as follows:

$$\begin{aligned} d(X_t\, Y_t) &= X_{t+dt}Y_{t+dt} - X_t Y_t \\ &= X_t(Y_{t+dt} - Y_t) + Y_t(X_{t+dt} - X_t) + (X_{t+dt} - X_t)(Y_{t+dt} - Y_t) \\ &= X_t\, dY_t + Y_t\, dX_t + dX_t\, dY_t, \end{aligned}$$

where the second identity is verified by direct computation.

If the process X_t is replaced by the deterministic function $f(t)$, then the aforementioned formula becomes

$$\boxed{d(f(t)Y_t) = f(t)\, dY_t + Y_t\, df(t) + df(t)\, dY_t.}$$

Since in most practical cases the process Y_t satisfies the equation

$$dY_t = a(t, W_t)dt + b(t, W_t)dW_t, \tag{6.1.1}$$

using relations $dt\, dW_t = dt^2 = 0$, the last term vanishes

$$df(t)\, dY_t = f'(t)dtdY_t = 0,$$

and hence

$$\boxed{d(f(t)Y_t) = f(t)\, dY_t + Y_t\, df(t).}$$

This relation looks like the usual product rule.

The quotient rule If X_t and Y_t are two stochastic processes, then

$$\boxed{d\Big(\frac{X_t}{Y_t}\Big) = \frac{Y_t dX_t - X_t dY_t - dX_t dY_t}{Y_t^2} + \frac{X_t}{Y_t^3}(dY_t)^2.}$$

The proof follows from Ito's formula and will be addressed in section 6.2.3.

When the process Y_t is replaced by the deterministic function $f(t)$, and X_t is a process satisfying an equation of type (6.1.1), then the previous formula becomes

$$\boxed{d\Big(\frac{X_t}{f(t)}\Big) = \frac{f(t)dX_t - X_t df(t)}{f(t)^2}.}$$

Example 6.1.1 *We shall show that*

$$\boxed{d(W_t^2) = 2W_t\, dW_t + dt.}$$

Applying the product rule and the fundamental relation $(dW_t)^2 = dt$, yields

$$d(W_t^2) = W_t\, dW_t + W_t\, dW_t + dW_t\, dW_t = 2W_t\, dW_t + dt.$$

Example 6.1.2 *Show that*

$$\boxed{d(W_t^3) = 3W_t^2\, dW_t + 3W_t dt.}$$

Applying the product rule and the previous exercise yields

$$\begin{aligned} d(W_t^3) &= d(W_t \cdot W_t^2) = W_t d(W_t^2) + W_t^2\, dW_t + d(W_t^2)\, dW_t \\ &= W_t(2W_t\, dW_t + dt) + W_t^2\, dW_t + dW_t(2W_t\, dW_t + dt) \\ &= 2W_t^2\, dW_t + W_t\, dt + W_t^2\, dW_t + 2W_t(dW_t)^2 + dt\, dW_t \\ &= 3W_t^2\, dW_t + 3W_t\, dt, \end{aligned}$$

where we used $(dW_t)^2 = dt$ and $dt\, dW_t = 0$.

Example 6.1.3 *Show that* $d(tW_t) = W_t\, dt + t\, dW_t$.

Using the product rule and $dt\, dW_t = 0$, we get

$$\begin{aligned} d(tW_t) &= W_t\, dt + t\, dW_t + dt\, dW_t \\ &= W_t\, dt + t\, dW_t. \end{aligned}$$

Example 6.1.4 *Let* $Z_t = \int_0^t W_u\, du$ *be the integrated Brownian motion. Show that*

$$\boxed{dZ_t = W_t\, dt.}$$

The infinitesimal change of Z_t is

$$dZ_t = Z_{t+dt} - Z_t = \int_t^{t+dt} W_s\, ds = W_t\, dt,$$

since W_s is a continuous function in s.

Example 6.1.5 *Let $A_t = \frac{1}{t}Z_t = \frac{1}{t}\int_0^t W_u\,du$ be the average of the Brownian motion on the time interval $[0,t]$. Show that*

$$\boxed{dA_t = \frac{1}{t}\Big(W_t - \frac{1}{t}Z_t\Big)\,dt.}$$

We have

$$\begin{aligned} dA_t &= d\Big(\frac{1}{t}\Big)\,Z_t + \frac{1}{t}\,dZ_t + d\Big(\frac{1}{t}\Big)\,dZ_t \\ &= \frac{-1}{t^2}Z_t\,dt + \frac{1}{t}W_t\,dt + \frac{-1}{t^2}W_t\underbrace{dt^2}_{=0} \\ &= \frac{1}{t}\Big(W_t - \frac{1}{t}Z_t\Big)\,dt. \end{aligned}$$

Exercise 6.1.6 *Let $G_t = \frac{1}{t}\int_0^t e^{W_u}\,du$ be the average of the geometric Brownian motion on $[0,t]$. Find dG_t.*

6.2 Ito's Formula

Ito's formula is the analog of the chain rule from elementary Calculus. We shall start by reviewing a few concepts regarding function approximations.

Let f be a twice continuously differentiable function of a real variable x. Let x_0 be fixed and consider the changes $\Delta x = x - x_0$ and $\Delta f(x) = f(x) - f(x_0)$. It is known from Calculus that the following second order Taylor approximation holds

$$\Delta f(x) = f'(x)\Delta x + \frac{1}{2}f''(x)(\Delta x)^2 + O(\Delta x)^3.$$

When x is infinitesimally close to x_0, we replace Δx by the differential dx and obtain

$$df(x) = f'(x)dx + \frac{1}{2}f''(x)(dx)^2 + O(dx)^3. \qquad (6.2.2)$$

In elementary Calculus, all terms involving terms of equal or higher order to dx^2 are neglected; then the aforementioned formula becomes

$$df(x) = f'(x)dx.$$

Now, if we consider $x = x(t)$ to be a differentiable function of t, substituting into the previous formula we obtain the differential form of the well known chain rule

$$df\big(x(t)\big) = f'\big(x(t)\big)dx(t) = f'\big(x(t)\big)x'(t)\,dt.$$

We shall present a similar formula for the stochastic environment. In this case the deterministic function $x(t)$ is replaced by a stochastic process X_t. The composition between the differentiable function f and the process X_t is a process denoted by $F_t = f(X_t)$.

Neglecting the increment powers higher than or equal to $(dX_t)^3$, the expression (6.2.2) becomes

$$dF_t = f'(X_t)dX_t + \frac{1}{2}f''(X_t)\big(dX_t\big)^2. \tag{6.2.3}$$

In the computation of dX_t we may take into the account stochastic relations such as $dW_t^2 = dt$, or $dt\,dW_t = 0$.

Theorem 6.2.1 (Ito's formula) *Let X_t be a stochastic process satisfying*

$$dX_t = b_t dt + \sigma_t dW_t,$$

with $b_t(\omega)$ and $\sigma_t(\omega)$ measurable processes. Let $F_t = f(X_t)$, with f twice continuously differentiable. Then

$$\boxed{dF_t = \Big[b_t f'(X_t) + \frac{\sigma_t^2}{2}f''(X_t)\Big]dt + \sigma_t f'(X_t)\,dW_t.} \tag{6.2.4}$$

Proof: We shall provide an informal proof. Using relations $dW_t^2 = dt$ and $dt^2 = dW_t\,dt = 0$, we have

$$\begin{aligned}
(dX_t)^2 &= \Big(b_t dt + \sigma_t dW_t\Big)^2 \\
&= b_t^2 dt^2 + 2b_t\sigma_t dW_t dt + \sigma_t^2 dW_t^2 \\
&= \sigma_t^2 dt.
\end{aligned}$$

Substituting into (6.2.3) yields

$$\begin{aligned}
dF_t &= f'(X_t)dX_t + \frac{1}{2}f''(X_t)\big(dX_t\big)^2 \\
&= f'(X_t)\Big(b_t dt + \sigma_t dW_t\Big) + \frac{1}{2}f''(X_t)\sigma_t^2 dt \\
&= \Big[b_t f'(X_t) + \frac{\sigma_t^2}{2}f''(X_t)\Big]dt + \sigma_t f'(X_t)\,dW_t.
\end{aligned}$$

■

Remark 6.2.2 Ito's formula can also be written under the following equivalent integral form

$$F_t = F_0 + \int_0^t \Big(b_s f'(X_s) + \frac{1}{2}\sigma_s^2 f''(X_s)\Big)\,ds + \int_0^t \sigma_s f'(X_s)\,dW_s.$$

In the case $X_t = W_t$ we obtain the following consequence:

Corollary 6.2.3 *Let $F_t = f(W_t)$. Then*

$$\boxed{dF_t = \frac{1}{2}f''(W_t)dt + f'(W_t)\,dW_t.} \tag{6.2.5}$$

Particular cases In the following we shall present the most often used cases:
1. If $f(x) = x^\alpha$, with α constant, then $f'(x) = \alpha x^{\alpha-1}$, $f''(x) = \alpha(\alpha-1)x^{\alpha-2}$. Then (6.2.5) becomes the following useful formula

$$\boxed{d(W_t^\alpha) = \frac{1}{2}\alpha(\alpha-1)W_t^{\alpha-2}dt + \alpha W_t^{\alpha-1}\,dW_t.}$$

A couple of useful cases easily follow:

$$\begin{aligned} d(W_t^2) &= 2W_t\,dW_t + dt \\ d(W_t^3) &= 3W_t^2\,dW_t + 3W_t dt. \end{aligned}$$

2. If $f(x) = e^{kx}$, with k constant, $f'(x) = ke^{kx}$, $f''(x) = k^2e^{kx}$. Therefore

$$\boxed{d(e^{kW_t}) = ke^{kW_t}dW_t + \frac{1}{2}k^2e^{kW_t}\,dt.}$$

In particular, for $k = 1$, we obtain the increments of a geometric Brownian motion

$$d(e^{W_t}) = e^{W_t}dW_t + \frac{1}{2}e^{W_t}\,dt.$$

3. If $f(x) = \sin x$, then

$$\boxed{d(\sin W_t) = \cos W_t\,dW_t - \frac{1}{2}\sin W_t\,dt.}$$

Exercise 6.2.4 *Use the previous rules to find the following increments*

(*a*) $d(W_te^{W_t})$

(*b*) $d(3W_t^2 + 2e^{5W_t})$

(*c*) $d(e^{t+W_t^2})$

(*d*) $d((t+W_t)^n)$

(*e*) $d\Big(\frac{1}{t}\int_0^t W_u\,du\Big)$

(*f*) $d\Big(\frac{1}{t^\alpha}\int_0^t e^{W_u}\,du\Big)$, *where α is a constant.*

In the case when the function $f = f(t,x)$ is also time dependent, the analog of (6.2.2) is given by

$$df(t,x) = \partial_t f(t,x)dt + \partial_x f(t,x)dx + \frac{1}{2}\partial_x^2 f(t,x)(dx)^2 + O(dx)^3 + O(dt)^2. \tag{6.2.6}$$

Substituting $x = X_t$ yields

$$df(t,X_t) = \partial_t f(t,X_t)dt + \partial_x f(t,X_t)dX_t + \frac{1}{2}\partial_x^2 f(t,X_t)(dX_t)^2. \tag{6.2.7}$$

If X_t is a process satisfying an equation of type (6.1.1), then we obtain an extra-term in formula (6.2.4)

$$\begin{aligned} dF_t &= \Big[\partial_t f(t,X_t) + a(W_t,t)\partial_x f(t,X_t) + \frac{b(W_t,t)^2}{2}\partial_x^2 f(t,X_t)\Big]dt \\ &\quad + b(W_t,t)\partial_x f(t,X_t)\, dW_t. \end{aligned} \tag{6.2.8}$$

Exercise 6.2.5 *Show that*

$$d(tW_t^2) = (t + W_t^2)dt + 2tW_t\, dW_t.$$

Exercise 6.2.6 *Find the following increments*

$$\begin{array}{ll} (a)\ \ d(tW_t) & (c)\ \ d(t^2\cos W_t) \\ (b)\ \ d(e^t W_t) & (d)\ \ d(\sin t\, W_t^2). \end{array}$$

6.2.1 Ito diffusions

Consider the process X_t given by

$$dX_t = b(X_t,t)dt + \sigma(X_t,t)dW_t. \tag{6.2.9}$$

A process $X_t = (X_t^i) \in \mathbb{R}^n$ satisfying this relation is called an *Ito diffusion* in $\mathbb{R}^n$. Equation (6.2.9) models the position of a small particle that moves under the influence of a drift force $b(X_t,t)$, and is subject to random deviations. This situation occurs in the physical world when a particle suspended in a moving liquid is subject to random molecular bombardments. The amount $\frac{1}{2}\sigma\sigma^T$ is called the *diffusion coefficient* and describes the difussion of the particle.

Exercise 6.2.7 *Consider the time-homogeneous Ito diffusion in* $\mathbb{R}^n$

$$dX_t = b(X_t)dt + \sigma(X_t)dW_t.$$

Show that:

$$(a)\ b_i(x) = \lim_{t\to 0} \frac{\mathbb{E}[(X_t^i - x_i)]}{t}.$$

$$(b)\ (\sigma\sigma^T)_{ij}(x) = \lim_{t\to 0} \frac{\mathbb{E}[(X_t^i - x_i)(X_t^j - x_j)]}{t}.$$

6.2.2 Ito's formula for Poisson processes

Consider the process $F_t = F(M_t)$, where $M_t = N_t - \lambda t$ is the compensated Poisson process. Ito's formula for the process F_t takes the following integral form. For a proof the reader can consult Kuo [30].

Proposition 6.2.8 *Let F be a twice differentiable function. Then for any $a < t$ we have*

$$F_t = F_a + \int_a^t F'(M_{s-})\,dM_s + \sum_{a<s\leq t}\Big(\Delta F(M_s) - F'(M_{s-})\Delta M_s\Big),$$

where $\Delta M_s = M_s - M_{s-}$ and $\Delta F(M_s) = F(M_s) - F(M_{s-})$.

We shall apply the aforementioned result for the case $F_t = F(M_t) = M_t^2$. We have

$$M_t^2 = M_a^2 + 2\int_a^t M_{s-}\,dM_s + \sum_{a<s\leq t}\Big(M_s^2 - M_{s-}^2 - 2M_{s-}(M_s - M_{s-})\Big). \tag{6.2.10}$$

Since the jumps in N_s are of size 1, we have $(\Delta N_s)^2 = \Delta N_s$. Since the difference of the processes M_s and N_s is continuous, then $\Delta M_s = \Delta N_s$. Using these formulas we have

$$\begin{aligned}
\Big(M_s^2 - M_{s-}^2 - 2M_{s-}(M_s - M_{s-})\Big) &= (M_s - M_{s-})\Big(M_s + M_{s-} - 2M_{s-}\Big)\\
&= (M_s - M_{s-})^2 = (\Delta M_s)^2 = (\Delta N_s)^2\\
&= \Delta N_s = N_s - N_{s-}.
\end{aligned}$$

Since the sum of the jumps between s and t is $\sum_{a<s\leq t}\Delta N_s = N_t - N_a$, formula (6.2.10) becomes

$$M_t^2 = M_a^2 + 2\int_a^t M_{s-}\,dM_s + N_t - N_a. \tag{6.2.11}$$

The differential form is

$$d(M_t^2) = 2M_{t-}\,dM_t + dN_t,$$

which is equivalent to

$$d(M_t^2) = (1 + 2M_{t-})\,dM_t + \lambda dt,$$

since $dN_t = dM_t + \lambda dt$.

Exercise 6.2.9 *Show that*

$$\int_0^T M_{t-}\,dM_t = \frac{1}{2}(M_T^2 - N_T).$$

Exercise 6.2.10 *Use Ito's formula for the Poisson process to find the conditional expectation $\mathbb{E}[M_t^2|\mathcal{F}_s]$ for $s < t$.*

6.2.3 Ito's multidimensional formula

If the process F_t depends on several Ito diffusions, say $F_t = f(t, X_t, Y_t)$, then a similar formula to (6.2.8) leads to

$$\begin{aligned} dF_t &= \frac{\partial f}{\partial t}(t, X_t, Y_t)dt + \frac{\partial f}{\partial x}(t, X_t, Y_t)dX_t + \frac{\partial f}{\partial y}(t, X_t, Y_t)dY_t \\ &\quad + \frac{1}{2}\frac{\partial^2 f}{\partial x^2}(t, X_t, Y_t)(dX_t)^2 + \frac{1}{2}\frac{\partial^2 f}{\partial y^2}(t, X_t, Y_t)(dY_t)^2 \\ &\quad + \frac{\partial^2 f}{\partial x \partial y}(t, X_t, Y_t)dX_t\, dY_t. \end{aligned}$$

Example 6.2.11 (Harmonic function of a Brownian motion) *In the case when $F_t = f(X_t, Y_t)$, with $X_t = W_t^1$, $Y_t = W_t^2$ independent Brownian motions, we have*

$$\begin{aligned} dF_t &= \frac{\partial f}{\partial x}dW_t^1 + \frac{\partial f}{\partial y}dW_t^2 + \frac{1}{2}\frac{\partial^2 f}{\partial x^2}(dW_t^1)^2 + \frac{1}{2}\frac{\partial^2 f}{\partial y^2}(dW_t^2)^2 \\ &\quad + \frac{\partial^2 f}{\partial x \partial y}dW_t^1\, dW_t^2 \\ &= \frac{\partial f}{\partial x}dW_t^1 + \frac{\partial f}{\partial y}dW_t^2 + \frac{1}{2}\Big(\frac{\partial^2 f}{\partial x^2} + \frac{\partial^2 f}{\partial y^2}\Big)dt. \end{aligned}$$

The expression

$$\Delta f = \frac{1}{2}\Big(\frac{\partial^2 f}{\partial x^2} + \frac{\partial^2 f}{\partial y^2}\Big)$$

is called the Laplacian of f. We can rewrite the previous formula as

$$dF_t = \frac{\partial f}{\partial x}dW_t^1 + \frac{\partial f}{\partial y}dW_t^2 + \Delta f\, dt.$$

A function f with $\Delta f = 0$ is called harmonic. The aforementioned formula in the case of harmonic functions takes the simple form

$$\boxed{dF_t = \frac{\partial f}{\partial x}dW_t^1 + \frac{\partial f}{\partial y}dW_t^2.} \tag{6.2.12}$$

Exercise 6.2.12 *Let W_t^1, W_t^2 be two independent Brownian motions. If the function f is harmonic, show that $F_t = f(W_t^1, W_t^2)$ is a martingale. Is the converse true?*

Exercise 6.2.13 *Use the previous formulas to find dF_t in the following cases*

(a) $F_t = (W_t^1)^2 + (W_t^2)^2$

(b) $F_t = \ln[(W_t^1)^2 + (W_t^2)^2]$.

Exercise 6.2.14 *Consider the Bessel process* $R_t = \sqrt{(W_t^1)^2 + (W_t^2)^2}$, *where* W_t^1 *and* W_t^2 *are two independent Brownian motions. Prove that*

$$dR_t = \frac{1}{2R_t}dt + \frac{W_t^1}{R_t}dW_t^1 + \frac{W_t^2}{R_t}dW_t^2.$$

Example 6.2.15 (The product rule) *Let* X_t *and* Y_t *be two Ito diffusions. Show that*

$$d(X_tY_t) = Y_tdX_t + X_tdY_t + dX_tdY_t.$$

Consider the function $f(x,y) = xy$. Since $\partial_x f = y$, $\partial_y f = x$, $\partial_x^2 f = \partial_y^2 f = 0$, $\partial_x\partial_y f = 1$, then Ito's multidimensional formula yields

$$\begin{aligned} d(X_tY_t) &= d\big(f(X_,Y_t)\big) = \partial_x f\, dX_t + \partial_y f\, dY_t \\ &\quad + \frac{1}{2}\partial_x^2 f(dX_t)^2 + \frac{1}{2}\partial_y^2 f(dY_t)^2 + \partial_x\partial_y f\, dX_tdY_t \\ &= Y_tdX_t + X_tdY_t + dX_tdY_t. \end{aligned}$$

Example 6.2.16 (The quotient rule) *Let* X_t *and* Y_t *be two Ito diffusions. Show that*

$$d\Big(\frac{X_t}{Y_t}\Big) = \frac{Y_tdX_t - X_tdY_t - dX_tdY_t}{Y_t^2} + \frac{X_t}{Y_t^3}(dY_t)^2.$$

Consider the function $f(x,y) = \frac{x}{y}$. Since $\partial_x f = \frac{1}{y}$, $\partial_y f = -\frac{x}{y^2}$, $\partial_x^2 f = 0$, $\partial_y f = -\frac{x}{y^2}$, $\partial_y^2 f = \frac{2x}{y^3}$, $\partial_x\partial_y = -\frac{1}{y^2}$, then applying Ito's multidimensional formula yields

$$\begin{aligned} d\Big(\frac{X_t}{Y_t}\Big) &= d\big(f(X_,Y_t)\big) = \partial_x f\, dX_t + \partial_y f\, dY_t \\ &\quad + \frac{1}{2}\partial_x^2 f(dX_t)^2 + \frac{1}{2}\partial_y^2 f(dY_t)^2 + \partial_x\partial_y f\, dX_tdY_t \\ &= \frac{Y_tdX_t - X_tdY_t - dX_tdY_t}{Y_t^2} + \frac{X_t}{Y_t^3}(dY_t)^2. \end{aligned}$$

Chapter 7

Stochastic Integration Techniques

Computing a stochastic integral starting from the definition of the Ito integral is not only difficult, but also rather inefficient. Like in elementary Calculus, several methods can be developed to compute stochastic integrals. We tried to keep the analogy with elementary Calculus as much as possible. The integration by substitution is more complicated in the stochastic environment and we have considered only a particular case of it, which we called *the method of heat equation.*

7.1 Notational Conventions

The intent of this section is to discuss some equivalent integral notations for a stochastic differential equation. Consider a process X_t whose increments satisfy the stochastic differential equation $dX_t = f(t, W_t)dW_t$. This can be written equivalently in the integral form as

$$\int_a^t dX_s = \int_a^t f(s, W_s)dW_s. \tag{7.1.1}$$

If we consider the partition $0 = t_0 < t_1 < \cdots < t_{n-1} < t_n = t$, then the left side becomes

$$\int_a^t dX_s = \text{ms-}\lim_{n\to\infty} \sum_{j=0}^{n-1}(X_{t_{j+1}} - X_{t_j}) = X_t - X_a,$$

after canceling the terms in pairs. Substituting into formula (7.1.1) yields the equivalent form

$$X_t = X_a + \int_a^t f(s, W_s)dW_s.$$

This can also be written as $dX_t = d\Big(\int_a^t f(s, W_s)dW_s\Big)$, since X_a is a constant. Using $dX_t = f(t, W_t)dW_t$, the previous formula can be written in the following two equivalent ways:

(i) For any $a < t$, we have

$$\boxed{d\Big(\int_a^t f(s, W_s)dW_s\Big) = f(t, W_t)dW_t.} \tag{7.1.2}$$

(ii) If Y_t is a stochastic process, such that $Y_t dW_t = dF_t$, then

$$\boxed{\int_a^b Y_t\, dW_t = F_b - F_a.}$$

These formulas are equivalent ways of writing the stochastic differential equation (7.1.1), and will be useful in future computations. A few applications follow.

Example 7.1.1 *Verify the stochastic formula*

$$\int_0^t W_s\, dW_s = \frac{W_t^2}{2} - \frac{t}{2}.$$

Let $X_t = \int_0^t W_s\, dW_s$ and $Y_t = \frac{W_t^2}{2} - \frac{t}{2}$. From Ito's formula

$$dY_t = d\Big(\frac{W_t^2}{2}\Big) - d\Big(\frac{t}{2}\Big) = \frac{1}{2}(2W_t\, dW_t + dt) - \frac{1}{2}dt = W_t\, dW_t,$$

and from formula (7.1.2) we get

$$dX_t = d\Big(\int_0^t W_s\, dW_s\Big) = W_t\, dW_t.$$

Hence $dX_t = dY_t$, or $d(X_t - Y_t) = 0$. Since the process $X_t - Y_t$ has zero increments, then $X_t - Y_t = c$, constant. Taking $t = 0$, yields

$$c = X_0 - Y_0 = \int_0^0 W_s\, dW_s - \Big(\frac{W_0^2}{2} - \frac{0}{2}\Big) = 0,$$

and hence $c = 0$. It follows that $X_t = Y_t$, which verifies the desired relation.

Example 7.1.2 *Verify the formula*

$$\int_0^t sW_s\, dW_s = \frac{t}{2}\Big(W_t^2 - \frac{t}{2}\Big) - \frac{1}{2}\int_0^t W_s^2\, ds.$$

Consider the stochastic processes $X_t = \int_0^t sW_s\,dW_s$, $Y_t = \frac{t}{2}\Big(W_t^2 - 1\Big)$, and $Z_t = \frac{1}{2}\int_0^t W_s^2\,ds$. Formula (7.1.2) yields

$$\begin{aligned} dX_t &= tW_t\,dW_t \\ dZ_t &= \frac{1}{2}W_t^2 dt. \end{aligned}$$

Applying Ito's formula, we get

$$\begin{aligned} dY_t &= d\Big(\frac{t}{2}\Big(W_t^2 - \frac{t}{2}\Big)\Big) = \frac{1}{2}d(tW_t^2) - d\Big(\frac{t^2}{4}\Big) \\ &= \frac{1}{2}\Big[(t + W_t^2)dt + 2tW_t\,dW_t\Big] - \frac{1}{2}tdt \\ &= \frac{1}{2}W_t^2 dt + tW_t\,dW_t. \end{aligned}$$

We can easily see that

$$dX_t = dY_t - dZ_t.$$

This implies $d(X_t - Y_t + Z_t) = 0$, i.e. $X_t - Y_t + Z_t = c$, constant. Since $X_0 = Y_0 = Z_0 = 0$, it follows that $c = 0$. This proves the desired relation.

Example 7.1.3 *Show that*

$$\int_0^t (W_s^2 - s)\,dW_s = \frac{1}{3}W_t^3 - tW_t.$$

Consider the function $f(t,x) = \frac{1}{3}x^3 - tx$, and let $F_t = f(t, W_t)$. Since $\partial_t f = -x$, $\partial_x f = x^2 - t$, and $\partial_x^2 f = 2x$, then Ito's formula provides

$$\begin{aligned} dF_t &= \partial_t f\,dt + \partial_x f\,dW_t + \frac{1}{2}\partial_x^2 f\,(dW_t)^2 \\ &= -W_t dt + (W_t^2 - t)\,dW_t + \frac{1}{2}2W_t\,dt \\ &= (W_t^2 - t)dW_t. \end{aligned}$$

From formula (7.1.2) we get

$$\int_0^t (W_s^2 - s)\,dW_s = \int_0^t dF_s = F_t - F_0 = F_t = \frac{1}{3}W_t^3 - tW_t.$$

Exercise 7.1.4 *Show that*

(*a*) $\displaystyle\int_0^t \ln W_s^2\,dW_s = W_t(\ln W_t^2 - 2) - \int_0^t \frac{1}{W_s}\,ds;$

(*b*) $\displaystyle\int_0^t e^{\frac{s}{2}}\cos W_s\,dW_s = e^{\frac{t}{2}}\sin W_t;$

(c) $\displaystyle\int_0^t e^{\frac{s}{2}} \sin W_s \, dW_s = 1 - e^{\frac{t}{2}} \cos W_t;$

(d) $\displaystyle\int_0^t e^{W_s - \frac{s}{2}} \, dW_s = e^{W_t - \frac{t}{2}} - 1;$

(e) $\displaystyle\int_0^t \cos W_s \, dW_s = \sin W_t + \frac{1}{2} \int_0^t \sin W_s \, ds;$

(f) $\displaystyle\int_0^t \sin W_s \, dW_s = 1 - \cos W_t - \frac{1}{2} \int_0^t \cos W_s \, ds.$

7.2 Stochastic Integration by Parts

Consider the process $F_t = f(t)g(W_t)$, with f and g differentiable. Using the product rule yields

$$\begin{aligned} dF_t &= df(t)\, g(W_t) + f(t)\, dg(W_t) \\ &= f'(t)g(W_t)dt + f(t)\big(g'(W_t)dW_t + \frac{1}{2}g''(W_t)dt\big) \\ &= f'(t)g(W_t)dt + \frac{1}{2}f(t)g''(W_t)dt + f(t)g'(W_t)dW_t. \end{aligned}$$

Writing the relation in the integral form, we obtain the first integration by parts formula:

$$\boxed{\int_a^b f(t)g'(W_t)\, dW_t = f(t)g(W_t)\Big|_a^b - \int_a^b f'(t)g(W_t)\, dt - \frac{1}{2}\int_a^b f(t)g''(W_t)\, dt.}$$

This formula is to be used when integrating a product between a function of t and a function of the Brownian motion W_t, for which an antiderivative is known. The following two particular cases are important and useful in applications.

1. If $g(W_t) = W_t$, the aforementioned formula takes the simple form

$$\boxed{\int_a^b f(t)\, dW_t = f(t)W_t\Big|_{t=a}^{t=b} - \int_a^b f'(t)W_t\, dt.} \tag{7.2.3}$$

It is worth noting that the left side is a Wiener integral.

2. If $f(t) = 1$, then the formula becomes

$$\boxed{\int_a^b g'(W_t)\, dW_t = g(W_t)\Big|_{t=a}^{t=b} - \frac{1}{2}\int_a^b g''(W_t)\, dt.} \tag{7.2.4}$$

Application 7.2.1 *Consider the Wiener integral* $I_T = \int_0^T t\,dW_t$*. From the general theory, see Proposition 5.6.1, it is known that* I *is a random variable normally distributed with mean 0 and variance*

$$Var[I_T] = \int_0^T t^2\,dt = \frac{T^3}{3}.$$

Recall the definition of integrated Brownian motion

$$Z_t = \int_0^t W_u\,du.$$

Formula (7.2.3) yields a relationship between I *and the integrated Brownian motion*

$$I_T = \int_0^T t\,dW_t = TW_T - \int_0^T W_t\,dt = TW_T - Z_T,$$

and hence $I_T + Z_T = TW_T$*. This relation can be used to compute the covariance between* I_T *and* Z_T*.*

$$\begin{aligned}
Cov(I_T + Z_T, I_T + Z_T) &= Var[TW_T] \Longleftrightarrow \\
Var[I_T] + Var[Z_T] + 2Cov(I_T, Z_T) &= T^2 Var[W_T] \Longleftrightarrow \\
T^3/3 + T^3/3 + 2Cov(I_T, Z_T) &= T^3 \Longleftrightarrow \\
Cov(I_T, Z_T) &= T^3/6,
\end{aligned}$$

where we used that $Var[Z_T] = T^3/3$*. The processes* I_t *and* Z_t *are not independent. Their correlation coefficient is 0.5 as the following calculation shows*

$$\begin{aligned}
Corr(I_T, Z_T) &= \frac{Cov(I_T, Z_T)}{\Big(Var[I_T]Var[Z_T]\Big)^{1/2}} = \frac{T^3/6}{T^3/3} \\
&= 1/2.
\end{aligned}$$

Application 7.2.2 *If we let* $g(x) = \frac{x^2}{2}$ *in formula (7.2.4), we get*

$$\boxed{\int_a^b W_t\,dW_t = \frac{W_b^2 - W_a^2}{2} - \frac{1}{2}(b-a).}$$

It is worth noting that letting $a = 0$ *and* $b = T$*, we retrieve a formula that was proved by direct methods in chapter 3*

$$\boxed{\int_0^T W_t\,dW_t = \frac{W_T^2}{2} - \frac{T}{2}.}$$

Similarly, if we let $g(x) = \frac{x^3}{3}$ *in (7.2.4) yields*

$$\boxed{\int_a^b W_t^2\, dW_t = \frac{W_t^3}{3}\Big|_a^b - \int_a^b W_t\, dt.}$$

Application 7.2.3 *Choosing* $f(t) = e^{\alpha t}$ *and* $g(x) = \sin x$*, we shall compute the stochastic integral* $\int_0^T e^{\alpha t} \cos W_t\, dW_t$ *using the formula of integration by parts*

$$\begin{aligned}
\int_0^T e^{\alpha t} \cos W_t\, dW_t &= \int_0^T e^{\alpha t} (\sin W_t)'\, dW_t \\
&= e^{\alpha t} \sin W_t \Big|_0^T - \int_0^T (e^{\alpha t})' \sin W_t\, dt - \frac{1}{2} \int_0^T e^{\alpha t} (\sin W_t)''\, dt \\
&= e^{\alpha T} \sin W_T - \alpha \int_0^T e^{\alpha t} \sin W_t\, dt + \frac{1}{2} \int_0^T e^{\alpha t} \sin W_t\, dt \\
&= e^{\alpha T} \sin W_T - \Big(\alpha - \frac{1}{2}\Big) \int_0^T e^{\alpha t} \sin W_t\, dt.
\end{aligned}$$

The particular case $\alpha = \frac{1}{2}$ *leads to the following exact formula of a stochastic integral*

$$\boxed{\int_0^T e^{\frac{t}{2}} \cos W_t\, dW_t = e^{\frac{T}{2}} \sin W_T.} \tag{7.2.5}$$

In a similar way, we can obtain an exact formula for the stochastic integral $\int_0^T e^{\beta t} \sin W_t\, dW_t$ *as follows*

$$\begin{aligned}
\int_0^T e^{\beta t} \sin W_t\, dW_t &= -\int_0^T e^{\beta t} (\cos W_t)'\, dW_t \\
&= -e^{\beta t} \cos W_t \Big|_0^T + \beta \int_0^T e^{\beta t} \cos W_t\, dt - \frac{1}{2} \int_0^T e^{\beta t} \cos W_t\, dt.
\end{aligned}$$

Taking $\beta = \frac{1}{2}$ *yields the closed form formula*

$$\boxed{\int_0^T e^{\frac{t}{2}} \sin W_t\, dW_t = 1 - e^{\frac{T}{2}} \cos W_T.} \tag{7.2.6}$$

A consequence of the last two formulas and of Euler's formula

$$e^{iW_t} = \cos W_t + i \sin W_t,$$

is

$$\boxed{\int_0^T e^{\frac{t}{2} + iW_t}\, dW_t = i(1 - e^{\frac{T}{2} + iW_T}).}$$

The proof details are left to the reader.

A general form of the integration by parts formula In general, if X_t and Y_t are two Ito diffusions, from the product formula

$$d(X_tY_t) = X_t dY_t + Y_t dX_t + dX_t\, dY_t.$$

Integrating between the limits a and b

$$\int_a^b d(X_tY_t) = \int_a^b X_t dY_t + \int_a^b Y_t dX_t + \int_a^b dX_t\, dY_t.$$

From the Fundamental Theorem

$$\int_a^b d(X_tY_t) = X_bY_b - X_aY_a,$$

so the previous formula takes the following form of integration by parts

$$\boxed{\int_a^b X_t dY_t = X_bY_b - X_aY_a - \int_a^b Y_t dX_t - \int_a^b dX_t\, dY_t.}$$

This formula is of theoretical value. In practice, the term $dX_t\, dY_t$ needs to be computed using the rules $dW_t^2 = dt$, and $dt\, dW_t = 0$.

Exercise 7.2.4 *(a) Use integration by parts to get*

$$\int_0^T \frac{1}{1+W_t^2} dW_t = \tan^{-1}(W_T) + \int_0^T \frac{W_t}{(1+W_t^2)^2}\, dt, \quad T > 0.$$

(b) Show that

$$\mathbb{E}[\tan^{-1}(W_T)] = -\int_0^T E\Big[\frac{W_t}{(1+W_t^2)^2}\Big]\, dt.$$

(c) Prove the double inequality

$$-\frac{3\sqrt{3}}{16} \le \frac{x}{(1+x^2)^2} \le \frac{3\sqrt{3}}{16}, \qquad \forall x \in \mathbb{R}.$$

(d) Use part (c) to obtain

$$-\frac{3\sqrt{3}}{16} T \le \int_0^T \frac{W_t}{(1+W_t^2)^2}\, dt \le \frac{3\sqrt{3}}{16} T.$$

(e) Use part (d) to get

$$-\frac{3\sqrt{3}}{16} T \le \mathbb{E}[\tan^{-1}(W_T)] \le \frac{3\sqrt{3}}{16} T.$$

(f) Does part (e) contradict the inequality

$$-\frac{\pi}{2} < \tan^{-1}(W_T) < \frac{\pi}{2}?$$

Exercise 7.2.5 *(a) Show the relation*

$$\int_0^T e^{W_t}\,dW_t = e^{W_T} - 1 - \frac{1}{2}\int_0^T e^{W_t}\,dt.$$

(b) Use part (a) to find $\mathbb{E}[e^{W_t}]$.

Exercise 7.2.6 *(a) Use integration by parts to show*

$$\int_0^T W_t e^{W_t}\,dW_t = 1 + W_T e^{W_T} - e^{W_T} - \frac{1}{2}\int_0^T e^{W_t}(1+W_t)\,dt;$$

(b) Use part (a) to find $\mathbb{E}[W_t e^{W_t}]$;
(c) Show that $Cov(W_t, e^{W_t}) = te^{t/2}$;

(d) Prove that $Corr(W_t, e^{W_t}) = \sqrt{\dfrac{t}{e^t - 1}}$, *and compute the limits as* $t \to 0$ *and* $t \to \infty$.

Exercise 7.2.7 *(a) Let* $T > 0$. *Show the following relation using integration by parts*

$$\int_0^T \frac{2W_t}{1+W_t^2}\,dW_t = \ln(1+W_T^2) - \int_0^T \frac{1-W_t^2}{(1+W_t^2)^2}\,dt.$$

(b) Show that for any real number x *the following double inequality holds*

$$-\frac{1}{8} \le \frac{1-x^2}{(1+x^2)^2} \le 1.$$

(c) Use part (b) to show that

$$-\frac{1}{8}T \le \int_0^T \frac{1-W_t^2}{(1+W_t^2)^2}\,dt \le T.$$

(d) Use parts (a) and (c) to get

$$-\frac{T}{8} \le \mathbb{E}[\ln(1+W_T^2)] \le T.$$

(e) Use Jensen's inequality to get

$$\mathbb{E}[\ln(1+W_T^2)] \le \ln(1+T).$$

Does this contradict the upper bound provided in (d)?

Exercise 7.2.8 *Use integration by parts to show*

$$\int_0^t \arctan W_s\,dW_s = W_t \arctan W_t - \frac{1}{2}\ln(1+W_t^2) - \frac{1}{2}\int_0^t \frac{1}{1+W_s^2}\,ds.$$

Exercise 7.2.9 *(a) Using integration by parts prove the identity*

$$\int_0^t W_s e^{W_s}\, dW_s = 1 + e^{W_t}(W_t - 1) - \frac{1}{2}\int_0^t (1 + W_s)e^{W_s}\, ds;$$

(b) Use part (a) to compute $\mathbb{E}[W_t e^{W_t}]$.

Exercise 7.2.10 *Check the following formulas using integration by parts*

(a) $\displaystyle\int_0^t W_s \sin W_s\, dW_s = \sin W_t - W_t \cos W_t - \frac{1}{2}\int_0^t (\sin W_s + W_s \cos W_s)\, ds;$

(b) $\displaystyle\int_0^t \frac{W_s}{\sqrt{1 + W_s^2}}\, dW_s = \sqrt{1 + W_t^2} - 1 - \frac{1}{2}\int_0^t (1 + W_s^2)^{-3/2}\, ds.$

7.3 The Heat Equation Method

In elementary Calculus, integration by substitution is the inverse application of the chain rule. In the stochastic environment, this will be the inverse application of Ito's formula. This is difficult to apply in general, but there is a particular case of great importance.

Let $\varphi(t,x)$ be a solution of the equation

$$\partial_t \varphi + \frac{1}{2}\partial_x^2 \varphi = 0. \tag{7.3.7}$$

This is called the *heat equation without sources.* The non-homogeneous equation

$$\partial_t \varphi + \frac{1}{2}\partial_x^2 \varphi = G(t,x) \tag{7.3.8}$$

is called the *heat equation with sources.* The function $G(t,x)$ represents the density of heat sources, while the function $\varphi(t,x)$ is the temperature at the point x at time t in a one-dimensional wire. If the heat source is time independent, then $G = G(x)$, i.e. G is a function of x only.

Example 7.3.1 *Find all solutions of the equation (7.3.7) of type* $\varphi(t,x) = a(t) + b(x)$.

Substituting into equation (7.3.7) yields

$$\frac{1}{2}b''(x) = -a'(t).$$

Since the left side is a function of x only, while the right side is a function of variable t, the only case where the previous equation is satisfied is when both sides are equal to the same constant C. This is called a separation constant. Therefore $a(t)$ and $b(x)$ satisfy the equations

$$a'(t) = -C, \qquad \frac{1}{2}b''(x) = C.$$

Integrating yields $a(t) = -Ct + C_0$ and $b(x) = Cx^2 + C_1x + C_2$. It follows that

$$\varphi(t,x) = C(x^2 - t) + C_1x + C_3,$$

with C_0, C_1, C_2, C_3 arbitrary constants.

Example 7.3.2 *Find all solutions of the equation (7.3.7) of the type* $\varphi(t,x) = a(t)b(x)$.

Substituting into the equation and dividing by $a(t)b(x)$ yields

$$\frac{a'(t)}{a(t)} + \frac{1}{2}\frac{b''(x)}{b(x)} = 0.$$

There is a separation constant C such that $\dfrac{a'(t)}{a(t)} = -C$ and $\dfrac{b''(x)}{b(x)} = 2C$. There are three distinct cases to discuss:

1. $C = 0$. In this case $a(t) = a_0$ and $b(x) = b_1x + b_0$, with a_0, a_1, b_0, b_1 real constants. Then

$$\varphi(t,x) = a(t)b(x) = c_1x + c_0, \qquad c_0, c_1 \in \mathbb{R}$$

is just a linear function in x.

2. $C > 0$. Let $\lambda > 0$ such that $2C = \lambda^2$. Then $a'(t) = -\frac{\lambda^2}{2}a(t)$ and $b''(x) = \lambda^2 b(x)$, with solutions

$$\begin{aligned} a(t) &= a_0e^{-\lambda^2t/2} \\ b(x) &= c_1e^{\lambda x} + c_2e^{-\lambda x}. \end{aligned}$$

The general solution of (7.3.7) is

$$\varphi(t,x) = e^{-\lambda^2t/2}(c_1e^{\lambda x} + c_2e^{-\lambda x}), \quad c_1, c_2 \in \mathbb{R}.$$

3. $C < 0$. Let $\lambda > 0$ such that $2C = -\lambda^2$. Then $a'(t) = \frac{\lambda^2}{2}a(t)$ and $b''(x) = -\lambda^2 b(x)$. Solving yields

$$\begin{aligned} a(t) &= a_0e^{\lambda^2t/2} \\ b(x) &= c_1\sin(\lambda x) + c_2\cos(\lambda x). \end{aligned}$$

The general solution of (7.3.7) in this case is

$$\varphi(t,x) = e^{\lambda^2 t/2}\big(c_1 \sin(\lambda x) + c_2 \cos(\lambda x)\big), \quad c_1, c_2 \in \mathbb{R}.$$

In particular, the functions x, $x^2 - t$, $e^{x-t/2}$, $e^{-x-t/2}$, $e^{t/2}\sin x$ and $e^{t/2}\cos x$, or any linear combination of them, are solutions of the heat equation (7.3.7). However, there are other solutions which are not of the previous type.

Exercise 7.3.3 *Prove that $\varphi(t,x) = \frac{1}{3}x^3 - tx$ is a solution of the heat equation (7.3.7).*

Exercise 7.3.4 *Show that $\varphi(t,x) = t^{-1/2}e^{-x^2/(2t)}$ is a solution of the heat equation (7.3.7) for $t > 0$.*

Exercise 7.3.5 *Let $\varphi = u(\lambda)$, with $\lambda = \dfrac{x}{2\sqrt{t}}$, $t > 0$. Show that φ satisfies the heat equation (7.3.7) if and only if $u'' + 2\lambda u' = 0$.*

Exercise 7.3.6 *Let $erfc(x) = \dfrac{2}{\sqrt{\pi}}\displaystyle\int_x^\infty e^{-r^2}\,dr$. Show that $\varphi = erfc(x/(2\sqrt{t}))$ is a solution of the equation (7.3.7).*

Exercise 7.3.7 (the fundamental solution) *Show that $\varphi(t,x) = \frac{1}{\sqrt{4\pi t}}e^{-\frac{x^2}{4t}}$, $t > 0$ satisfies the equation (7.3.7).*

Sometimes it is useful to generate new solutions for the heat equation from other solutions. Below we present a few ways to accomplish this:

(i) by linear combination: if φ_1 and φ_2 are solutions, then $a_1\varphi_1 + a_1\varphi_2$ is a solution, where a_1, a_2 are constants.

(ii) by translation: if $\varphi(t,x)$ is a solution, then $\varphi(t-\tau, x-\xi)$ is a solution, where (τ, ξ) is a translation vector.

(iii) by affine transforms: if $\varphi(t,x)$ is a solution, then $\varphi(\lambda t, \lambda^2 x)$ is a solution, for any constant λ.

(iv) by differentiation: if $\varphi(t,x)$ is a solution, then $\dfrac{\partial^{n+m}}{\partial^n x \partial^m t}\varphi(t,x)$ is a solution.

(v) by convolution: if $\varphi(t,x)$ is a solution, then so are

$$\int_a^b \varphi(t, x-\xi) f(\xi)\,d\xi$$

$$\int_a^b \varphi(t-\tau, x) g(t)\,dt.$$

For more detail on the subject the reader can consult Widder [46] and Cannon [11].

Theorem 7.3.8 *Let $\varphi(t,x)$ be a solution of the heat equation (7.3.7) and denote $f(t,x) = \partial_x \varphi(t,x)$. Then*

$$\boxed{\int_a^b f(t,W_t)\,dW_t = \varphi(b,W_b) - \varphi(a,W_a).}$$

Proof: Let $F_t = \varphi(t,W_t)$. Applying Ito's formula we get

$$dF_t = \partial_x \varphi(t,W_t)\,dW_t + \Big(\partial_t \varphi + \frac{1}{2}\partial_x^2 \varphi\Big)dt.$$

Since $\partial_t \varphi + \frac{1}{2}\partial_x^2 \varphi = 0$ and $\partial_x \varphi(t,W_t) = f(t,W_t)$, we have

$$dF_t = f(t,W_t)\,dW_t.$$

Writing in the integral form, yields

$$\int_a^b f(t,W_t)\,dW_t = \int_a^b dF_t = F_b - F_a = \varphi(b,W_b) - \varphi(a,W_a).$$

■

Application 7.3.9 *Show that*

$$\int_0^T W_t\,dW_t = \frac{1}{2}W_T^2 - \frac{1}{2}T.$$

Choose the solution of the heat equation (7.3.7) given by $\varphi(t,x) = x^2 - t$. Then $f(t,x) = \partial_x \varphi(t,x) = 2x$. Theorem 7.3.8 yields

$$\int_0^T 2W_t\,dW_t = \int_0^T f(t,W_t)\,dW_t = \varphi(t,x)\Big|_0^T = W_T^2 - T.$$

Dividing by 2 leads to the desired result.

Application 7.3.10 *Show that*

$$\int_0^T (W_t^2 - t)\,dW_t = \frac{1}{3}W_T^3 - TW_T.$$

Consider the function $\varphi(t,x) = \frac{1}{3}x^3 - tx$, which is a solution of the heat equation (7.3.7), see Exercise 7.3.3. Then $f(t,x) = \partial_x \varphi(t,x) = x^2 - t$. Applying Theorem 7.3.8 yields

$$\int_0^T (W_t^2 - t)\,dW_t = \int_0^T f(t,W_t)\,dW_t = \varphi(t,W_t)\Big|_0^T = \frac{1}{3}W_T^3 - TW_T.$$

Application 7.3.11 *Let $\lambda > 0$. Prove the identities*

$$\boxed{\int_0^T e^{-\frac{\lambda^2 t}{2} \pm \lambda W_t}\, dW_t = \frac{1}{\pm\lambda}\Big(e^{-\frac{\lambda^2 T}{2} \pm \lambda W_T} - 1\Big).}$$

Consider the function $\varphi(t,x) = e^{-\frac{\lambda^2 t}{2} \pm \lambda x}$, which is a solution of the homogeneous heat equation (7.3.7), see Example 7.3.2. Then $f(t,x) = \partial_x \varphi(t,x) = \pm\lambda e^{-\frac{\lambda^2 t}{2} \pm \lambda x}$. Apply Theorem 7.3.8 to get

$$\int_0^T \pm\lambda e^{-\frac{\lambda^2 t}{2} \pm \lambda x}\, dW_t = \int_0^T f(t, W_t)\, dW_t = \varphi(t, W_t)\Big|_0^T = e^{-\frac{\lambda^2 T}{2} \pm \lambda W_T} - 1.$$

Dividing by the constant $\pm\lambda$ ends the proof.

In particular, for $\lambda = 1$ the aforementioned formula becomes

$$\boxed{\int_0^T e^{-\frac{t}{2} + W_t}\, dW_t = e^{-\frac{T}{2} + W_T} - 1.} \tag{7.3.9}$$

Application 7.3.12 *Let $\lambda > 0$. Prove the identity*

$$\boxed{\int_0^T e^{\frac{\lambda^2 t}{2}} \cos(\lambda W_t)\, dW_t = \frac{1}{\lambda} e^{\frac{\lambda^2 T}{2}} \sin(\lambda W_T).}$$

From the Example 7.3.2 we know that $\varphi(t,x) = e^{\frac{\lambda^2 t}{2}} \sin(\lambda x)$ is a solution of the heat equation. Applying Theorem 7.3.8 to the function $f(t,x) = \partial_x \varphi(t,x) = \lambda e^{\frac{\lambda^2 t}{2}} \cos(\lambda x)$, yields

$$\begin{aligned}
\int_0^T \lambda e^{\frac{\lambda^2 t}{2}} \cos(\lambda W_t)\, dW_t &= \int_0^T f(t, W_t)\, dW_t = \varphi(t, W_t)\Big|_0^T \\
&= e^{\frac{\lambda^2 t}{2}} \sin(\lambda W_t)\Big|_0^T = e^{\frac{\lambda^2 T}{2}} \sin(\lambda W_T).
\end{aligned}$$

Divide by λ to end the proof.

If we choose $\lambda = 1$ we recover a result already familiar to the reader from section 7.2

$$\boxed{\int_0^T e^{\frac{t}{2}} \cos(W_t)\, dW_t = e^{\frac{T}{2}} \sin W_T.} \tag{7.3.10}$$

Application 7.3.13 *Let $\lambda > 0$. Show that*

$$\boxed{\int_0^T e^{\frac{\lambda^2 t}{2}} \sin(\lambda W_t)\, dW_t = \frac{1}{\lambda}\Big(1 - e^{\frac{\lambda^2 T}{2}} \cos(\lambda W_T)\Big).}$$

Choose $\varphi(t,x) = e^{\frac{\lambda^2 t}{2}}\cos(\lambda x)$ to be a solution of the heat equation. Apply Theorem 7.3.8 for the function $f(t,x) = \partial_x \varphi(t,x) = -\lambda e^{\frac{\lambda^2 t}{2}}\sin(\lambda x)$ to get

$$\begin{aligned}\int_0^T (-\lambda) e^{\frac{\lambda^2 t}{2}} \sin(\lambda W_t)\,dW_t &= \varphi(t,W_t)\Big|_0^T \\ &= e^{\frac{\lambda^2 T}{2}}\cos(\lambda W_t)\Big|_0^T = e^{\frac{\lambda^2 T}{2}}\cos(\lambda W_T) - 1,\end{aligned}$$

and then divide by $-\lambda$.

Application 7.3.14 *Let $0 < a < b$. Show that*

$$\boxed{\int_a^b t^{-\frac{3}{2}} W_t e^{-\frac{W_t^2}{2t}}\,dW_t = a^{-\frac{1}{2}} e^{-\frac{W_a^2}{2a}} - b^{-\frac{1}{2}} e^{-\frac{W_b^2}{2b}}.} \tag{7.3.11}$$

From Exercise 7.3.4 we have that $\varphi(t,x) = t^{-1/2}e^{-x^2/(2t)}$ is a solution of the homogeneous heat equation. Since $f(t,x) = \partial_x\varphi(t,x) = -t^{-3/2}xe^{-x^2/(2t)}$, applying Theorem 7.3.8 yields the desired result. The reader can easily fill in the details.

Integration techniques will be used when solving stochastic differential equations in the next chapter.

Exercise 7.3.15 *Find the value of the following stochastic integrals*

(a) $\displaystyle\int_0^1 e^t \cos(\sqrt{2}W_t)\,dW_t$

(b) $\displaystyle\int_0^3 e^{2t}\cos(2W_t)\,dW_t$

(c) $\displaystyle\int_0^4 e^{-t+\sqrt{2}W_t}\,dW_t.$

Exercise 7.3.16 *Let $\varphi(t,x)$ be a solution of the following non-homogeneous heat equation with time-dependent and uniform heat source $G(t)$*

$$\partial_t\varphi + \frac{1}{2}\partial_x^2\varphi = G(t).$$

Denote $f(t,x) = \partial_x\varphi(t,x)$. Show that

$$\boxed{\int_a^b f(t,W_t)\,dW_t = \varphi(b,W_b) - \varphi(a,W_a) - \int_a^b G(t)\,dt.}$$

How does the formula change if the heat source G is constant?

7.4 Table of Usual Stochastic Integrals

Now we present a user-friendly table, which enlists integral identities developed in this chapter. This table is far too complicated to be memorized in full. However, the first couple of identities in this table are the most memorable, and should be remembered.

Let $a < b$ and $0 < T$. Then we have:

1. $\displaystyle\int_a^b dW_t = W_b - W_a;$

2. $\displaystyle\int_0^T W_t\,dW_t = \frac{W_T^2}{2} - \frac{T}{2};$

3. $\displaystyle\int_0^T (W_t^2 - t)\,dW_t = \frac{W_T^3}{3} - TW_T;$

4. $\displaystyle\int_0^T t\,dW_t = TW_T - \int_0^T W_t\,dt, \quad 0 < T;$

5. $\displaystyle\int_0^T W_t^2\,dW_t = \frac{W_T^3}{3} - \int_0^T W_t\,dt;$

6. $\displaystyle\int_0^T e^{\frac{t}{2}} \cos W_t\,dW_t = e^{\frac{T}{2}} \sin W_T;$

7. $\displaystyle\int_0^T e^{\frac{t}{2}} \sin W_t\,dW_t = 1 - e^{\frac{T}{2}} \cos W_T;$

8. $\displaystyle\int_0^T e^{-\frac{t}{2}+W_t}\,dW_t = e^{-\frac{T}{2}+W_T} - 1;$

9. $\displaystyle\int_0^T e^{\frac{\lambda^2 t}{2}} \cos(\lambda W_t)\,dW_t = \frac{1}{\lambda} e^{\frac{\lambda^2 T}{2}} \sin(\lambda W_T);$

10. $\displaystyle\int_0^T e^{\frac{\lambda^2 t}{2}} \sin(\lambda W_t)\,dW_t = \frac{1}{\lambda}\Big(1 - e^{\frac{\lambda^2 T}{2}} \cos(\lambda W_T)\Big);$

11. $\displaystyle\int_0^T e^{-\frac{\lambda^2 t}{2} \pm \lambda W_t}\,dW_t = \frac{1}{\pm\lambda}\Big(e^{-\frac{\lambda^2 T}{2} \pm \lambda W_T} - 1\Big);$

12. $\displaystyle\int_a^b t^{-\frac{3}{2}} W_t e^{-\frac{W_t^2}{2t}}\,dW_t = a^{-\frac{1}{2}} e^{-\frac{W_a^2}{2a}} - b^{-\frac{1}{2}} e^{-\frac{W_b^2}{2b}};$

13. $\displaystyle\int_0^T \cos W_t\,dW_t = \sin W_T + \frac{1}{2}\int_0^T \sin W_t\,dt;$

14. $\displaystyle\int_0^T \sin W_t\, dW_t = 1 - \cos W_T - \frac{1}{2}\int_0^T \cos W_t\, dt;$

15. $\displaystyle d\Big(\int_a^t f(s, W_s)\, dW_s\Big) = f(t, W_t)\, dW_t;$

16. $\displaystyle\int_a^b Y_t\, dW_t = F_b - F_a$, if $Y_t dW_t = dF_t$;

17. $\displaystyle\int_a^b f(t)\, dW_t = f(t)W_t|_a^b - \int_a^b f'(t)W_t\, dt;$

18. $\displaystyle\int_a^b g'(W_t)\, dW_t = g(W_t)\Big|_a^b - \frac{1}{2}\int_a^b g''(W_t)\, dt.$

Chapter 8

Stochastic Differential Equations

If deterministic Calculus was developed mainly to put into differential equations form the fundamental principles which govern all evolution phenomena, then Stochastic Calculus plays a similar role for the case of noisy evolution systems, which provide a more realistic description of the real world.

This chapter deals with several analytic techniques of solving stochastic differential equations. The number of these techniques is limited and follows quite closely the methods used in the ordinary differential equations treated, for instance, in classical books of Arnold [3] or Boyce and DiPrima [7].

8.1 Definitions and Examples

Let X_t be a continuous stochastic process. If small changes in the process X_t can be written as a linear combination of small changes in t and small increments of the Brownian motion W_t, we may write

$$\boxed{dX_t = a(t, W_t, X_t)dt + b(t, W_t, X_t)\, dW_t} \tag{8.1.1}$$

and call it a *stochastic differential equation*. In fact, this differential relation has the following integral meaning:

$$\boxed{X_t = X_0 + \int_0^t a(s, W_s, X_s)\, ds + \int_0^t b(s, W_s, X_s)\, dW_s,} \tag{8.1.2}$$

where the last integral is taken in the Ito sense. Relation (8.1.2) is taken as the definition for the stochastic differential equation (8.1.1). However, since it is convenient to use stochastic differentials informally, we shall approach stochastic differential equations by analogy with the ordinary differential equations, and try to present the same methods of solving equations in the new stochastic environment.

Most of the stochastic differential equations considered here describe diffusions, and are of the type

$$dX_t = a(t, X_t)dt + b(t, X_t)dW_t, \qquad X_0 = \zeta, \tag{8.1.3}$$

with $a(t, x)$ and $b(t, x)$ measurable functions. The functions $a(t, x)$ and $b(t, x)$ are called the *drift rate* and the *volatility* of the process X_t, respectively. Given these two functions as input, one may seek for the solution X_t of the stochastic differential equation as an output. The desired outputs X_t are the so-called *strong solutions*. The precise definition of this concept is given in the following. The beginner can skip this definition; all solutions in this book will be solutions in the strong sense anyway.

Definition 8.1.1 *A process X_t is a strong solution for the stochastic equation (8.1.3) on the probability space $(\Omega, \mathcal{F}, P)$ if it satisfies the following properties:*

(i) X_t is adapted to the augmented filtration $\mathcal{F}_t$ generated by the Brownian motion W_t and the initial condition ζ;

(ii) $P(X_0 = \zeta) = 1$;

(iii) For any $0 \le t < \infty$ we have

$$\int_0^t \Big(|a(s, X_s)| + b^2(s, X_s) \Big)\, ds < \infty;$$

(iv) The formula

$$X_t = X_0 + \int_0^t a(s, X_s)\, ds + \int_0^t b(s, X_s)\, dW_s$$

holds almost surely.

A few comments regarding the previous definition. Part (*i*) states that given the information induced by ζ and the history of the Brownian motion until time t, one can determine the value X_t. Part (*ii*) states that X_0 takes the value ζ with probability 1. Part (*iii*) deals with a non-explosive condition for the coefficients. Part (*iv*) states that X_t verifies the associated integral equation.

We shall start with an example.

Example 8.1.2 (The Brownian bridge) *Let $a, b \in \mathbb{R}$. Show that the process*

$$X_t = a(1-t) + bt + (1-t)\int_0^t \frac{1}{1-s}\, dW_s, \quad 0 \le t < 1$$

is a solution of the stochastic differential equation

$$dX_t = \frac{b - X_t}{1-t}dt + dW_t, \qquad 0 \le t < 1, X_0 = a.$$

We shall perform a routine verification to show that X_t is a solution. First we compute the quotient $\dfrac{b - X_t}{1-t}$:

$$\begin{aligned} b - X_t &= b - a(1-t) - bt - (1-t)\int_0^t \frac{1}{1-s}\,dW_s \\ &= (b-a)(1-t) - (1-t)\int_0^t \frac{1}{1-s}\,dW_s, \end{aligned}$$

and dividing by $1 - t$ yields

$$\frac{b - X_t}{1-t} = b - a - \int_0^t \frac{1}{1-s}\,dW_s. \tag{8.1.4}$$

Using

$$d\Big(\int_0^t \frac{1}{1-s}\,dW_s\Big) = \frac{1}{1-t}\,dW_t,$$

the product rule provides

$$\begin{aligned} dX_t &= a\,d(1-t) + bdt + d(1-t)\int_0^t \frac{1}{1-s}\,dW_s + (1-t)d\Big(\int_0^t \frac{1}{1-s}\,dW_s\Big) \\ &= \Big(b - a - \int_0^t \frac{1}{1-s}\,dW_s\Big)dt + dW_t \\ &= \frac{b - X_t}{1-t}dt + dW_t, \end{aligned}$$

where the last identity comes from (8.1.4). We just verified that the process X_t is a solution of the given stochastic equation. The question of *how this solution was obtained in the first place*, is the subject of study for the next few sections.

8.2 The Integration Technique

We shall start with the simple case when both the drift and the volatility are just functions of time t.

Proposition 8.2.1 *The solution X_t of the stochastic differential equation*

$$dX_t = a(t)dt + b(t)dW_t$$

is Gaussian distributed with the mean $X_0 + \int_0^t a(s)\,ds$ and the variance $\int_0^t b^2(s)\,ds$.

Proof: Integrating in the equation yields

$$X_t - X_0 = \int_0^t dX_s = \int_0^t a(s)\, ds + \int_0^t b(s)\, dW_s.$$

Using the property of Wiener integrals, $\int_0^t b(s)\, dW_s$ is Gaussian distributed with mean 0 and variance $\int_0^t b^2(s)\, ds$. Then X_t is Gaussian (as a sum between a deterministic function and a Gaussian), with

$$\begin{aligned}
\mathbb{E}[X_t] &= \mathbb{E}[X_0 + \int_0^t a(s)\, ds + \int_0^t b(s)\, dW_s] \\
&= X_0 + \int_0^t a(s)\, ds + \mathbb{E}\Big[\int_0^t b(s)\, dW_s\Big] \\
&= X_0 + \int_0^t a(s)\, ds,
\end{aligned}$$

$$\begin{aligned}
Var[X_t] &= Var[X_0 + \int_0^t a(s)\, ds + \int_0^t b(s)\, dW_s] \\
&= Var\Big[\int_0^t b(s)\, dW_s\Big] \\
&= \int_0^t b^2(s)\, ds,
\end{aligned}$$

which ends the proof. ■

Exercise 8.2.2 *Solve the following stochastic differential equations for* $t \geq 0$ *and determine the mean and the variance of the solution:*

(a) $dX_t = \cos t\, dt - \sin t\, dW_t$, $X_0 = 1$.

(b) $dX_t = e^t\, dt + \sqrt{t}\, dW_t$, $X_0 = 0$.

(c) $dX_t = \frac{t}{1+t^2}\, dt + t^{3/2}\, dW_t$, $X_0 = 1$.

If the drift and the volatility depend only on variables t and W_t, the stochastic differential equation

$$dX_t = a(t, W_t)dt + b(t, W_t)dW_t, \qquad t \geq 0$$

defines a stochastic process that can be expressed in terms of Ito integrals

$$X_t = X_0 + \int_0^t a(s, W_s)\, ds + \int_0^t b(s, W_s)\, dW_s.$$

There are several cases when both integrals can be computed explicitly. In order to compute the second integral we shall often use the table of usual stochastic integrals provided in section 7.4.

Example 8.2.3 *Find the solution of the stochastic differential equation*

$$dX_t = dt + W_t\,dW_t, \qquad X_0 = 1.$$

Integrate between 0 and t and get

$$\begin{aligned} X_t &= 1 + \int_0^t ds + \int_0^t W_s\,dW_s = 1 + t + \frac{W_t^2}{2} - \frac{t}{2} \\ &= \frac{1}{2}(W_t^2 + t) + 1. \end{aligned}$$

Example 8.2.4 *Solve the stochastic differential equation*

$$dX_t = (W_t - 1)dt + W_t^2\,dW_t, \qquad X_0 = 0.$$

Let $Z_t = \int_0^t W_s\,ds$ denote the integrated Brownian motion process. Integrating the equation between 0 and t yields

$$\begin{aligned} X_t &= \int_0^t dX_s = \int_0^t (W_s - 1)ds + \int_0^t W_s^2\,dW_s \\ &= Z_t - t + \frac{1}{3}W_t^3 - Z_t \\ &= \frac{1}{3}W_t^3 - t, \end{aligned}$$

where we used that $\int_0^t W_s^2\,dW_s = \frac{1}{3}W_t^3 - Z_t$.

Example 8.2.5 *Solve the stochastic differential equation*

$$dX_t = t^2 dt + e^{t/2}\cos W_t\,dW_t, \qquad X_0 = 0,$$

and find $\mathbb{E}[X_t]$ *and* $Var(X_t)$.

Integrating yields

$$\begin{aligned} X_t &= \int_0^t s^2\,ds + \int_0^t e^{s/2}\cos W_s\,dW_s \\ &= \frac{t^3}{3} + e^{t/2}\sin W_t, \end{aligned} \tag{8.2.5}$$

where we used (7.3.10). Even if the process X_t is not Gaussian, we can still compute its mean and variance. Since Ito integrals have zero expectation,

$$\mathbb{E}[X_t] = \mathbb{E}\Big[\int_0^t s^2\,ds + \int_0^t e^{s/2}\cos W_s\,dW_s\Big] = \int_0^t s^2\,ds = \frac{t^3}{3}.$$

Another variant of computation is using Ito's formula

$$d(\sin W_t) = \cos W_t\, dW_t - \frac{1}{2}\sin W_t\, dt$$

Integrating between 0 and t yields

$$\sin W_t = \int_0^t \cos W_s\, dW_s - \frac{1}{2}\int_0^t \sin W_s\, ds,$$

where we used that $\sin W_0 = \sin 0 = 0$. Taking the expectation in the previous relation yields

$$\mathbb{E}[\sin W_t] = E\Big[\int_0^t \cos W_s\, dW_s\Big] - \frac{1}{2}\int_0^t \mathbb{E}[\sin W_s]\, ds.$$

From the properties of the Ito integral, the first expectation on the right side is zero. Denoting $\mu(t) = \mathbb{E}[\sin W_t]$, we obtain the integral equation

$$\mu(t) = -\frac{1}{2}\int_0^t \mu(s)\, ds.$$

Differentiating yields the differential equation

$$\mu'(t) = -\frac{1}{2}\mu(t)$$

with the solution $\mu(t) = ke^{-t/2}$. Since $k = \mu(0) = \mathbb{E}[\sin W_0] = 0$, it follows that $\mu(t) = 0$. Hence

$$\mathbb{E}[\sin W_t] = 0.$$

Taking expectation in (8.2.5) leads to

$$\mathbb{E}[X_t] = E\Big[\frac{t^3}{3}\Big] + e^{t/2}\mathbb{E}[\sin W_t] = \frac{t^3}{3}.$$

Since the variance of deterministic functions is zero,

$$\begin{aligned} Var[X_t] &= Var\Big[\frac{t^3}{3} + e^{t/2}\sin W_t\Big] = (e^{t/2})^2 Var[\sin W_t] \\ &= e^t\mathbb{E}[\sin^2 W_t] = \frac{e^t}{2}(1 - \mathbb{E}[\cos 2W_t]). \end{aligned} \tag{8.2.6}$$

In order to compute the last expectation we use Ito's formula

$$d(\cos 2W_t) = -2\sin 2W_t\, dW_t - 2\cos 2W_t\, dt$$

and integrate to get

$$\cos 2W_t = \cos 2W_0 - 2\int_0^t \sin 2W_s \, dW_s - 2\int_0^t \cos 2W_s \, ds.$$

Taking the expectation and using that Ito integrals have zero expectation, yields

$$\mathbb{E}[\cos 2W_t] = 1 - 2\int_0^t \mathbb{E}[\cos 2W_s] \, ds.$$

If we denote $m(t) = \mathbb{E}[\cos 2W_t]$, the previous relation becomes an integral equation

$$m(t) = 1 - 2\int_0^t m(s) \, ds.$$

Differentiate and get

$$m'(t) = -2m(t),$$

with the solution $m(t) = ke^{-2t}$. Since $k = m(0) = \mathbb{E}[\cos 2W_0] = 1$, we have $m(t) = e^{-2t}$. Substituting into (8.2.6) yields

$$Var[X_t] = \frac{e^t}{2}(1 - e^{-2t}) = \frac{e^t - e^{-t}}{2} = \sinh t.$$

In conclusion, the solution X_t has the mean and the variance given by

$$\mathbb{E}[X_t] = \frac{t^3}{3}, \qquad Var[X_t] = \sinh t.$$

Example 8.2.6 *Solve the following stochastic differential equation*

$$e^{t/2} dX_t = dt + e^{W_t} \, dW_t, \qquad X_0 = 0,$$

and find the distribution of the solution X_t and its mean and variance.

Dividing by $e^{t/2}$, integrating between 0 and t, and using formula (7.3.9) yields

$$\begin{aligned} X_t &= \int_0^t e^{-s/2} \, ds + \int_0^t e^{-s/2 + W_s} \, dW_s \\ &= 2(1 - e^{-t/2}) + e^{-t/2} e^{W_t} - 1 \\ &= 1 + e^{-t/2}(e^{W_t} - 2). \end{aligned}$$

Since e^{W_t} is a geometric Brownian motion, using Proposition 3.2.2 yields

$$\begin{aligned} \mathbb{E}[X_t] &= \mathbb{E}[1 + e^{-t/2}(e^{W_t} - 2)] = 1 - 2e^{-t/2} + e^{-t/2}\mathbb{E}[e^{W_t}] \\ &= 2 - 2e^{-t/2}. \\ Var(X_t) &= Var[1 + e^{-t/2}(e^{W_t} - 2)] = Var[e^{-t/2} e^{W_t}] = e^{-t} Var[e^{W_t}] \\ &= e^{-t}(e^{2t} - e^t) = e^t - 1. \end{aligned}$$

The process X_t has the following distribution:

$$\begin{aligned}
F(y) &= P(X_t \le y) = P\big(1 + e^{-t/2}(e^{W_t} - 2) \le y\big) \\
&= P\Big(W_t \le \ln\big(2 + e^{t/2}(y-1)\big)\Big) = P\Big(\frac{W_t}{\sqrt{t}} \le \frac{1}{\sqrt{t}}\ln\big(2 + e^{t/2}(y-1)\big)\Big) \\
&= N\Big(\frac{1}{\sqrt{t}}\ln\big(2 + e^{t/2}(y-1)\big)\Big),
\end{aligned}$$

where $N(u) = \dfrac{1}{\sqrt{2\pi}}\displaystyle\int_{-\infty}^{u} e^{-s^2/2}\,ds$ is the distribution function of a standard normal distributed random variable.

Example 8.2.7 *Solve the stochastic differential equation*

$$dX_t = dt + t^{-3/2} W_t e^{-W_t^2/(2t)}\,dW_t, \qquad X_1 = 1.$$

Integrating between 1 and t and applying formula (7.3.11) yields

$$\begin{aligned}
X_t &= X_1 + \int_1^t ds + \int_1^t s^{-3/2} W_s e^{-W_s^2/(2s)}\,dW_s \\
&= 1 + t - 1 - e^{-W_1^2/2} - \frac{1}{t^{1/2}} e^{-W_t^2/(2t)} \\
&= t - e^{-W_1^2/2} - \frac{1}{t^{1/2}} e^{-W_t^2/(2t)}, \qquad \forall t \ge 1.
\end{aligned}$$

Exercise 8.2.8 *Solve the following stochastic differential equations by the method of integration*

(*a*) $dX_t = (t - \frac{1}{2}\sin W_t)dt + (\cos W_t)dW_t, \;\; X_0 = 0;$

(*b*) $dX_t = (\frac{1}{2}\cos W_t - 1)dt + (\sin W_t)dW_t, \;\; X_0 = 0;$

(*c*) $dX_t = \frac{1}{2}(\sin W_t + W_t\cos W_t)dt + (W_t \sin W_t)dW_t, \;\; X_0 = 0.$

8.3 Exact Stochastic Equations

The stochastic differential equation

$$dX_t = a(t, W_t)dt + b(t, W_t)dW_t \tag{8.3.7}$$

is called *exact* if there is a differentiable function $f(t, x)$ such that

$$a(t,x) = \partial_t f(t,x) + \frac{1}{2}\partial_x^2 f(t,x) \tag{8.3.8}$$

$$b(t,x) = \partial_x f(t,x). \tag{8.3.9}$$

Assume the equation is exact. Then substituting in (8.3.7) yields

$$dX_t = \Big(\partial_t f(t, W_t) + \frac{1}{2}\partial_x^2 f(t, W_t)\Big)dt + \partial_x f(t, W_t)dW_t.$$

Applying Ito's formula, the previous equation becomes

$$dX_t = d\big(f(t, W_t)\big),$$

which implies $X_t = f(t, W_t) + c$, with c constant.

Solving the partial differential equations system (8.3.8)–(8.3.9) requires the following steps:

1. Integrating partially with respect to x in the second equation to obtain $f(t, x)$ up to an additive function $T(t)$;

2. Substitute into the first equation and determine the function $T(t)$;

3. The solution is $X_t = f(t, W_t) + c$, with c determined from the initial condition on X_t.

Example 8.3.1 *Solve the stochastic differential equation as an exact equation*

$$dX_t = e^t(1 + W_t^2)dt + (1 + 2e^t W_t)dW_t, \quad X_0 = 0.$$

In this case $a(t, x) = e^t(1 + x^2)$ and $b(t, x) = 1 + 2e^t x$. The associated system is

$$\begin{aligned} e^t(1 + x^2) &= \partial_t f(t, x) + \frac{1}{2}\partial_x^2 f(t, x) \\ 1 + 2e^t x &= \partial_x f(t, x). \end{aligned}$$

Integrating partially in x in the second equation yields

$$f(t, x) = \int (1 + 2e^t x)\, dx = x + e^t x^2 + T(t).$$

Then $\partial_t f = e^t x^2 + T'(t)$ and $\partial_x^2 f = 2e^t$. Substituting in the first equation yields

$$e^t(1 + x^2) = e^t x^2 + T'(t) + e^t.$$

This implies $T'(t) = 0$, or $T = c$ constant. Hence $f(t, x) = x + e^t x^2 + c$, and $X_t = f(t, W_t) = W_t + e^t W_t^2 + c$. Since $X_0 = 0$, it follows that $c = 0$. The solution is $X_t = W_t + e^t W_t^2$.

Example 8.3.2 *Find the solution of*

$$dX_t = \big(2tW_t^3 + 3t^2(1 + W_t)\big)dt + (3t^2 W_t^2 + 1)dW_t, \quad X_0 = 0.$$

The coefficient functions are $a(t,x) = 2tx^3 + 3t^2(1+x)$ and $b(t,x) = 3t^2x^2 + 1$. The associated system is given by

$$\begin{aligned} 2tx^3 + 3t^2(1+x) &= \partial_t f(t,x) + \frac{1}{2}\partial_x^2 f(t,x) \\ 3t^2x^2 + 1 &= \partial_x f(t,x). \end{aligned}$$

Integrating partially in the second equation yields

$$f(t,x) = \int (3t^2x^2 + 1)\,dx = t^2x^3 + x + T(t).$$

Then $\partial_t f = 2tx^3 + T'(t)$ and $\partial_x^2 f = 6t^2x$, and substituting into the first equation we get

$$2tx^3 + 3t^2(1+x) = 2tx^3 + T'(t) + \frac{1}{2}6t^2x.$$

After cancelations we get $T'(t) = 3t^2$, so $T(t) = t^3 + c$. Then

$$f(t,x) = t^2x^3 + x + t^3 + c.$$

The solution process is given by $X_t = f(t,W_t) = t^2W_t^3 + W_t + t^3 + c$. Using $X_0 = 0$ we get $c = 0$. Hence the solution is $X_t = t^2W_t^3 + W_t + t^3$.

The next result deals with a condition regarding the *closeness* of the stochastic differential equation.

Theorem 8.3.3 *If the stochastic differential equation (8.3.7) is exact, then the coefficient functions $a(t,x)$ and $b(t,x)$ satisfy the condition*

$$\boxed{\partial_x a = \partial_t b + \frac{1}{2}\partial_x^2 b.} \tag{8.3.10}$$

Proof: If the stochastic equation is exact, there is a function $f(t,x)$ satisfying the system (8.3.8)–(8.3.9). Differentiating the first equation of the system with respect to x yields

$$\partial_x a = \partial_t \partial_x f + \frac{1}{2}\partial_x^2 \partial_x f.$$

Substituting $b = \partial_x f$ yields the desired relation. ∎

Remark 8.3.4 The equation (8.3.10) has the meaning of a heat equation. The function $b(t,x)$ represents the temperature measured at x at the instance t, while $\partial_x a$ is the density of heat sources. The function $a(t,x)$ can be regarded as the potential from which the density of heat sources is derived by taking the gradient in x.

It is worth noting that equation (8.3.10) is just a necessary condition for exactness. This means that if this condition is not satisfied, then the equation is not exact. In that case we need to try a different method to solve the equation.

Example 8.3.5 *Is the stochastic differential equation*

$$dX_t = (1 + W_t^2)dt + (t^4 + W_t^2)dW_t$$

exact?

Collecting the coefficients, we have $a(t, x) = 1 + x^2$, $b(t, x) = t^4 + x^2$. Since $\partial_x a = 2x$, $\partial_t b = 4t^3$, and $\partial_x^2 b = 2$, the condition (8.3.10) is not satisfied, and hence the equation is not exact.

Exercise 8.3.6 *Solve the following exact stochastic differential equations*

(a) $dX_t = e^t dt + (W_t^2 - t)dW_t, \qquad X_0 = 1;$

(b) $dX_t = (\sin t)dt + (W_t^2 - t)dW_t, \qquad X_0 = -1;$

(c) $dX_t = t^2 dt + e^{W_t - \frac{t}{2}} dW_t, \qquad X_0 = 0;$

(d) $dX_t = t dt + e^{t/2}(\cos W_t)\, dW_t, \qquad X_0 = 1.$

Exercise 8.3.7 *Verify the closeness condition and then solve the following exact stochastic differential equations*

(a) $dX_t = \left(W_t + \frac{3}{2}W_t^2\right)dt + (t + W_t^3)dW_t, \qquad X_0 = 0;$

(b) $dX_t = 2tW_t dt + (t^2 + W_t)dW_t, \qquad X_0 = 0;$

(c) $dX_t = \left(e^t W_t + \frac{1}{2}\cos W_t\right)dt + (e^t + \sin W_t)dW_t, \qquad X_0 = 0;$

(d) $dX_t = e^{W_t}(1 + \frac{t}{2})dt + te^{W_t}\, dW_t, \qquad X_0 = 2.$

8.4 Integration by Inspection

When solving a stochastic differential equation by inspection we look for opportunities to apply the product or the quotient formulas:

$$\boxed{d(f(t)Y_t) = f(t)\, dY_t + Y_t\, df(t)}$$

$$\boxed{d\Big(\frac{X_t}{f(t)}\Big) = \frac{f(t)dX_t - X_t df(t)}{f(t)^2}.}$$

For instance, if a stochastic differential equation can be written as

$$dX_t = f'(t)W_t dt + f(t)dW_t,$$

the product rule brings the equation into the exact form

$$dX_t = d\Big(f(t)W_t\Big),$$

which after integration leads to the solution

$$X_t = X_0 + f(t)W_t.$$

Example 8.4.1 *Solve*

$$dX_t = (t + W_t^2)dt + 2tW_t dW_t, \qquad X_0 = a.$$

We can write the equation as

$$dX_t = W_t^2 dt + t(2W_t dW_t + dt),$$

which can be contracted to

$$dX_t = W_t^2 dt + td(W_t^2).$$

Using the product rule we can bring it to the exact form

$$dX_t = d(tW_t^2),$$

with the solution $X_t = tW_t^2 + a$.

Example 8.4.2 *Solve the stochastic differential equation*

$$dX_t = (W_t + 3t^2)dt + tdW_t.$$

If we rewrite the equation as

$$dX_t = 3t^2 dt + (W_t dt + tdW_t),$$

we note the exact expression formed by the last two terms $W_t dt + tdW_t = d(tW_t)$. Then

$$dX_t = d(t^3) + d(tW_t),$$

which is equivalent to $d(X_t) = d(t^3 + tW_t)$. Hence $X_t = t^3 + tW_t + c$, $c \in \mathbb{R}$.

Example 8.4.3 *Solve the stochastic differential equation*

$$e^{-2t}dX_t = (1 + 2W_t^2)dt + 2W_t dW_t.$$

Multiply by e^{2t} to get

$$dX_t = e^{2t}(1 + 2W_t^2)dt + 2e^{2t}W_t dW_t.$$

After regrouping, this becomes

$$dX_t = (2e^{2t}dt)W_t^2 + e^{2t}(2W_t dW_t + dt).$$

Since $d(e^{2t}) = 2e^{2t}dt$ and $d(W_t^2) = 2W_t dW_t + dt$, the previous relation becomes

$$dX_t = d(e^{2t})W_t^2 + e^{2t}d(W_t^2).$$

By the product rule, the right side becomes exact

$$dX_t = d(e^{2t}W_t^2),$$

and hence the solution is $X_t = e^{2t}W_t^2 + c$, $c \in \mathbb{R}$.

Example 8.4.4 *Solve the equation*

$$t^3 dX_t = (3t^2 X_t + t)dt + t^6 dW_t, \qquad X_1 = 0.$$

The equation can be written as

$$t^3 dX_t - 3X_t t^2 dt = tdt + t^6 dW_t.$$

Divide by t^6

$$\frac{t^3 dX_t - X_t d(t^3)}{(t^3)^2} = t^{-5}dt + dW_t.$$

Applying the quotient rule yields

$$d\Big(\frac{X_t}{t^3}\Big) = -d\Big(\frac{t^{-4}}{4}\Big) + dW_t.$$

Integrating between 1 and t, yields

$$\frac{X_t}{t^3} = -\frac{t^{-4}}{4} + W_t - W_1 + c$$

so

$$X_t = ct^3 - \frac{1}{4t} + t^3(W_t - W_1), \quad c \in \mathbb{R}.$$

Using $X_1 = 0$ yields $c = 1/4$ and hence the solution is

$$X_t = \frac{1}{4}\Big(t^3 - \frac{1}{t}\Big) + t^3(W_t - W_1), \quad c \in \mathbb{R}.$$

Exercise 8.4.5 *Solve the following stochastic differential equations by the inspection method*

(a) $dX_t = (1 + W_t)dt + (t + 2W_t)dW_t, \qquad X_0 = 0;$

(b) $t^2 dX_t = (2t^3 - W_t)dt + tdW_t, \qquad X_1 = 0;$

(c) $e^{-t/2}dX_t = \frac{1}{2}W_t dt + dW_t, \qquad X_0 = 0;$

(d) $dX_t = 2tW_t dW_t + W_t^2 dt, \qquad X_0 = 0;$

(e) $dX_t = \Big(1 + \frac{1}{2\sqrt{t}}W_t\Big)dt + \sqrt{t}\,dW_t, \qquad X_1 = 0.$

8.5 Linear Stochastic Differential Equations

Consider the stochastic differential equation with the drift term linear in X_t

$$dX_t = \big(\alpha(t)X_t + \beta(t)\big)dt + b(t, W_t)dW_t, \qquad t \geq 0.$$

This can also be written as

$$dX_t - \alpha(t)X_t dt = \beta(t)dt + b(t, W_t)dW_t.$$

Let $A(t) = \int_0^t \alpha(s)\,ds$. Multiplying by the integrating factor $e^{-A(t)}$, the left side of the previous equation becomes an exact expression

$$\begin{aligned} e^{-A(t)}\Big(dX_t - \alpha(t)X_t dt\Big) &= e^{-A(t)}\beta(t)dt + e^{-A(t)}b(t, W_t)dW_t \\ d\Big(e^{-A(t)}X_t\Big) &= e^{-A(t)}\beta(t)dt + e^{-A(t)}b(t, W_t)dW_t. \end{aligned}$$

Integrating yields

$$\begin{aligned} e^{-A(t)}X_t &= X_0 + \int_0^t e^{-A(s)}\beta(s)\,ds + \int_0^t e^{-A(s)}b(s, W_s)\,dW_s \\ X_t &= X_0 e^{A(t)} + e^{A(t)}\Big(\int_0^t e^{-A(s)}\beta(s)\,ds + \int_0^t e^{-A(s)}b(s, W_s)\,dW_s\Big). \end{aligned}$$

The first integral within the previous parentheses is a Riemann integral, and the latter one is an Ito stochastic integral. Sometimes, in practical applications these integrals can be computed explicitly.

When $b(t, W_t) = b(t)$, the latter integral becomes a Wiener integral. In this case the solution X_t is Gaussian with mean and variance given by

$$\begin{aligned} \mathbb{E}[X_t] &= X_0 e^{A(t)} + e^{A(t)}\int_0^t e^{-A(s)}\beta(s)\,ds \\ Var[X_t] &= e^{2A(t)}\int_0^t e^{-2A(s)}b(s)^2\,ds. \end{aligned}$$

Another important particular case is when $\alpha(t) = \alpha \neq 0$, $\beta(t) = \beta$ are constants and $b(t, W_t) = b(t)$. The equation in this case is

$$dX_t = \big(\alpha X_t + \beta\big)dt + b(t)dW_t, \qquad t \geq 0,$$

and the solution takes the form

$$X_t = X_0 e^{\alpha t} + \frac{\beta}{\alpha}(e^{\alpha t} - 1) + \int_0^t e^{\alpha(t-s)}b(s)\,dW_s.$$

Example 8.5.1 *Solve the linear stochastic differential equation*

$$dX_t = (2X_t + 1)dt + e^{2t}dW_t.$$

Write the equation as

$$dX_t - 2X_t dt = dt + e^{2t}dW_t$$

and multiply by the integrating factor e^{-2t} to get

$$d\Big(e^{-2t}X_t\Big) = e^{-2t}dt + dW_t.$$

Integrate between 0 and t and multiply by e^{2t}, to obtain

$$\begin{aligned} X_t &= X_0 e^{2t} + e^{2t}\int_0^t e^{-2s}ds + e^{2t}\int_0^t dW_s \\ &= X_0 e^{2t} + \frac{1}{2}(e^{2t} - 1) + e^{2t}W_t. \end{aligned}$$

Example 8.5.2 *Solve the linear stochastic differential equation*

$$dX_t = (2 - X_t)dt + e^{-t}W_t dW_t.$$

Multiplying by the integrating factor e^t yields

$$e^t(dX_t + X_t dt) = 2e^t dt + W_t\, dW_t.$$

Since $e^t(dX_t + X_t dt) = d(e^t X_t)$, integrating between 0 and t we get

$$e^t X_t = X_0 + \int_0^t 2e^t\, dt + \int_0^t W_s\, dW_s.$$

Dividing by e^t and performing the integration yields

$$X_t = X_0 e^{-t} + 2(1 - e^{-t}) + \frac{1}{2}e^{-t}(W_t^2 - t).$$

Example 8.5.3 *Solve the linear stochastic differential equation*

$$dX_t = (\frac{1}{2}X_t + 1)dt + e^t \cos W_t\, dW_t.$$

Write the equation as

$$dX_t - \frac{1}{2}X_t dt = dt + e^t \cos W_t\, dW_t$$

and multiply by the integrating factor $e^{-t/2}$ to get

$$d(e^{-t/2}X_t) = e^{-t/2}dt + e^{t/2}\cos W_t\, dW_t.$$

Integrating yields

$$e^{-t/2}X_t = X_0 + \int_0^t e^{-s/2}\, ds + \int_0^t e^{s/2}\cos W_s\, dW_s.$$

Multiply by $e^{t/2}$ and use formula (7.3.10) to obtain the solution

$$X_t = X_0 e^{t/2} + 2(e^{t/2} - 1) + e^t \sin W_t.$$

Exercise 8.5.4 *Solve the following linear stochastic differential equations*

(a) $dX_t = (4X_t - 1)dt + 2dW_t$;
(b) $dX_t = (3X_t - 2)dt + e^{3t}dW_t$;
(c) $dX_t = (1 + X_t)dt + e^t W_t dW_t$;
(d) $dX_t = (4X_t + t)dt + e^{4t}dW_t$;
(e) $dX_t = \left(t + \frac{1}{2}X_t\right)dt + e^t \sin W_t\, dW_t$;
(f) $dX_t = -X_t dt + e^{-t}\, dW_t$.

In the following we present an important example of stochastic differential equation, which can be solved by the method presented in this section.

Proposition 8.5.5 (The mean-reverting Ornstein-Uhlenbeck process) *Let m and α be two constants. Then the solution X_t of the stochastic equation*

$$\boxed{dX_t = (m - X_t)dt + \alpha dW_t} \qquad (8.5.11)$$

is given by

$$\boxed{X_t = m + (X_0 - m)e^{-t} + \alpha \int_0^t e^{s-t}\, dW_s.} \qquad (8.5.12)$$

X_t is Gaussian with mean and variance given by

$$\begin{aligned} \mathbb{E}[X_t] &= m + (X_0 - m)e^{-t} \\ Var(X_t) &= \frac{\alpha^2}{2}(1 - e^{-2t}). \end{aligned}$$

Proof: Adding $X_t dt$ to both sides and multiplying by the integrating factor e^t we get

$$d(e^t X_t) = me^t dt + \alpha e^t dW_t,$$

which after integration yields

$$e^t X_t = X_0 + m(e^t - 1) + \alpha \int_0^t e^s \, dW_s.$$

Hence

$$\begin{aligned} X_t &= X_0 e^{-t} + m - e^{-t} + \alpha e^{-t} \int_0^t e^s \, dW_s \\ &= m + (X_0 - m)e^{-t} + \alpha \int_0^t e^{s-t} \, dW_s. \end{aligned}$$

Since X_t is the sum between a predictable function and a Wiener integral, then we can use Proposition 5.6.1 and it follows that X_t is Gaussian, with

$$\begin{aligned} \mathbb{E}[X_t] &= m + (X_0 - m)e^{-t} + E\Big[\alpha \int_0^t e^{s-t} \, dW_s\Big] = m + (X_0 - m)e^{-t} \\ Var(X_t) &= Var\Big[\alpha \int_0^t e^{s-t} \, dW_s\Big] = \alpha^2 e^{-2t} \int_0^t e^{2s} \, ds \\ &= \alpha^2 e^{-2t} \frac{e^{2t} - 1}{2} = \frac{1}{2}\alpha^2 (1 - e^{-2t}). \end{aligned}$$

■

The name *mean-reverting* comes from the fact that

$$\lim_{t \to \infty} \mathbb{E}[X_t] = m.$$

The variance also tends to zero exponentially, $\lim_{t \to \infty} Var[X_t] = 0$. According to Proposition 4.9.1, the process X_t tends to m in the mean square sense.

Proposition 8.5.6 (The Brownian bridge) *For $a, b \in \mathbb{R}$ fixed, the stochastic differential equation*

$$dX_t = \frac{b - X_t}{1 - t} dt + dW_t, \quad 0 \le t < 1, X_0 = a$$

has the solution

$$\boxed{X_t = a(1 - t) + bt + (1 - t) \int_0^t \frac{1}{1 - s} dW_s, \; 0 \le t < 1.} \tag{8.5.13}$$

The solution has the property $\lim_{t \to 1} X_t = b$, almost certainly.

Proof: If we let $Y_t = b - X_t$ the equation becomes linear in Y_t

$$dY_t + \frac{1}{1-t} Y_t dt = -dW_t.$$

Multiplying by the integrating factor $\rho(t) = \frac{1}{1-t}$ yields

$$d\Big(\frac{Y_t}{1-t}\Big) = -\frac{1}{1-t} dW_t,$$

which leads by integration to

$$\frac{Y_t}{1-t} = c - \int_0^t \frac{1}{1-s}\, dW_s.$$

Making $t = 0$ yields $c = a - b$, so

$$\frac{b - X_t}{1-t} = a - b - \int_0^t \frac{1}{1-s}\, dW_s.$$

Solving for X_t yields

$$X_t = a(1-t) + bt + (1-t)\int_0^t \frac{1}{1-s} dW_s, \ 0 \le t < 1.$$

Let $U_t = (1-t)\int_0^t \frac{1}{1-s}\, dW_s$. First we notice that

$$\begin{aligned}
\mathbb{E}[U_t] &= (1-t)E\Big[\int_0^t \frac{1}{1-s}\, dW_s\Big] = 0, \\
Var(U_t) &= (1-t)^2 Var\Big[\int_0^t \frac{1}{1-s}\, dW_s\Big] = (1-t)^2 \int_0^t \frac{1}{(1-s)^2}\, ds \\
&= (1-t)^2\Big(\frac{1}{1-t} - 1\Big) = t(1-t).
\end{aligned}$$

In order to show as-$\lim_{t\to 1} X_t = b$, we need to prove

$$P\big(\omega; \lim_{t\to 1} X_t(\omega) = b\big) = 1.$$

Since $X_t = a(1-t) + bt + U_t$, it suffices to show that

$$P\big(\omega; \lim_{t\to 1} U_t(\omega) = 0\big) = 1. \tag{8.5.14}$$

We evaluate the probability of the complementary event

$$P\big(\omega; \lim_{t\to 1} U_t(\omega) \neq 0\big) = P\big(\omega; |U_t(\omega)| > \epsilon, \forall t\big),$$

for some $\epsilon > 0$. Since by Markov's inequality

$$P\big(\omega; |U_t(\omega)| > \epsilon\big) < \frac{Var(U_t)}{\epsilon^2} = \frac{t(1-t)}{\epsilon^2}$$

holds for any $0 \le t < 1$, choosing $t \to 1$ implies that

$$P\big(\omega; |U_t(\omega)| > \epsilon, \forall t\big) = 0,$$

which implies (8.5.14). ∎

The process (8.5.13) is called the *Brownian bridge* because it joins $X_0 = a$ with $X_1 = b$. Since X_t is the sum between a deterministic linear function in t and a Wiener integral, it follows that it is a Gaussian process, with mean and variance

$$\begin{aligned} \mathbb{E}[X_t] &= a(1-t) + bt \\ Var(X_t) &= Var(U_t) = t(1-t). \end{aligned}$$

It is worth noting that the variance is maximum at the midpoint $t = (b-a)/2$ and zero at the end points a and b.

Exercise 8.5.7 *Show that the Brownian bridge* (8.5.13) *satisfies* $X_t \xrightarrow{ms} b$ *as* $t \to 1$.

Exercise 8.5.8 *Find* $Cov(X_s, X_t)$, $0 < s < t$ *for the following cases:*

(*a*) X_t *is a mean reverting Ornstein-Uhlenbeck process;*

(*b*) X_t *is a Brownian bridge process.*

8.6 Stochastic Equations with respect to a Poisson Process

Similar techniques can be applied in the case when the Brownian motion process W_t is replaced by a Poisson process N_t with constant rate λ. For instance, the stochastic differential equation

$$\begin{aligned} dX_t &= 3X_t dt + e^{3t} dN_t \\ X_0 &= 1 \end{aligned}$$

can be solved multiplying by the integrating factor e^{-3t} to obtain

$$d(e^{-3t}X_t) = dN_t.$$

Integrating yields $e^{-3t}X_t = N_t + 1$, so the solution is $X_t = e^{3t}(1 + N_t)$.

The following equation

$$dX_t = (m - X_t)dt + \alpha dN_t$$

is similar to the equation defining the mean-reverting Ornstein-Uhlenbeck process. As we shall see, in this case, the process is no more mean-reverting, but it reverts to a certain constant. A similar method yields the solution

$$X_t = m + (X_0 - m)e^{-t} + \alpha e^{-t}\int_0^t e^s\,dN_s.$$

Since from Proposition 5.8.5 and Exercise 5.8.8 we have

$$\begin{aligned}
\mathbb{E}[\int_0^t e^s\,dN_s] &= \lambda\int_0^t e^s\,ds = \lambda(e^t - 1)\\
Var(\int_0^t e^s\,dN_s) &= \lambda\int_0^t e^{2s}\,ds = \frac{\lambda}{2}(e^{2t} - 1),
\end{aligned}$$

it follows that

$$\begin{aligned}
\mathbb{E}[X_t] &= m + (X_0 - m)e^{-t} + \alpha\lambda(1 - e^{-t}) \to m + \alpha\lambda\\
Var(X_t) &= \frac{\lambda\alpha^2}{2}(1 - e^{-2t}).
\end{aligned}$$

It is worth noting that in this case the process X_t is not Gaussian any more.

8.7 The Method of Variation of Parameters

Let us start by considering the following stochastic equation

$$dX_t = \alpha X_t dW_t, \tag{8.7.15}$$

with α constant. This is the equation which, in physics, is known to model the *linear noise*. Dividing by X_t yields

$$\frac{dX_t}{X_t} = \alpha\,dW_t.$$

Switch to the integral form

$$\int\frac{dX_t}{X_t} = \int\alpha\,dW_t,$$

and integrate "blindly" to get $\ln X_t = \alpha W_t + c$, with c an integration constant. This leads to the "pseudo-solution"

$$X_t = e^{\alpha W_t + c}.$$

The nomination "pseudo" stands for the fact that X_t does not satisfy the initial equation. We shall find a correct solution by letting the parameter c be a function of t. In other words, we are looking for a solution of the following type:

$$X_t = e^{\alpha W_t + c(t)}, \tag{8.7.16}$$

where the function $c(t)$ is subject to be determined. Using Ito's formula we get

$$\begin{aligned} dX_t &= d(e^{\alpha W_t + c(t)}) = e^{\alpha W_t + c(t)}(c'(t) + \alpha^2/2)dt + \alpha e^{\alpha W_t + c(t)} dW_t \\ &= X_t(c'(t) + \alpha^2/2)dt + \alpha X_t \, dW_t. \end{aligned}$$

Substituting the last term from the initial equation (8.7.15) yields

$$dX_t = X_t(c'(t) + \alpha^2/2)dt + dX_t,$$

which leads to the equation

$$c'(t) + \alpha^2/2 = 0$$

with the solution $c(t) = -\frac{\alpha^2}{2}t + k$. Substituting into (8.7.16) yields

$$X_t = e^{\alpha W_t - \frac{\alpha^2}{2}t + k}.$$

The value of the constant k is determined by taking $t = 0$. This leads to $X_0 = e^k$. Hence we have obtained the solution of the equation (8.7.15)

$$X_t = X_0 e^{\alpha W_t - \frac{\alpha^2}{2}t}.$$

Example 8.7.1 *Use the method of variation of parameters to solve the stochastic differential equation*

$$dX_t = \mu X_t dt + \sigma X_t dW_t,$$

with μ and σ constants.

After dividing by X_t we bring the equation into the equivalent integral form

$$\int \frac{dX_t}{X_t} = \int \mu \, dt + \int \sigma \, dW_t.$$

Integrate on the left "blindly" and get

$$\ln X_t = \mu t + \sigma W_t + c,$$

where c is an integration constant. We arrive at the following "pseudo-solution"

$$X_t = e^{\mu t + \sigma W_t + c}.$$

Assume the constant c is replaced by a function $c(t)$, so we are looking for a solution of the form

$$X_t = e^{\mu t + \sigma W_t + c(t)}. \tag{8.7.17}$$

Apply Ito's formula and get

$$dX_t = X_t(\mu + c'(t) + \frac{\sigma^2}{2})dt + \sigma X_t dW_t.$$

Subtracting the initial equation yields

$$(c'(t) + \frac{\sigma^2}{2})dt = 0,$$

which is satisfied for $c'(t) = -\frac{\sigma^2}{2}$, with the solution $c(t) = -\frac{\sigma^2}{2}t + k$, $k \in \mathbb{R}$. Substituting into (8.7.17) yields the solution

$$X_t = e^{\mu t + \sigma W_t - \frac{\sigma^2}{2}t + k} = e^{(\mu - \frac{\sigma^2}{2})t + \sigma W_t + k} = X_0 e^{(\mu - \frac{\sigma^2}{2})t + \sigma W_t}.$$

Exercise 8.7.2 *Use the method of variation of parameters to solve the equation*

$$dX_t = X_t W_t dW_t$$

by following the next two steps:

(a) Divide by X_t and integrate "blindly" to get the "pseudo-solution"

$$X_t = e^{\frac{W_t^2}{2} - \frac{t}{2} + c},$$

with c constant.

(b) Consider $c = c(t, W_t)$ and find a solution of type

$$X_t = e^{\frac{W_t^2}{2} - \frac{t}{2} + c(t, W_t)}.$$

Example 8.7.3 (Langevin equation) *Solve $dX_t = -qX_t dt + \sigma dW_t$, with q constant.*

We start solving the associated deterministic equation

$$dX_t = -qX_t dt,$$

which has the solution $X_t = Ce^{-qt}$, with C constant. Now, look for a solution of the type $X_t = C(t)e^{-qt}$, and determine the function $C(t)$ such that the initial equation is satisfied. Comparing

$$\begin{aligned} dX_t &= d\Big(C(t)e^{-qt}\Big) = -qC(t)e^{-qt}dt + e^{-qt}dC(t) \\ &= -qX_t dt + e^{-qt}dC(t) \end{aligned}$$

with the initial stochastic differential equation of X_t implies that $C(t)$ satisfies

$$dC(t) = \sigma e^{qt} dW_t.$$

Integrating we obtain $C(t) = C(0) + \sigma \int_0^t e^{qs}\, dW_s$, and hence $X_t = C(0)e^{-qt} + \sigma e^{-qt} \int_0^t e^{qs}\, dW_s$. It is not hard to see that $C(0) = X_0$, which enables us to write the final solution as

$$X_t = X_0 e^{-qt} + \sigma e^{-qt} \int_0^t e^{qs}\, dW_s.$$

Exercise 8.7.4 (the mean reverting Orstein-Uhlenbeck process) *Use the method of variation of constants to solve*

$$dX_t = \lambda(\mu - X_r)dt + \sigma dW_t,$$

where λ and μ are constants.

8.8 Integrating Factors

The method of integrating factors can be applied to a class of stochastic differential equations of the type

$$dX_t = f(t, X_t)dt + g(t)X_t dW_t, \tag{8.8.18}$$

where f and g are continuous deterministic functions. The integrating factor is given by

$$\rho_t = e^{-\int_0^t g(s)\, dW_s + \frac{1}{2}\int_0^t g^2(s)\, ds}.$$

The equation can be brought into the following exact form

$$d(\rho_t X_t) = \rho_t f(t, X_t)dt.$$

Substituting $Y_t = \rho_t X_t$, we obtain that Y_t satisfies the deterministic differential equation

$$dY_t = \rho_t f(t, Y_t/\rho_t)dt,$$

which can be solved by either integration or as an exact equation. We shall exemplify the method of integrating factors with a few examples.

Example 8.8.1 *Solve the stochastic differential equation*

$$dX_t = r dt + \alpha X_t dW_t, \tag{8.8.19}$$

with r and α constants.

The integrating factor is given by $\rho_t = e^{\frac{1}{2}\alpha^2 t - \alpha W_t}$. Using Ito's formula, we can easily check that

$$d\rho_t = \rho_t(\alpha^2 dt - \alpha dW_t).$$

Using $dt^2 = dt\, dW_t = 0$, $(dW_t)^2 = dt$ we obtain

$$dX_t d\rho_t = -\alpha^2 \rho_t X_t dt.$$

Multiplying by ρ_t, the initial equation becomes

$$\rho_t dX_t - \alpha \rho_t X_t dW_t = r\rho_t dt,$$

and adding and subtracting $\alpha^2 \rho_t X_t dt$ from the left side yields

$$\rho_t dX_t - \alpha \rho_t X_t dW_t + \alpha^2 \rho_t X_t dt - \alpha^2 \rho_t X_t dt = r\rho_t dt.$$

This can be written as

$$\rho_t dX_t + X_t d\rho_t + d\rho_t dX_t = r\rho_t dt,$$

which, by virtue of the product rule, becomes

$$d(\rho_t X_t) = r\rho_t dt.$$

Integrating yields

$$\rho_t X_t = \rho_0 X_0 + r\int_0^t \rho_s\, ds$$

and hence the solution is

$$\begin{aligned} X_t &= \frac{1}{\rho_t} X_0 + \frac{r}{\rho_t}\int_0^t \rho_s\, ds \\ &= X_0 e^{\alpha W_t - \frac{1}{2}\alpha^2 t} + r\int_0^t e^{-\frac{1}{2}\alpha^2(t-s) + \alpha(W_t - W_s)}\, ds. \end{aligned}$$

Exercise 8.8.2 *Let α be a constant. Solve the following stochastic differential equations by the method of integrating factors*

(a) $dX_t = \alpha X_t dW_t$;

(b) $dX_t = X_t dt + \alpha X_t dW_t$;

(c) $dX_t = \dfrac{1}{X_t} dt + \alpha X_t dW_t$, $X_0 > 0$.

Exercise 8.8.3 *Let X_t be the solution of the stochastic equation $dX_t = \sigma X_t\, dW_t$, with σ constant. Let $A_t = \frac{1}{t}\int_0^t X_s\, dW_s$ be the stochastic average of X_t. Find the stochastic equation satisfied by A_t, the mean and variance of A_t.*

8.9 Existence and Uniqueness

The following theorem is the analog of Picard's result from ordinary differential equations. It states the existence and uniqueness of strong solutions of stochastic differential equations. The proof of the next theorem can be found in Øksendal [37]. For more results regarding existence and uniqueness of strong solutions the reader is referred to Krylov and Zvonkin [29].

Theorem 8.9.1 (Existence and Uniqueness) *Consider the stochastic differential equation*

$$dX_t = b(t, X_t)dt + \sigma(t, X_t)dW_t, \qquad X_0 = c$$

where c is a constant and b and σ are continuous functions on $[0,T] \times \mathbb{R}$ satisfying

1. $|b(t,x)| + |\sigma(t,x)| \le C(1 + |x|); \qquad x \in \mathbb{R}, t \in [0,T]$
2. $|b(t,x) - b(t,y)| + |\sigma(t,x) - \sigma(t,y)| \le K|x - y|, \qquad x, y \in \mathbb{R}, t \in [0,T]$

with C, K positive constants. Let $\mathcal{F}_t = \sigma\{W_s; s \le t\}$. Then there is a unique solution process X_t that is continuous and $\mathcal{F}_t$-adapted and satisfies

$$E\Big[\int_0^T X_t^2\, dt\Big] < \infty.$$

The first condition says that the drift and volatility increase no faster than a linear function in x. This condition ensures that the solution X_t does not explode in finite time, i.e. does not tend to ∞ for finite t. The second conditions states that the functions are Lipschitz in the second argument; this condition guarantees the solution uniqueness.

The following example deals with an exploding solution. Consider the nonlinear stochastic differential equation

$$dX_t = X_t^3 dt + X_t^2\, dW_t, \qquad X_0 = 1/a, \tag{8.9.20}$$

where a is a nonzero constant. It is clear that condition 1. does not hold, since the drift increases cubically.

We shall look for a solution of the type $X_t = f(W_t)$. Ito's formula yields

$$dX_t = f'(W_t)dW_t + \frac{1}{2}f''(W_t)dt.$$

Equating the coefficients of dt and dW_t in the last two equations yields

$$f'(W_t) = X_t^2 \Longrightarrow f'(W_t) = f(W_t)^2 \tag{8.9.21}$$

$$\frac{1}{2}f''(W_t) = X_t^3 \Longrightarrow f''(W_t) = 2f(W_t)^3. \tag{8.9.22}$$

We note that equation (8.9.21) implies (8.9.22) by differentiation. So it suffices to solve only the ordinary differential equation

$$f'(x) = f(x)^2, \qquad f(0) = 1/a.$$

Separating and integrating we have

$$\int \frac{df}{f(x)^2} = \int ds \Longrightarrow f(x) = \frac{1}{a-x}.$$

Hence a solution of equation (8.9.20) is

$$X_t = \frac{1}{a - W_t}.$$

Let T_a be the first time the Brownian motion W_t hits a. Then the process X_t is defined only for $0 \leq t < T_a$. T_a is a random variable with $P(T_a < \infty) = 1$ and $\mathbb{E}[T_a] = \infty$, see section 4.3.

Example 8.9.2 *Show that that the following stochastic differential equations have a unique (strong) solution, without solving the equations explicitly:*

(a) $dX_t = \mu X_t\, dt + \sigma\, dW_t$ *(Langevin equation);*

(b) $dX_t = (m - X_t)\, dt + \sigma\, dW_t$ *(Mean reverting Ornstein-Uhlenbeck process);*

(c) $dX_t = \sigma X_t\, dW_t$ *(Linear noise);*

(d) $dX_t = \mu\, dt + \sqrt{\frac{X_t}{2}}\, dW_t$ *(Squared Bessel process);*

(e) $dX_t = \mu X_t\, dt + \sigma X_t\, dW_t$ *(Geometric Brownian motion);*

(f) $dX_t = (\mu_0 + \mu_1 X_t)\, dt + \sqrt{\frac{X_t}{2}}\, dW_t$ *(CIR process),*
with m, μ, μ_0, μ_1 *and* σ *positive constants.*

Example 8.9.3 *Consider the stochastic differential equation*

$$dX_t = \Big(\sqrt{1 + X_t^2} + \frac{1}{2}X_t\Big)dt + \sqrt{1 + X_t^2}\, dW_t, \quad X_0 = x_0.$$

(a) *Solve the equation;*

(b) *Show that there is a unique solution.*

8.10 Finding Mean and Variance

For most practical purposes, the most important information one needs to know about a process is its mean and variance. These can be found directly from the stochastic equation in some particular cases without solving explicitly the equation. We shall deal with this problem in the present section.

Consider the process X_t satisfying (6.1.1). Then taking the expectation in (8.1.2) and using the property of the Ito integral as a zero mean random variable yields

$$\mathbb{E}[X_t] = X_0 + \int_0^t \mathbb{E}[a(s, W_s, X_s)]\, ds. \tag{8.10.23}$$

Applying the Fundamental Theorem of Calculus we obtain

$$\frac{d}{dt}\mathbb{E}[X_t] = \mathbb{E}[a(t, W_t, X_t)].$$

We note that X_t is not differentiable, but its expectation $\mathbb{E}[X_t]$ is. This equation can be solved exactly in a few particular cases.

1. If $a(t, W_t, X_t) = a(t)$, then $\frac{d}{dt}\mathbb{E}[X_t] = a(t)$ with the exact solution $\mathbb{E}[X_t] = X_0 + \int_0^t a(s)\, ds$.

2. If $a(t, W_t, X_t) = \alpha(t)X_t + \beta(t)$, with $\alpha(t)$ and $\beta(t)$ continuous deterministic function, then

$$\frac{d}{dt}\mathbb{E}[X_t] = \alpha(t)\mathbb{E}[X_t] + \beta(t),$$

which is a linear differential equation in $\mathbb{E}[X_t]$. Its solution is given by

$$\mathbb{E}[X_t] = e^{A(t)}\Big(X_0 + \int_0^t e^{-A(s)}\beta(s)\, ds\Big), \tag{8.10.24}$$

where $A(t) = \int_0^t \alpha(s)\, ds$. It is worth noting that the expectation $\mathbb{E}[X_t]$ does not depend on the volatility term $b(t, W_t, X_t)$.

Exercise 8.10.1 *If $dX_t = (2X_t + e^{2t})dt + b(t, W_t, X_t)dW_t$, then show that*

$$\mathbb{E}[X_t] = e^{2t}(X_0 + t).$$

Proposition 8.10.2 *Let X_t be a process satisfying the stochastic equation*

$$dX_t = \alpha(t)X_t dt + b(t)dW_t.$$

Then the mean and variance of X_t are given by

$$\begin{aligned} \mathbb{E}[X_t] &= e^{A(t)}X_0 \\ Var[X_t] &= e^{2A(t)}\int_0^t e^{-A(s)}b^2(s)\, ds, \end{aligned}$$

where $A(t) = \int_0^t \alpha(s)\, ds$.

Proof: The expression of $\mathbb{E}[X_t]$ follows directly from formula (8.10.24) with $\beta = 0$. In order to compute the second moment we first compute

$$\begin{aligned}
(dX_t)^2 &= b^2(t)\,dt; \\
d(X_t^2) &= 2X_t dX_t + (dX_t)^2 \\
&= 2X_t\big(\alpha(t)X_t dt + b(t)dW_t\big) + b^2(t)dt \\
&= \big(2\alpha(t)X_t^2 + b^2(t)\big)dt + 2b(t)X_t dW_t,
\end{aligned}$$

where we used Ito's formula. If we let $Y_t = X_t^2$, the previous equation becomes

$$dY_t = \big(2\alpha(t)Y_t + b^2(t)\big)dt + 2b(t)\sqrt{Y_t}\,dW_t.$$

Applying formula (8.10.24) with $\alpha(t)$ replaced by $2\alpha(t)$ and $\beta(t)$ by $b^2(t)$, yields

$$\mathbb{E}[Y_t] = e^{2A(t)}\Big(Y_0 + \int_0^t e^{-2A(s)}b^2(s)\,ds\Big),$$

which is equivalent to

$$\mathbb{E}[X_t^2] = e^{2A(t)}\Big(X_0^2 + \int_0^t e^{-2A(s)}b^2(s)\,ds\Big).$$

It follows that the variance is

$$Var[X_t] = \mathbb{E}[X_t^2] - (\mathbb{E}[X_t])^2 = e^{2A(t)}\int_0^t e^{-2A(s)}b^2(s)\,ds.$$

■

Remark 8.10.3 We note that the previous equation is of linear type. This shall be solved explicitly in a future section.

The mean and variance for a given stochastic process can be computed by working out the associated stochastic equation. We shall provide next a few examples.

Example 8.10.4 *Find the mean and variance of e^{kW_t}, with k constant.*

From Ito's formula

$$d(e^{kW_t}) = ke^{kW_t}dW_t + \frac{1}{2}k^2e^{kW_t}dt,$$

and integrating yields

$$e^{kW_t} = 1 + k\int_0^t e^{kW_s}\,dW_s + \frac{1}{2}k^2\int_0^t e^{kW_s}\,ds.$$

Taking the expectations we have

$$\mathbb{E}[e^{kW_t}] = 1 + \frac{1}{2}k^2 \int_0^t \mathbb{E}[e^{kW_s}]\, ds.$$

If we let $f(t) = \mathbb{E}[e^{kW_t}]$, then differentiating the previous relations yields the differential equation

$$f'(t) = \frac{1}{2}k^2 f(t)$$

with the initial condition $f(0) = \mathbb{E}[e^{kW_0}] = 1$. The solution is $f(t) = e^{k^2 t/2}$, and hence

$$\boxed{\mathbb{E}[e^{kW_t}] = e^{k^2 t/2}.}$$

The variance is

$$\begin{aligned} Var(e^{kW_t}) &= \mathbb{E}[e^{2kW_t}] - (\mathbb{E}[e^{kW_t}])^2 = e^{4k^2 t/2} - e^{k^2 t} \\ &= e^{k^2 t}(e^{k^2 t} - 1). \end{aligned}$$

Example 8.10.5 *Find the mean of the process $W_t e^{W_t}$.*

We shall set up a stochastic differential equation for $W_t e^{W_t}$. Using the product formula and Ito's formula yields

$$\begin{aligned} d(W_t e^{W_t}) &= e^{W_t} dW_t + W_t d(e^{W_t}) + dW_t\, d(e^{W_t}) \\ &= e^{W_t} dW_t + (W_t + dW_t)(e^{W_t} dW_t + \frac{1}{2} e^{W_t} dt) \\ &= (\frac{1}{2} W_t e^{W_t} + e^{W_t}) dt + (e^{W_t} + W_t e^{W_t}) dW_t. \end{aligned}$$

Integrating and using that $W_0 e^{W_0} = 0$ yields

$$W_t e^{W_t} = \int_0^t (\frac{1}{2} W_s e^{W_s} + e^{W_s})\, ds + \int_0^t (e^{W_s} + W_s e^{W_s})\, dW_s.$$

Since the expectation of an Ito integral is zero, we have

$$\mathbb{E}[W_t e^{W_t}] = \int_0^t \big(\frac{1}{2}\mathbb{E}[W_s e^{W_s}] + \mathbb{E}[e^{W_s}]\big)\, ds.$$

Let $f(t) = \mathbb{E}[W_t e^{W_t}]$. Using $\mathbb{E}[e^{W_s}] = e^{s/2}$, the previous integral equation becomes

$$f(t) = \int_0^t (\frac{1}{2} f(s) + e^{s/2})\, ds.$$

Differentiating yields the following linear differential equation

$$f'(t) = \frac{1}{2} f(t) + e^{t/2}$$

with the initial condition $f(0) = 0$. Multiplying by $e^{-t/2}$ yields the following exact equation

$$(e^{-t/2} f(t))' = 1.$$

The solution is $f(t) = te^{t/2}$. Hence we obtained that

$$\mathbb{E}[W_t e^{W_t}] = te^{t/2}.$$

Exercise 8.10.6 *Find (a)* $\mathbb{E}[W_t^2 e^{W_t}]$; *(b)* $\mathbb{E}[W_t e^{kW_t}]$.

Example 8.10.7 *Show that for any integer* $k \geq 0$ *we have*

$$\boxed{\mathbb{E}[W_t^{2k}] = \frac{(2k)!}{2^k k!} t^k, \qquad \mathbb{E}[W_t^{2k+1}] = 0.}$$

In particular, $\mathbb{E}[W_t^4] = 3t^2$, $\mathbb{E}[W_t^6] = 15t^3$.

From Ito's formula we have

$$d(W_t^n) = nW_t^{n-1} dW_t + \frac{n(n-1)}{2} W_t^{n-2} dt.$$

Integrate and get

$$W_t^n = n \int_0^t W_s^{n-1} \, dW_s + \frac{n(n-1)}{2} \int_0^t W_s^{n-2} \, ds.$$

Since the expectation of the first integral on the right side is zero, taking the expectation yields the following recursive relation

$$\mathbb{E}[W_t^n] = \frac{n(n-1)}{2} \int_0^t \mathbb{E}[W_s^{n-2}] \, ds.$$

Using the initial values $\mathbb{E}[W_t] = 0$ and $\mathbb{E}[W_t^2] = t$, the method of mathematical induction implies that $\mathbb{E}[W_t^{2k+1}] = 0$ and $\mathbb{E}[W_t^{2k}] = \frac{(2k)!}{2^k k!} t^k$. The details are left to the reader.

Exercise 8.10.8 *(a) Is* $W_t^4 - 3t^2$ *an* $\mathcal{F}_t$*-martingale?*
(b) What about W_t^3*?*

Example 8.10.9 *Find* $\mathbb{E}[\sin W_t]$.

From Ito's formula

$$d(\sin W_t) = \cos W_t \, dW_t - \frac{1}{2} \sin W_t \, dt,$$

then integrating yields

$$\sin W_t = \int_0^t \cos W_s \, dW_s - \frac{1}{2}\int_0^t \sin W_s \, ds.$$

Taking expectations we arrive at the integral equation

$$\mathbb{E}[\sin W_t] = -\frac{1}{2}\int_0^t \mathbb{E}[\sin W_s] \, ds.$$

Let $f(t) = \mathbb{E}[\sin W_t]$. Differentiating yields the equation $f'(t) = -\frac{1}{2}f(t)$ with $f(0) = \mathbb{E}[\sin W_0] = 0$. The unique solution is $f(t) = 0$. Hence

$$\mathbb{E}[\sin W_t] = 0.$$

Exercise 8.10.10 *Let σ be a constant. Show that*

(*a*) $\mathbb{E}[\sin(\sigma W_t)] = 0$;

(*b*) $\mathbb{E}[\cos(\sigma W_t)] = e^{-\sigma^2 t/2}$;

(*c*) $\mathbb{E}[\sin(t + \sigma W_t)] = e^{-\sigma^2 t/2} \sin t$;

(*d*) $\mathbb{E}[\cos(t + \sigma W_t)] = e^{-\sigma^2 t/2} \cos t$.

Exercise 8.10.11 *Use the previous exercise and the definition of expectation to show that*

(*a*) $\displaystyle\int_{-\infty}^{\infty} e^{-x^2} \cos x \, dx = \frac{\pi^{1/2}}{e^{1/4}}$;

(*b*) $\displaystyle\int_{-\infty}^{\infty} e^{-x^2/2} \cos x \, dx = \sqrt{\frac{2\pi}{e}}$.

Exercise 8.10.12 *Using expectations show that*

(*a*) $\displaystyle\int_{-\infty}^{\infty} x e^{-ax^2+bx} \, dx = \sqrt{\frac{\pi}{a}}\Big(\frac{b}{2a}\Big) e^{b^2/(4a)}$;

(*b*) $\displaystyle\int_{-\infty}^{\infty} x^2 e^{-ax^2+bx} \, dx = \sqrt{\frac{\pi}{a}}\frac{1}{2a}\Big(1 + \frac{b^2}{2a}\Big) e^{b^2/(4a)}$;

(*c*) *Can you apply a similar method to find a closed form expression for the integral*

$$\int_{-\infty}^{\infty} x^n e^{-ax^2+bx} \, dx?$$

Exercise 8.10.13 *Using the result given by Example 8.10.7 show that*

(*a*) $\mathbb{E}[\cos(tW_t)] = e^{-t^3/2}$;

(*b*) $\mathbb{E}[\sin(tW_t)] = 0$;

(*c*) $\mathbb{E}[e^{tW_t}] = 0$.

For general drift rates we cannot find the mean, but in the case of concave drift rates we can find an upper bound for the expectation $\mathbb{E}[X_t]$. The following classical result will be used for this purpose.

Lemma 8.10.14 (Gronwall's inequality) *Let $f(t)$ be a non-negative continuous function satisfying the inequality*

$$f(t) \leq C + M \int_0^t f(s)\, ds$$

for $0 \leq t \leq T$, with C, M constants. Then

$$f(t) \leq Ce^{Mt}, \qquad 0 \leq t \leq T.$$

Proof: The proof follows Revuz and Yor [41]. Iterating the integral inequality one gets

$$\begin{aligned} f(t) &\leq C + M \int_0^t f(s)\, ds \\ &\leq C + M \int_0^t \Big(C + M \int_0^s f(u)\, du \Big) ds \\ &= C + MCt + M^2 \int_0^t \int_0^s f(u)\, du\, ds \\ &= C + MCt + M^2 t \int_0^t f(u)\, du. \end{aligned}$$

Working inductively, we obtain the following inequality

$$\begin{aligned} f(t) &\leq C + MCt + M^2 C \frac{t^2}{2} + \cdots + M^n C \frac{t^n}{n!} \\ &\quad + \frac{M^{n+1} t^n}{n!} \int_0^t f(u)\, du. \end{aligned} \tag{8.10.25}$$

The last term tends to 0 as $n \to \infty$, since

$$0 \leq \int_0^t f(u)\, du \leq t \max_{0 \leq u \leq t} f(u),$$

$$\lim_{n \to \infty} \frac{M^{n+1} t^n}{n!} = 0.$$

Taking the limit in (8.10.25) it is not hard to obtain

$$f(t) \;\leq\; C \sum_{k=0}^{\infty} \frac{M^n t^n}{n!} = Ce^{Mt}.$$

■

Proposition 8.10.15 *Let X_t be a continuous stochastic process such that*

$$dX_t = a(X_t)dt + b(t, W_t, X_t)\, dW_t,$$

with the function $a(\cdot)$ satisfying the following conditions

1. $a(x) \geq 0$, *for* $0 \leq x \leq T$;
2. $a''(x) < 0$, *for* $0 \leq x \leq T$;
3. $a'(0) = M$.

Then $\mathbb{E}[X_t] \leq X_0 e^{Mt}$, *for* $0 \leq X_t \leq T$.

Proof: From the mean value theorem there is $\xi \in (0, x)$ such that

$$a(x) = a(x) - a(0) = (x - 0)a'(\xi) \leq xa'(0) = Mx, \tag{8.10.26}$$

where we used that $a'(x)$ is a decreasing function. Applying Jensen's inequality for concave functions yields

$$\mathbb{E}[a(X_t)] \leq a(\mathbb{E}[X_t]).$$

Combining with (8.10.26) we obtain $\mathbb{E}[a(X_t)] \leq M\mathbb{E}[X_t]$. Substituting in the identity (8.10.23) implies

$$\mathbb{E}[X_t] \leq X_0 + M \int_0^t \mathbb{E}[X_s]\, ds.$$

Applying Gronwall's inequality we obtain $\mathbb{E}[X_t] \leq X_0 e^{Mt}$. ∎

Exercise 8.10.16 *State the previous result in the particular case when* $a(x) = \sin x$, *with* $0 \leq x \leq \pi$.

Not in all cases can the mean and the variance be obtained directly from the stochastic equation. In these cases one may try to produce closed form solutions. Some of these techniques were developed in the previous sections of the current chapter.

Chapter 9

Applications of Brownian Motion

This chapter deals with a surprising relation between stochastic differential equations and second order partial differential equations. This will provide a way of computing solutions of parabolic differential equations, which is a deterministic problem, by means of studying the transition probability density of the underlying stochastic process.

9.1 The Directional Derivative

Consider a smooth curve $x : [0, \infty) \to \mathbb{R}^n$, starting at $x(0) = x_0$ with the initial velocity $v = x'(0)$. Then the derivative of a function $f : \mathbb{R}^n \to \mathbb{R}$ in the direction v is defined by

$$D_v f(x_0) = \lim_{t \searrow 0} \frac{f\big(x(t)\big) - f(x)}{t} = \frac{d}{dt} f\big(x(t)\big)\Big|_{t=0+}.$$

Applying the chain rule yields

$$D_v f(x_0) = \sum_{k=1}^{n} \frac{\partial f}{\partial x_k}(x(t)) \frac{dx_k(t)}{dt}\Big|_{t=0+} = \sum_{k=1}^{n} \frac{\partial f}{\partial x_k}(x_0) v_k(0) = \langle \nabla f(x_0), v \rangle,$$

where ∇f stands for the gradient of f and $\langle \, , \rangle$ denotes the scalar product. The linear differential operator D_v is called *the directional derivative* with respect to the vector v. In the next section we shall extend this definition to the case when the curve $x(t)$ is replaced by an Ito diffusion X_t; in this case the corresponding "directional derivative" will be a second order partial differential operator.

9.2 The Generator of an Ito Diffusion

Let $(X_t)_{t\geq 0}$ be a stochastic process with $X_0 = x_0$. We shall consider an operator that describes infinitesimally the rate of change of a function which depends smoothly on X_t.

More precisely, the *generator* of the stochastic process X_t is the second order partial differential operator A defined by

$$Af(x) = \lim_{t\searrow 0} \frac{\mathbb{E}^x[f(X_t)] - f(x)}{t},$$

for any smooth function (at least of class C^2) with compact support, i.e. $f : \mathbb{R}^n \to \mathbb{R}$, $f \in C_0^2(\mathbb{R}^n)$. Here $\mathbb{E}^x$ stands for the expectation operator given the initial condition $X_0 = x$, i.e.,

$$\mathbb{E}^x[f(X_t)] = \mathbb{E}[f(X_t)|X_0 = x] = \int_{\mathbb{R}^n} f(y)p_t(x, y)\,dy,$$

where $p_t(x, y) = p(x, y; t, 0)$ is the transition density of X_t, given $X_0 = x$ (the initial value X_0 is a deterministic value x).

In the following we shall find the generator associated with the Ito diffusion

$$dX_t = b(X_t)dt + \sigma(X_t)dW(t), \qquad t \geq 0, X_0 = x, \tag{9.2.1}$$

where $W(t) = \big(W_1(t), \ldots, W_m(t)\big)$ is an m-dimensional Brownian motion, with $b : \mathbb{R}^n \to \mathbb{R}^n$ and $\sigma : \mathbb{R}^n \to \mathbb{R}^{n\times m}$ measurable functions.

The main tool used in deriving the formula for the generator A is Ito's formula in several variables. If $F_t = f(X_t)$, then using Ito's formula we have

$$dF_t = \sum_i \frac{\partial f}{\partial x_i}(X_t)\,dX_t^i + \frac{1}{2}\sum_{i,j} \frac{\partial^2 f}{\partial x_i \partial x_j}(X_t)\,dX_t^i\,dX_t^j, \tag{9.2.2}$$

where $X_t = (X_t^1, \cdots, X_t^n)$ satisfies the Ito diffusion (9.2.1) on components, i.e.,

$$\begin{aligned} dX_t^i &= b_i(X_t)dt + [\sigma(X_t)\,dW(t)]_i \\ &= b_i(X_t)dt + \sum_k \sigma_{ik}\,dW_k(t). \end{aligned} \tag{9.2.3}$$

Using the stochastic relations $dt^2 = dt\,dW_k(t) = 0$ and $dW_k(t)\,dW_r(t) = \delta_{kr}dt$, a computation provides

$$\begin{aligned}
dX_t^i\,dX_t^j &= \Big(b_i dt + \sum_k \sigma_{ik} dW_k(t)\Big)\Big(b_j dt + \sum_k \sigma_{jk} dW_k(t)\Big) \\
&= \Big(\sum_k \sigma_{ik} dW_k(t)\Big)\Big(\sum_r \sigma_{jr} dW_r(t)\Big) \\
&= \sum_{k,r} \sigma_{ik}\sigma_{jr}\, dW_k(t) dW_r(t) = \sum_k \sigma_{ik}\sigma_{jk}\, dt \\
&= (\sigma\sigma^T)_{ij}\, dt.
\end{aligned}$$

Therefore

$$dX_t^i\,dX_t^j = (\sigma\sigma^T)_{ij}\, dt. \tag{9.2.4}$$

Substituting (9.2.3) and (9.2.4) into (9.2.2) yields

$$\begin{aligned}
dF_t &= \Big[\frac{1}{2}\sum_{i,j} \frac{\partial^2 f}{\partial x_i \partial x_j}(X_t)(\sigma\sigma^T)_{ij} + \sum_i b_i(X_t)\frac{\partial f}{\partial x_i}(X_t)\Big] dt \\
&\quad + \sum_{i,k} \frac{\partial f}{\partial x_i}(X_t)\sigma_{ik}(X_t)\, dW_k(t).
\end{aligned}$$

Integrate and obtain

$$\begin{aligned}
F_t &= F_0 + \int_0^t \Big[\frac{1}{2}\sum_{i,j} \frac{\partial^2 f}{\partial x_i \partial x_j}(\sigma\sigma^T)_{ij} + \sum_i b_i \frac{\partial f}{\partial x_i}\Big](X_s)\, ds \\
&\quad + \sum_k \int_0^t \sum_i \sigma_{ik} \frac{\partial f}{\partial x_i}(X_s)\, dW_k(s).
\end{aligned}$$

Since $F_0 = f(X_0) = f(x)$ and $\mathbb{E}^x(f(x)) = f(x)$, applying the expectation operator in the previous relation we obtain

$$\mathbb{E}^x[F_t] = f(x) + \mathbb{E}^x\Big[\int_0^t \Big(\frac{1}{2}\sum_{i,j}(\sigma\sigma^T)_{ij}\frac{\partial^2 f}{\partial x_i \partial x_j} + \sum_i b_i \frac{\partial f}{\partial x_i}\Big)(X_s)\, ds\Big]. \tag{9.2.5}$$

Using the commutativity between the operator $\mathbb{E}^x$ and the integral $\int_0^t$, applying l'Hospital rule (see Exercise 9.2.7), yields

$$\lim_{t\searrow 0} \frac{\mathbb{E}^x[F_t] - f(x)}{t} = \frac{1}{2}\sum_{i,j}(\sigma\sigma^T)_{ij}\frac{\partial^2 f(x)}{\partial x_i \partial x_j} + \sum_k b_k \frac{\partial f(x)}{\partial x_k}.$$

We conclude the previous computations with the following result.

Theorem 9.2.1 *The generator of the Ito diffusion (9.2.1) is given by*

$$A = \frac{1}{2}\sum_{i,j}(\sigma\sigma^T)_{ij}\frac{\partial^2}{\partial x_i \partial x_j} + \sum_k b_k \frac{\partial}{\partial x_k}. \tag{9.2.6}$$

The matrix σ is called *dispersion* and the product $\sigma\sigma^T$ is called *diffusion* matrix. These names are related with their physical significance. Substituting (9.2.6) in (9.2.5) we obtain the following formula

$$\mathbb{E}^x[f(X_t)] = f(x) + \mathbb{E}^x\Big[\int_0^t Af(X_s)\,ds\Big], \tag{9.2.7}$$

for any $f \in C_0^2(\mathbb{R}^n)$.

Exercise 9.2.2 *Find the generator operator associated with the n-dimensional Brownian motion.*

Exercise 9.2.3 *Find the Ito diffusion corresponding to the generator $Af(x) = f''(x) + f'(x)$.*

Exercise 9.2.4 *Let $\Delta_G = \frac{1}{2}(\partial_{x_1}^2 + x_1^2\partial_{x_2}^2)$ be the Grushin's operator.*

(a) Find the diffusion process associated with the generator Δ_G.

(b) Find the diffusion and dispersion matrices and show that they are degenerate.

Exercise 9.2.5 *Let X_t and Y_t be two one-dimensional independent Ito diffusions with infinitesimal generators A_X and A_Y. Let $Z_t = (X_t, Y_t)$ with the infinitesimal generator A_Z. Show that $A_Z = A_X + A_Y$.*

Exercise 9.2.6 *Let X_t be an Ito diffusion with infinitesimal generator A_X. Consider the process $Y_t = (t, X_t)$. Show that the infinitesimal generator of Y_t is given by $A_Y = \partial_t + A_X$.*

Exercise 9.2.7 *Let X_t be an Ito diffusion with $X_0 = x$, and φ a smooth function. Using l'Hospital rule, show that*

$$\lim_{t\to 0+}\frac{1}{t}\mathbb{E}^x\Big[\int_0^t \varphi(X_s)\,ds\Big] = \varphi(x).$$

9.3 Dynkin's Formula

Formula (9.2.7) holds under more general conditions, when t is a stopping time. First we need the following result, which deals with a continuity-type property in the upper limit of an Ito integral.

Lemma 9.3.1 *Let g be a bounded measurable function and τ be a stopping time for X_t with $\mathbb{E}[\tau] < \infty$. Then*

$$\lim_{k\to\infty} \mathbb{E}\Big[\int_0^{\tau\wedge k} g(X_s)\,dW_s\Big] = \mathbb{E}\Big[\int_0^{\tau} g(X_s)\,dW_s\Big]; \tag{9.3.8}$$

$$\lim_{k\to\infty} \mathbb{E}\Big[\int_0^{\tau\wedge k} g(X_s)\,ds\Big] = \mathbb{E}\Big[\int_0^{\tau} g(X_s)\,ds\Big]. \tag{9.3.9}$$

Proof: Let $|g| < K$. Using the properties of Ito integrals, we have

$$\mathbb{E}\Big[\Big(\int_0^{\tau} g(X_s)\,dW_s - \int_0^{\tau\wedge k} g(X_t)\,dW_s\Big)^2\Big] = \mathbb{E}\Big[\Big(\int_{\tau\wedge k}^{\tau} g(X_s)\,dW_s\Big)^2\Big]$$
$$= \mathbb{E}\Big[\int_{\tau\wedge k}^{\tau} g^2(X_s)\,ds\Big] \le K^2\mathbb{E}[\tau - \tau\wedge k] \to 0, \qquad k\to\infty.$$

Since $\mathbb{E}[X^2] \le \mathbb{E}[X]^2$, it follows that

$$\mathbb{E}\Big[\int_0^{\tau} g(X_s)\,dW_s - \int_0^{\tau\wedge k} g(X_t)\,dW_s\Big] \to 0, \qquad k\to\infty,$$

which is equivalent to relation (9.3.8).

The second relation can be proved similarly and is left as an exercise for the reader. ■

Exercise 9.3.2 *Assume the hypothesis of the previous lemma. Let $\mathbf{1}_{\{s<\tau\}}$ be the characteristic function of the interval $(-\infty,\tau)$*

$$\mathbf{1}_{\{s<\tau\}}(u) = \begin{cases} 1, & \textit{if}\, u<\tau \\ 0, & \textit{otherwise.}\end{cases}$$

Show that

(a) $\displaystyle\int_0^{\tau\wedge k} g(X_s)\,dW_s = \int_0^{k} \mathbf{1}_{\{s<\tau\}}\, g(X_s)\,dW_s,$

(b) $\displaystyle\int_0^{\tau\wedge k} g(X_s)\,ds = \int_0^{k} \mathbf{1}_{\{s<\tau\}}\, g(X_s)\,ds.$

Theorem 9.3.3 (Dynkin's formula) *Let $f \in C_0^2(\mathbb{R}^n)$, and X_t be an Ito diffusion starting at x. If τ is a stopping time with $\mathbb{E}[\tau] < \infty$, then*

$$\mathbb{E}^x[f(X_\tau)] = f(x) + \mathbb{E}^x\Big[\int_0^\tau Af(X_s)\,ds\Big], \tag{9.3.10}$$

where A is the infinitesimal generator of X_t.

Proof: Replace t by k and f by $\mathbf{1}_{\{s<\tau\}}f$ in (9.2.7) and obtain

$$\mathbb{E}[1_{s<\tau}f(X_k)] = \mathbf{1}_{\{s<\tau\}}f(x) + \mathbb{E}\Big[\int_0^k A(\mathbf{1}_{\{s<\tau\}}f)(X_s)\,ds\Big],$$

which can be written as

$$\begin{aligned}\mathbb{E}[f(X_{k\wedge\tau})] &= \mathbf{1}_{\{s<\tau\}}f(x) + \mathbb{E}\Big[\int_0^k \mathbf{1}_{\{s<\tau\}}(s)A(f)(X_s)\,ds\Big] \\ &= \mathbf{1}_{\{s<\tau\}}f(x) + \mathbb{E}\Big[\int_0^{k\wedge\tau} A(f)(X_s)\,ds\Big]. \end{aligned} \tag{9.3.11}$$

Since by Lemma 9.3.1

$$\mathbb{E}[f(X_{k\wedge\tau})] \to \mathbb{E}[f(X_\tau)], \qquad k \to \infty$$

$$\mathbb{E}\Big[\int_0^{k\wedge\tau} A(f)(X_s)\,ds\Big] \to \mathbb{E}\Big[\int_0^\tau A(f)(X_s)\,ds\Big], \qquad k \to \infty,$$

using Exercise 9.3.2 and relation (9.3.11) yields (9.3.10). ■

Exercise 9.3.4 *Write Dynkin's formula for the case of a function $f(t, X_t)$. Use Exercise 9.2.6.*

More details in this direction can be found in Dynkin [16]. In the following sections we shall present a few important results of stochastic calculus that can be obtained as direct consequences of Dynkin's formula.

9.4 Kolmogorov's Backward Equation

For any function $f \in C_0^2(\mathbb{R}^n)$ let $v(t, x) = \mathbb{E}^x[f(X_t)]$, given that $X_0 = x$. The operator E denotes the expectation given the initial condition $X_0 = x$. Then $v(0, x) = f(x)$, and differentiating in Dynkin's formula (9.2.7)

$$v(t, x) = f(x) + \int_0^t \mathbb{E}^x[Af(X_s)]\,ds$$

provides

$$\frac{\partial v}{\partial t} = \mathbb{E}^x[Af(X_t)].$$

Since we are allowed to differentiate inside of an integral, the operators A and $\mathbb{E}^x$ commute

$$\mathbb{E}^x[Af(X_t)] = A\mathbb{E}^x[f(X_t)].$$

Therefore

$$\frac{\partial v}{\partial t} = \mathbb{E}^x[Af(X_t)] = A\mathbb{E}^x[f(X_t)] = Av(t,x).$$

Hence, we arrive at the following result.

Theorem 9.4.1 (Kolmogorov's backward equation) *For any $f \in C_0^2(\mathbb{R}^n)$ the function $v(t,x) = \mathbb{E}^x[f(X_t)]$ satisfies the following Cauchy's problem*

$$\begin{aligned} \frac{\partial v}{\partial t} &= Av, \qquad t > 0 \\ v(0,x) &= f(x), \end{aligned}$$

where A denotes the generator of the Ito's diffusion (9.2.1).

Solving Kolmogorov's backward equation is a problem of partial differential equations. The reader interested in several methods for solving this equation can consult the book of Calin et al. [10].

9.5 Exit Time from an Interval

Let $X_t = x_0 + W_t$ be a one-dimensional Brownian motion starting at x_0, with $x_0 \in (a,b)$. Consider the exit time of the process X_t from the strip (a,b)

$$\tau = \inf\{t > 0; X_t \notin (a,b)\}.$$

Assuming $\mathbb{E}[\tau] < 0$, applying Dynkin's formula yields

$$\mathbb{E}\Big[f(X_\tau)\Big] = f(x_0) + \mathbb{E}\Big[\int_0^\tau \frac{1}{2}\frac{d^2}{dx^2}f(X_s)\,ds\Big]. \tag{9.5.12}$$

Choosing $f(x) = x$ in (9.5.12) we obtain

$$\mathbb{E}[X_\tau] = x_0. \tag{9.5.13}$$

Exercise 9.5.1 *Prove relation (9.5.13) using the Optional Stopping Theorem for the martingale X_t.*

Let $p_a = P(X_\tau = a)$ and $p_b = P(X_\tau = b)$ be the exit probabilities of X_t from the interval (a, b). Obviously, $p_a + p_b = 1$, since the probability that the Brownian motion will stay forever inside the bounded interval is zero. Using the expectation definition, relation (9.5.13) yields

$$ap_a + b(1 - p_a) = x_0.$$

Solving for p_a and p_b we get the following exit probabilities

$$p_a = \frac{b - x_0}{b - a} \tag{9.5.14}$$

$$p_b = 1 - p_a = \frac{x_0 - a}{b - a}. \tag{9.5.15}$$

It is worth noting that if $b \to \infty$ then $p_a \to 1$ and if $a \to -\infty$ then $p_b \to 1$. This can be stated by saying that a Brownian motion starting at x_0 reaches any level (below or above x_0) with probability 1.

Next we shall compute the mean of the exit time, $\mathbb{E}[\tau]$. Choosing $f(x) = x^2$ in (9.5.12) yields

$$\mathbb{E}[(X_\tau)^2] = x_0^2 + \mathbb{E}[\tau].$$

From the definition of the mean and formulas (9.5.14)-(9.5.15) we obtain

$$\begin{aligned}
\mathbb{E}[\tau] &= a^2 p_a + b^2 p_b - x_0^2 = a^2 \frac{b - x_0}{b - a} + b^2 \frac{x_0 - a}{b - a} - x_0 \\
&= \frac{1}{b - a}\Big[ba^2 - ab^2 + x_0(b - a)(b + a)\Big] - x_0^2 \\
&= -ab + x_0(b + a) - x_0^2 \\
&= (b - x_0)(x_0 - a). \qquad (9.5.16)
\end{aligned}$$

Exercise 9.5.2 (*a*) *Show that the equation* $x^2 - (b - a)x + \mathbb{E}[\tau] = 0$ *cannot have complex roots;*

(*b*) *Prove that* $\mathbb{E}[\tau] \leq \dfrac{(b - a)^2}{4}$;

(*c*) *Find the point* $x_0 \in (a, b)$ *such that the expectation of the exit time,* $\mathbb{E}[\tau]$, *is maximum.*

9.6 Transience and Recurrence of Brownian Motion

We shall consider first the expectation of the exit time from a ball. Then we shall extend it to an annulus and compute the transience probabilities.

1. Consider the process $X_t = a + W(t)$, where $W(t) = \big(W_1(t), \dots, W_n(t)\big)$ is an n-dimensional Brownian motion, and $a = (a_1, \dots, a_n) \in \mathbb{R}^n$ is a fixed

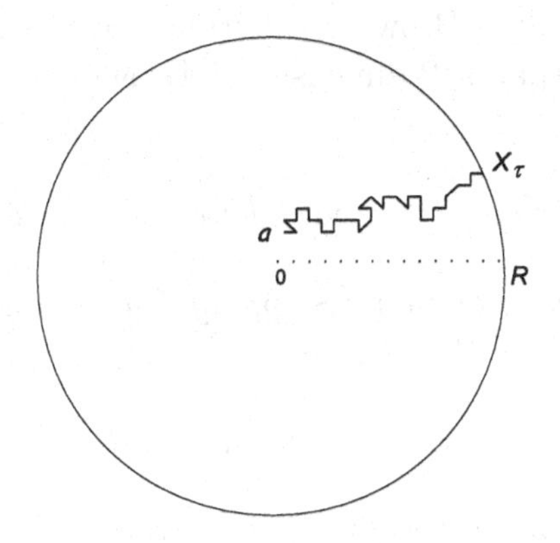

Figure 9.1: *The Brownian motion X_t in the ball $B(0, R)$.*

vector, see Fig. 9.1. Let $R > 0$ be such that $R > |a|$. Consider the exit time of the process X_t from the ball $B(0, R)$

$$\tau = \inf\{t > 0; |X_t| \geq R\}. \tag{9.6.17}$$

Assuming $\mathbb{E}[\tau] < \infty$ and letting $f(x) = |x|^2 = x_1^2 + \cdots + x_n^2$ in Dynkin's formula

$$\mathbb{E}\big[f(X_\tau)\big] = f(x) + \mathbb{E}\Big[\int_0^\tau \frac{1}{2}\Delta f(X_s)\,ds\Big]$$

yields

$$R^2 = |a|^2 + \mathbb{E}\Big[\int_0^\tau n\,ds\Big],$$

and hence

$$\mathbb{E}[\tau] = \frac{R^2 - |a|^2}{n}. \tag{9.6.18}$$

In particular, if the Brownian motion starts from the center, i.e. $a = 0$, the expectation of the exit time is

$$\mathbb{E}[\tau] = \frac{R^2}{n}.$$

We make a few remarks:
(*i*) Since $R^2/2 > R^2/3$, the previous relation implies that it takes longer for a Brownian motion to exit a disk of radius R rather than a ball of the same radius.

(*ii*) The probability that a Brownian motion leaves the interval $(-R, R)$ is twice the probability that a 2-dimensional Brownian motion exits the disk $B(0, R)$.

Exercise 9.6.1 *Prove that* $\mathbb{E}[\tau] < \infty$, *where* τ *is given by (9.6.17).*

Exercise 9.6.2 *Apply the Optional Stopping Theorem for the martingale* $W_t = W_t^2 - t$ *to show that* $\mathbb{E}[\tau] = R^2$, *where*

$$\tau = \inf\{t > 0; |W_t| \geq R\}$$

is the first exit time of the Brownian motion from $(-R, R)$.

2. Let $b \in \mathbb{R}^n$ such that $b \notin B(0, R)$, i.e. $|b| \geq R$, and consider the annulus

$$A_k = \{x; R < |x| < kR\}$$

where $k > 0$ such that $b \in A_k$. Consider the process $X_t = b + W(t)$ and let

$$\tau_k = \inf\{t > 0; X_t \notin A_k\}$$

be the first exit time of X_t from the annulus A_k. Let $f : A_k \to \mathbb{R}$ be defined by

$$f(x) = \begin{cases} -\ln |x|, & \text{if } n = 2 \\ \frac{1}{|x|^{n-2}}, & \text{if } n > 2. \end{cases}$$

A straightforward computation shows that $\Delta f = 0$. Substituting into Dynkin's formula

$$\mathbb{E}\big[f(X_{\tau_k})\big] = f(b) + \mathbb{E}\Big[\int_0^{\tau_k} \Big(\frac{1}{2}\Delta f\Big)(X_s)\, ds\Big]$$

yields

$$\mathbb{E}\big[f(X_{\tau_k})\big] = f(b). \qquad (9.6.19)$$

This can be stated by saying that the value of f at a point b in the annulus is equal to the expected value of f at the first exit time of a Brownian motion starting at b.

Since $|X_{\tau_k}|$ is a random variable with two outcomes, we have

$$\mathbb{E}\big[f(X_{\tau_k})\big] = p_k f(R) + q_k f(kR),$$

where $p_k = P(|X_{\tau_k}| = R)$, $q_k = P(|X_{X_{\tau_k}}|) = kR$ and $p_k + q_k = 1$. Substituting in (9.6.19) yields

$$p_k f(R) + q_k f(kR) = f(b). \qquad (9.6.20)$$

There are two distinguished cases:

(*i*) If $n = 2$ we obtain

$$-p_k \ln R - q_k(\ln k + \ln R) = -\ln b.$$

Using $p_k = 1 - q_k$, solving for p_k yields

$$p_k = 1 - \frac{\ln(\frac{b}{R})}{\ln k}.$$

Hence

$$P(\tau < \infty) = \lim_{k\to\infty} p_k = 1,$$

where $\tau = \inf\{t > 0; |X_t| \leq R\}$ is the first time X_t hits the ball $B(0, R)$. Hence in $\mathbb{R}^2$ a Brownian motion hits with probability 1 any ball. This is stated equivalently by saying that the Brownian motion is *recurrent* in $\mathbb{R}^2$.

(*ii*) If $n > 2$ the equation (9.6.20) becomes

$$\frac{p_k}{R^{n-2}} + \frac{q_k}{k^{n-2}R^{n-2}} = \frac{1}{b^{n-2}}.$$

Taking the limit $k \to \infty$ yields

$$\lim_{k\to\infty} p_k = \left(\frac{R}{b}\right)^{n-2} < 1.$$

Then in $\mathbb{R}^n$, $n > 2$, a Brownian motion starting outside of a ball hits it with a probability less than 1. This is usually stated by saying that the Brownian motion is *transient.*

3. We shall recover the previous results using the n-dimensional Bessel process

$$\mathcal{R}_t = dist\big(0, W(t)\big) = \sqrt{W_1(t)^2 + \cdots + W_n(t)^2}.$$

Consider the process $Y_t = \alpha + \mathcal{R}_t$, with $0 \leq \alpha < R$, see section 3.7. It can be shown that the generator of Y_t is the Bessel operator of order n, see Example 10.2.4

$$A = \frac{1}{2}\frac{d^2}{dx^2} + \frac{n-1}{2x}\frac{d}{dx}.$$

Consider the exit time

$$\tau = \{t > 0; Y_t \geq R\}.$$

Applying Dynkin's formula

$$\mathbb{E}\big[f(Y_\tau)\big] = f(Y_0) + \mathbb{E}\Big[\int_0^\tau (Af)(Y_s)\, ds\Big]$$

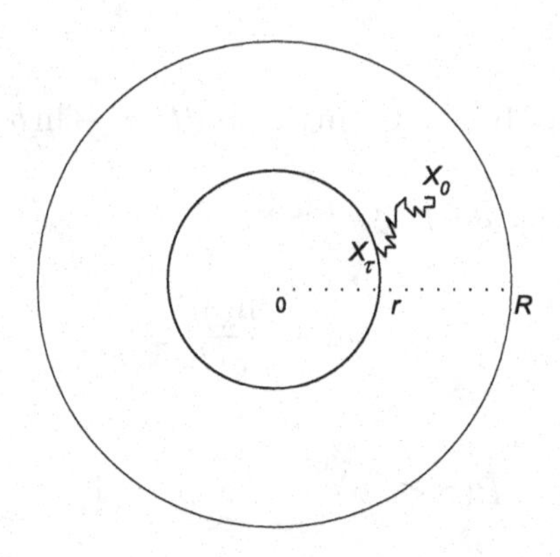

Figure 9.2: *The Brownian motion X_t in the annulus $A_{r,R}$.*

for $f(x) = x^2$ yields $R^2 = \alpha^2 + \mathbb{E}\big[\int_0^\tau n\,ds\big]$. This leads to

$$\mathbb{E}[\tau] = \frac{R^2 - \alpha^2}{n}$$

which recovers (9.6.18) with $\alpha = |a|$.

In the following assume $n \geq 3$ and consider the annulus

$$A_{r,R} = \{x \in \mathbb{R}^n; r < |x| < R\}.$$

Consider the stopping time $\tau = \inf\{t > 0; X_t \notin A_{r,R}\} = \inf\{t > 0; Y_t \notin (r, R)\}$, where $|Y_0| = \alpha \in (r, R)$. Applying Dynkin's formula for $f(x) = x^{2-n}$ yields $E\big[f(Y_\tau\big] = f(\alpha)$. This can be written as

$$p_r r^{2-n} + p_R R^{2-n} = \alpha^{2-n},$$

where

$$p_r = P(|X_t| = r), \qquad p_R = P(|X_t| = R), \qquad p_r + p_R = 1.$$

Solving for p_r and p_R yields

$$p_r = \frac{\left(\frac{R}{\alpha}\right)^{n-2} - 1}{\left(\frac{R}{r}\right)^{n-2} - 1}, \qquad p_R = \frac{\left(\frac{r}{\alpha}\right)^{n-2} - 1}{\left(\frac{r}{R}\right)^{n-2} - 1}.$$

The transience probability is obtained by taking the limit to infinity

$$p_r = \lim_{R\to\infty} p_{r,R} = \lim_{R\to\infty} \frac{\alpha^{2-n}R^{n-2}-1}{r^{2-n}R^{n-2}-1} = \Big(\frac{r}{\alpha}\Big)^{n-2},$$

where p_r is the probability that a Brownian motion starting outside the ball of radius r will hit the ball, see Fig. 9.2.

Exercise 9.6.3 *Solve the equation* $\frac{1}{2}f''(x) + \frac{n-1}{2x}f'(x) = 0$ *by looking for a solution of monomial type* $f(x) = x^k$.

9.7 Application to Parabolic Equations

This section deals with solving first and second order parabolic equations using the integral of the cost function along a certain *characteristic* solution. The first order equations are related to predictable characteristic curves, while the second order equations depend on stochastic characteristic curves.

9.7.1 Deterministic characteristics

Let $\varphi(s)$ be the solution of the following one-dimensional ODE

$$\begin{aligned} \frac{dX(s)}{ds} &= a\big(s, X(s)\big), \qquad t \le s \le T \\ X(t) &= x, \end{aligned}$$

and define the *cumulative cost* between t and T along the solution φ

$$u(t,x) = \int_t^T c\big(s,\varphi(s)\big)\,ds, \tag{9.7.21}$$

where c denotes a continuous *cost function.* Differentiate both sides with respect to t

$$\begin{aligned} \frac{\partial}{\partial t}u\big(t,\varphi(t)\big) &= \frac{\partial}{\partial t}\int_t^T c\big(s,\varphi(s)\big)\,ds \\ \partial_t u + \partial_x u\,\varphi'(t) &= -c\big(t,\varphi(t)\big). \end{aligned}$$

Hence (9.7.21) is a solution of the following final value problem

$$\begin{aligned} \partial_t u(t,x) + a(t,x)\partial_x u(t,x) &= -c(t,x) \\ u(T,x) &= 0. \end{aligned}$$

It is worth mentioning that this is a variant of the method of characteristics.[1] The curve given by the solution $\varphi(s)$ is called a *characteristic curve.*

Exercise 9.7.1 *Using the previous method solve the following final boundary problems:*

(a)

$$\begin{aligned} \partial_t u + x\partial_x u &= -x \\ u(T,x) &= 0. \end{aligned}$$

(b)

$$\begin{aligned} \partial_t u + tx\partial_x u &= \ln x, \qquad x > 0 \\ u(T,x) &= 0. \end{aligned}$$

9.7.2 Stochastic characteristics

Consider the Ito diffusion

$$\begin{aligned} dX_s &= a(s,X_s)ds + b(s,X_s)dW_s, \qquad t \le s \le T \\ X_t &= x, \end{aligned}$$

and define the stochastic cumulative cost function

$$u(t,X_t) = \int_t^T c(s,X_s)\,ds, \tag{9.7.22}$$

with the conditional expectation

$$\begin{aligned} u(t,x) &= \mathbb{E}\Big[u(t,X_t)|X_t = x\Big] \\ &= \mathbb{E}\Big[\int_t^T c(s,X_s)\,ds|X_t = x\Big]. \end{aligned}$$

Taking increments on both sides of (9.7.22) yields

$$du(t,X_t) = d\int_t^T c(s,X_s)\,ds.$$

Applying Ito's formula on one side and the Fundamental Theorem of Calculus on the other, we obtain

$$\partial_t u(t,x)dt + \partial_x u(t,X_t)dX_t + \frac{1}{2}\partial_x^2 u(t,t,X_t)dX_t^2 = -c(t,X_t)dt.$$

[1]This is a well known method of solving linear partial differential equations.

Taking the expectation $\mathbb{E}[\,\cdot\,|X_t = x]$ on both sides yields

$$\partial_t u(t,x)dt + \partial_x u(t,x)a(t,x)dt + \frac{1}{2}\partial_x^2 u(t,x)b^2(t,x)dt = -c(t,x)dt.$$

Hence, the expected cost

$$u(t,x) = \mathbb{E}\Big[\int_t^T c(s,X_s)\,ds|X_t = x\Big]$$

is a solution of the following second order parabolic equation

$$\begin{aligned} \partial_t u + a(t,x)\partial_x u + \frac{1}{2}b^2(t,x)\partial_x^2 u(t,x) &= -c(t,x) \\ u(T,x) &= 0. \end{aligned}$$

This represents the probabilistic interpretation of the solution of a parabolic equation.

Exercise 9.7.2 *Solve the following final boundary problems:*

(a)

$$\begin{aligned} \partial_t u + \partial_x u + \frac{1}{2}\partial_x^2 u &= -x \\ u(T,x) &= 0. \end{aligned}$$

(b)

$$\begin{aligned} \partial_t u + \partial_x u + \frac{1}{2}\partial_x^2 u &= e^x, \\ u(T,x) &= 0. \end{aligned}$$

(c)

$$\begin{aligned} \partial_t u + \mu x\partial_x u + \frac{1}{2}\sigma^2 x^2\partial_x^2 u &= -x, \\ u(T,x) &= 0. \end{aligned}$$

Chapter 10

Girsanov's Theorem and Brownian Motion

After setting the basis in martingales, we shall prove Girsanov's theorem, which is the main tool used in practice to eliminate drift. Then we present Lévy's theorem with applications as well as as the time change for Brownian motions.

10.1 Examples of Martingales

In this section we shall use the knowledge acquired in the previous chapters to present a few important examples of martingales and some of their particular cases. Some of these results will be useful later in the proof of Girsanov's theorem.

Example 10.1.1 *If $v(s)$ is a continuous function on $[0, T]$, then*

$$\boxed{X_t = \int_0^t v(s)\, dW_s}$$

is an $\mathcal{F}_t$-martingale.

The integrability of X_t follows from

$$\begin{aligned} \mathbb{E}[|X_t|]^2 &\leq \mathbb{E}[X_t^2] = \mathbb{E}\Big[\Big(\int_0^t v(s)\, dW_s\Big)^2\Big] \\ &= \mathbb{E}\Big[\int_0^t v(s)^2\, ds\Big] = \int_0^t v(s)^2\, ds < \infty. \end{aligned}$$

We note the continuity of $v(s)$ can be replaced by the weaker condition $v \in L^2[0, T]$.

X_t is obviously adapted to the information set $\mathcal{F}_t$ induced by the Brownian motion W_t. Taking out the predictable part leads to

$$\begin{aligned}\mathbb{E}[X_t|\mathcal{F}_s] &= \mathbb{E}\Big[\int_0^s v(\tau)\,dW_\tau + \int_s^t v(\tau)\,dW_\tau\Big|\mathcal{F}_s\Big] \\ &= X_s + \mathbb{E}\Big[\int_s^t v(\tau)\,dW_\tau\Big|\mathcal{F}_s\Big] = X_s,\end{aligned}$$

where we used that $\int_s^t v(\tau)\,dW_\tau$ is independent of $\mathcal{F}_s$ and the conditional expectation equals the usual expectation

$$\mathbb{E}\Big[\int_s^t v(\tau)\,dW_\tau\Big|\mathcal{F}_s\Big] = \mathbb{E}\Big[\int_s^t v(\tau)\,dW_\tau\Big] = 0.$$

Example 10.1.2 *Let* $X_t = \int_0^t v(s)\,dW_s$ *be a process as in Example 10.1.1. Then*

$$\boxed{M_t = X_t^2 - \int_0^t v^2(s)\,ds}$$

is an $\mathcal{F}_t$*-martingale.*

The process X_t satisfies the stochastic equation $dX_t = v(t)dW_t$. By Ito's formula

$$d(X_t^2) = 2X_t dX_t + (dX_t)^2 = 2v(t)X_t dW_t + v^2(t)dt. \tag{10.1.1}$$

Integrating between s and t yields

$$X_t^2 - X_s^2 = 2\int_s^t X_\tau v(\tau)\,dW_\tau + \int_s^t v^2(\tau)\,d\tau.$$

Then separating the deterministic from the random part, we have

$$\begin{aligned}\mathbb{E}[M_t|\mathcal{F}_s] &= \mathbb{E}\Big[X_t^2 - \int_0^t v^2(\tau)\,d\tau\Big|\mathcal{F}_s\Big] \\ &= \mathbb{E}\Big[X_t^2 - X_s^2 - \int_s^t v^2(\tau)\,d\tau + X_s^2 - \int_0^s v^2(\tau)\,d\tau\Big|\mathcal{F}_s\Big] \\ &= X_s^2 - \int_0^s v^2(\tau)\,d\tau + \mathbb{E}\Big[X_t^2 - X_s^2 - \int_s^t v^2(\tau)\,d\tau\Big|\mathcal{F}_s\Big] \\ &= M_s + 2\mathbb{E}\Big[\int_s^t X_\tau v(\tau)\,dW_\tau\Big|\mathcal{F}_s\Big] = M_s,\end{aligned}$$

where we used relation (10.1.1) and that $\int_s^t X_\tau v(\tau)\,dW_\tau$ is independent of the information set $\mathcal{F}_s$.

The integrability of M_t can be inferred from the following computation. Since taking the expectation in

$$X_t^2 = 2\int_0^t v(\tau)X_\tau \, dW_\tau + \int_0^t v^2(\tau)\, d\tau$$

yields

$$\mathbb{E}[X_t^2] = \int_0^t v^2(\tau)\, d\tau,$$

which leads to the following estimation

$$\mathbb{E}[|M_t|] \leq \mathbb{E}[X_t^2] + \int_0^t v^2(\tau)\, d\tau \leq 2\int_0^t v^2(\tau)\, d\tau < \infty.$$

In the following we shall mention a few particular cases.

1. If $v(s) = 1$, then $X_t = W_t$. In this case $M_t = W_t^2 - t$ is an $\mathcal{F}_t$-martingale.
2. If $v(s) = s$, then $X_t = \int_0^t s\, dW_s$, and hence

$$M_t = \Big(\int_0^t s\, dW_s\Big)^2 - \frac{t^3}{3}$$

is an $\mathcal{F}_t$-martingale.

Example 10.1.3 *Let $u : [0, T] \to \mathbb{R}$ be a continuous function. Then*

$$\boxed{M_t = e^{\int_0^t u(s)\, dW_s - \frac{1}{2}\int_0^t u^2(s)\, ds}}$$

is an $\mathcal{F}_t$-martingale for $0 \leq t \leq T$.

Using Exercise 10.1.8 we obtain $\mathbb{E}[M_t] = 1$, so M_t is integrable. Consider now the process $U_t = \displaystyle\int_0^t u(s)\, dW_s - \frac{1}{2}\int_0^t u^2(s)\, ds$. Then

$$\begin{aligned} dU_t &= u(t)dW_t - \frac{1}{2}u^2(t)dt \\ (dU_t)^2 &= u(t)dt. \end{aligned}$$

Then Ito's formula yields

$$\begin{aligned} dM_t &= d(e^{U_t}) = e^{U_t}dU_t + \frac{1}{2}e^{U_t}(dU_t)^2 \\ &= e^{U_t}\Big(u(t)dW_t - \frac{1}{2}u^2(t)dt + \frac{1}{2}u^2(t)dt\Big) \\ &= u(t)M_t dW_t. \end{aligned}$$

Integrating between s and t yields

$$M_t = M_s + \int_s^t u(\tau) M_\tau \, dW_\tau.$$

Since $\int_s^t u(\tau) M_\tau \, dW_\tau$ is independent of $\mathcal{F}_s$, then

$$\mathbb{E}\Big[\int_s^t u(\tau) M_\tau \, dW_\tau | \mathcal{F}_s\Big] = \mathbb{E}\Big[\int_s^t u(\tau) M_\tau \, dW_\tau\Big] = 0,$$

and hence

$$\mathbb{E}[M_t|\mathcal{F}_s] = \mathbb{E}[M_s + \int_s^t u(\tau) M_\tau \, dW_\tau | \mathcal{F}_s] = M_s.$$

Remark 10.1.4 The condition that $u(s)$ is continuous on $[0, T]$ can be relaxed by asking only

$$u \in L^2[0, T] = \{u : [0, T] \to \mathbb{R}; \text{measurable and } \int_0^t |u(s)|^2 \, ds < \infty\}.$$

It is worth noting that the conclusion still holds if the function $u(s)$ is replaced by a stochastic process $u(t, \omega)$ satisfying Novikov's condition

$$\mathbb{E}\big[e^{\frac{1}{2}\int_0^T u^2(s,\omega)\, ds}\big] < \infty.$$

The previous process has a distinguished importance in the theory of martingales and will be useful in the proof of Girsanov theorem.

Definition 10.1.5 *Let $u \in L^2[0, T]$ be a deterministic function. Then the stochastic process*

$$M_t = e^{\int_0^t u(s)\, dW_s - \frac{1}{2}\int_0^t u^2(s)\, ds}$$

is called the exponential process induced by u.

Particular cases of exponential processes In the following we shall consider a few cases of particular interest:

1. Let $u(s) = \sigma$, constant, then $M_t = e^{\sigma W_t - \frac{\sigma^2}{2} t}$ is an $\mathcal{F}_t$-martingale.
2. Let $u(s) = s$. Integrating in $d(tW_t) = t dW_t - W_t dt$ yields

$$\int_0^t s \, dW_s = tW_t - \int_0^t W_s \, ds.$$

Let $Z_t = \int_0^t W_s \, ds$ be the integrated Brownian motion. Then

$$\begin{aligned} M_t &= e^{\int_0^t s\, dW_s - \frac{1}{2}\int_0^t s^2\, ds} \\ &= e^{tW_t - \frac{t^3}{6} - Z_t} \end{aligned}$$

is an $\mathcal{F}_t$-martingale.

Example 10.1.6 *Let X_t be a solution of $dX_t = u(t)dt + dW_t$, with $u(s)$ bounded function. Consider the exponential process*

$$M_t = e^{-\int_0^t u(s)\,dW_s - \frac{1}{2}\int_0^t u^2(s)\,ds}. \tag{10.1.2}$$

Then $Y_t = M_t X_t$ is an $\mathcal{F}_t$-martingale.

Applying Ito's formula we obtain $dM_t = -u(t)M_t dW_t$. Then

$$dM_t\,dX_t = -u(t)M_t dt.$$

The product rule yields

$$\begin{aligned} dY_t &= M_t dX_t + X_t dM_t + dM_t\,dX_t \\ &= M_t\big(u(t)dt + dW_t\big) - X_t u(t)M_t dW_t - u(t)M_t dt \\ &= M_t\big(1 - u(t)X_t\big)dW_t. \end{aligned}$$

Integrating between s and t leads to

$$Y_t = Y_s + \int_s^t M_\tau\big(1 - u(\tau)X_\tau\big)dW_\tau.$$

Since $\int_s^t M_\tau\big(1 - u(\tau)X_\tau\big)dW_\tau$ is independent of $\mathcal{F}_s$, we have

$$\mathbb{E}\Big[\int_s^t M_\tau\big(1 - u(\tau)X_\tau\big)dW_\tau|\mathcal{F}_s\Big] = \mathbb{E}\Big[\int_s^t M_\tau\big(1 - u(\tau)X_\tau\big)dW_\tau\Big] = 0,$$

and hence

$$\mathbb{E}[Y_t|\mathcal{F}_s] = Y_s.$$

Exercise 10.1.7 *Prove that $(W_t + t)e^{-W_t - \frac{1}{2}t}$ is an $\mathcal{F}_t$-martingale.*

Exercise 10.1.8 *Let h be a continuous function. Using the properties of the Wiener integral and log-normal random variables, show that*

$$\mathbb{E}\Big[e^{\int_0^t h(s)\,dW_s}\Big] = e^{\frac{1}{2}\int_0^t h(s)^2\,ds}.$$

Exercise 10.1.9 *Let M_t be the exponential process (10.1.2). Use the previous exercise to show that for any $t > 0$*

$$(a)\ \mathbb{E}[M_t] = 1 \qquad (b)\ \mathbb{E}[M_t^2] = e^{\int_0^t u(s)^2\,ds}.$$

Exercise 10.1.10 *Let $\mathcal{F}_t = \sigma\{W_u; u \le t\}$. Show that the following processes are $\mathcal{F}_t$-martingales:*

(a) $e^{t/2}\cos W_t$;

(b) $e^{t/2}\sin W_t$.

Recall that the Laplacian of a twice differentiable function f is defined by $\Delta f(x) = \sum_{j=1}^n \partial_{x_j}^2 f$.

Example 10.1.11 *Consider the smooth function $f : \mathbb{R}^n \to R$, such that*
(i) $\Delta f = 0$;
(ii) $\mathbb{E}\big[|f(W_t)|\big] < \infty$, $\forall t > 0$ *and* $x \in R$.
Then the process $X_t = f(W_t)$ is an $\mathcal{F}_t$-martingale.

Exercise 10.1.12 *Let $W_1(t)$ and $W_2(t)$ be two independent Brownian motions. Show that $X_t = e^{W_1(t)} \cos W_2(t)$ is a martingale.*

Proposition 10.1.13 *Let $f : \mathbb{R}^n \to \mathbb{R}$ be a smooth function such that*
(i) $\mathbb{E}\big[|f(W_t)|\big] < \infty$;
(ii) $\mathbb{E}\Big[\int_0^t |\Delta f(W_s)|\, ds\Big] < \infty$.
Then the process $X_t = f(W_t) - \frac{1}{2}\int_0^t \Delta f(W_s)\, ds$ is a martingale.

Proof: For $0 \le s < t$ we have

$$\begin{aligned}
\mathbb{E}\big[X_t|\mathcal{F}_s\big] &= \mathbb{E}\big[f(W_t)|\mathcal{F}_s\big] - \mathbb{E}\Big[\frac{1}{2}\int_0^t \Delta f(W_u)\, du|\mathcal{F}_s\Big] \\
&= \mathbb{E}\big[f(W_t)|\mathcal{F}_s\big] - \frac{1}{2}\int_0^s \Delta f(W_u)\, du - \int_s^t \mathbb{E}\Big[\frac{1}{2}\Delta f(W_u)|\mathcal{F}_s\Big]\, du.
\end{aligned}$$

Let $p(t, y, x)$ be the probability density function of W_t. Integrating by parts and using that p satisfies the Kolmogorov's backward equation, we have

$$\begin{aligned}
\mathbb{E}\Big[\frac{1}{2}\Delta f(W_u)|\mathcal{F}_s\Big] &= \frac{1}{2}\int p(u-s, W_s, x)\, \Delta f(x)\, dx \\
&= \frac{1}{2}\int \Delta_x p(u-s, W_s, x) f(x)\, dx \\
&= \int \frac{\partial}{\partial u} p(u-s, W_s, x) f(x)\, dx.
\end{aligned}$$

Then, using the Fundamental Theorem of Calculus, we obtain

$$\begin{aligned}
\int_s^t \mathbb{E}\Big[\frac{1}{2}\Delta f(W_u)|\mathcal{F}_s\Big]\, du &= \int_s^t \Big(\frac{\partial}{\partial u}\int p(u-s, W_s, x) f(x)\, dx\Big)\, du \\
&= \int p(t-s, W_s, x) f(x)\, dx - \lim_{\epsilon \searrow 0}\int p(\epsilon, W_s, x) f(x)\, dx \\
&= \mathbb{E}\big[f(W_t)|\mathcal{F}_s\big] - \int \delta(x = W_s) f(x)\, dx \\
&= \mathbb{E}\big[f(W_t)|\mathcal{F}_s\big] - f(W_s).
\end{aligned}$$

Substituting in (10.1.3) yields

$$\begin{aligned}
\mathbb{E}[X_t|\mathcal{F}_s] &= \mathbb{E}[f(W_t)|\mathcal{F}_s] - \frac{1}{2}\int_0^s \Delta f(W_u)\,du - \mathbb{E}[f(W_t)|\mathcal{F}_s] + f(W_s) \\
&= f(W_s) - \frac{1}{2}\int_0^s \Delta f(W_u)\,du \\
&= X_s.
\end{aligned}$$

Hence X_t is an $\mathcal{F}_t$-martingale. ■

Exercise 10.1.14 *Use Proposition 10.1.13 to show that the following processes are martingales:*

(*a*) $X_t = W_t^2 - t$;

(*b*) $X_t = W_t^3 - 3\int_0^t W_s\,ds$;

(*c*) $X_t = \frac{1}{n(n-1)}W_t^n - \frac{1}{2}\int_0^t W_s^{n-2}\,ds$;

(*d*) $X_t = e^{cW_t} - \frac{1}{2}c^2\int_0^t e^{W_s}\,ds$, *with c constant;*

(*e*) $X_t = \sin(cW_t) + \frac{1}{2}c^2\int_0^t \sin(cW_s)\,ds$, *with c constant.*

Exercise 10.1.15 *Let* $f : \mathbb{R}^n \to \mathbb{R}$ *be a function such that*

(*i*) $\mathbb{E}[|f(W_t)|] < \infty$;

(*ii*) $\Delta f = \lambda f$, λ *constant.*

Show that the process $X_t = f(W_t) - \frac{\lambda}{2}\int_0^t f(W_s)\,ds$ *is a martingale.*

10.2 How to Recognize a Brownian Motion

Many processes are disguised Brownian motions. How can we recognize them? We already know that a Brownian motion B_t has the following properties:

(*i*) Is a continuous martingale with respect to $\sigma\{B_s; s \le t\}$;

(*ii*) Has the quadratic variation $\langle B, B\rangle_t = t$, for $t \ge 0$.

We state, without proof, a classical theorem due to Paul Lévy, which is a reciprocal of the foregoing result. The following theorem is a useful tool to show that a one-dimensional process is a Brownian motion. For a more general result and a proof, the reader can consult Durrett [15].

Theorem 10.2.1 (Lévy) *If* X_t *is a continuous martingale with respect to the filtration* $\mathcal{F}_t$*, with* $X_0 = 0$ *and* $\langle X, X\rangle_t = t$*, for all* $t \ge 0$*, then* X_t *is an* $\mathcal{F}_t$*-Brownian motion.*

Example 10.2.2 *Let B_t be a Brownian motion and consider the process*

$$X_t = \int_0^t sgn(B_s)\, dB_s,$$

where

$$sgn(x) = \begin{cases} 1, & if\, x > 0 \\ -1, & if\, x \leq 0. \end{cases}$$

We note that $dX_t = sgn(B_t)\, dB_t$ and hence $(dX_t)^2 = dt$. Since X_t is a continuous martingale (because it is an Ito integral) and its quadratic variation is given by

$$\langle X, X \rangle_t = \int_0^t (dX_s)^2 = \int_0^t ds = t,$$

then Lévy's theorem implies that X_t is a Brownian motion.

Example 10.2.3 (The squared Bessel process) *Let $W_1(t), \cdots, W_n(t)$ be n one-dimensional independent Brownian motions, and consider the n-dimensional Bessel process*

$$R_t = \sqrt{W_1(t)^2 + \cdots + W_n(t)^2}, \qquad n \geq 2.$$

Define the process

$$\beta_t = \sum_{i=1}^n \int_0^t \Big(\frac{W_i(s)}{R_s}\Big)\, dW_i(s).$$

Since the set $\{\omega; R_t(\omega) = 0\}$ has probability zero, the division by R_s does not cause any problems almost surely. As a sum of Ito integrals, β_t is an $\mathcal{F}_t$-martingale, with the quadratic variation given by

$$\langle \beta, \beta \rangle_t = \sum_{i=1}^n \int_0^t \frac{W_i(s)^2}{R_s^2}\, ds = \int_0^t \frac{R_s^2}{R_s^2}\, ds = t.$$

By Levy's theorem, β_t is an $\mathcal{F}_t$-Brownian motion. It satisfies the following equation

$$d\beta_t = \sum_{i=1}^n \Big(\frac{W_i(t)}{R_t}\Big)\, dW_i(t). \tag{10.2.3}$$

From Ito's formula and an application of (10.2.3) we get

$$\begin{aligned} d(R_t^2) &= \sum_{i=1}^n 2W_i(t)\, dW_i(t) + n\, dt \\ &= 2R_t \sum_{i=1}^n \frac{W_i(t)}{R_t}\, dW_i(t) + n\, dt \\ &= 2R_t\, d\beta_t + n\, dt. \end{aligned}$$

Hence, the squared Bessel process, $Z_t = R_t^2$, satisfies the following stochastic differential equation

$$dZ_t = 2\sqrt{Z_t}d\beta_t + n\,dt, \tag{10.2.4}$$

with β_t Brownian motion.

Example 10.2.4 (The Bessel process) *From Example 10.2.3 we recall that $(dZ_t)^2 = 4R_t^2\,dt$. Then, using $R_t = Z_t^{1/2}$, Ito's formula yields*

$$\begin{aligned} dR_t &= \frac{1}{2}Z_t^{1/2}\,dZ_t - \frac{1}{8}Z_t^{-3/2}(dZ_t)^2 \\ &= \frac{1}{2}\frac{1}{R_t}(2R_t d\beta_t + ndt) - \frac{1}{2}\frac{1}{R_t}dt \\ &= d\beta_t + \frac{n-1}{2R_t}\,dt. \end{aligned}$$

Hence the Bessel process satisfies the following stochastic differential equation

$$dR_t = d\beta_t + \frac{n-1}{2R_t}\,dt, \tag{10.2.5}$$

where β_t is a Brownian motion. It is worth noting that the infinitesimal generator of R_t is the operator

$$\mathcal{A} = \frac{1}{2}\partial_x^2 + \frac{n-1}{2x}\partial_x, \tag{10.2.6}$$

which is the Bessel operator of order n.

Example 10.2.5 *Let $f : \mathbb{R}^2 \to \mathbb{R}$ be a continuous twice differentiable function, and consider the process $X_t = f(W_1(t), W_2(t))$, with $W_1(t)$, $W_2(t)$ independent one-dimensional Brownian motions. Ito's formula implies*

$$dX_t = \frac{\partial f}{\partial x_1}dW_1(t) + \frac{\partial f}{\partial x_2}dW_2(t) + \frac{1}{2}\Big(\frac{\partial^2 f}{\partial x_1^2} + \frac{\partial^2 f}{\partial x_2^2}\Big)\,dt.$$

Then X_t is a continuous martingale if f is harmonic, i.e.

$$\frac{\partial^2 f}{\partial x_1^2} + \frac{\partial^2 f}{\partial x_2^2} = 0. \tag{10.2.7}$$

Then

$$(dX_t)^2 = \Big[\Big(\frac{\partial f}{\partial x_1}\Big)^2 + \Big(\frac{\partial f}{\partial x_2}\Big)^2\Big]\,dt = |\nabla f|^2\,dt,$$

so we have

$$\langle X, X\rangle_t = \int_0^t (dX_t)^2 = \int_0^t |\nabla f|^2\, ds.$$

Then the condition $\langle X, X\rangle_t = t$, *for any* $t \geq 0$, *implies* $|\nabla f| = 1$, *i.e.*

$$\Big(\frac{\partial f}{\partial x_1}\Big)^2 + \Big(\frac{\partial f}{\partial x_2}\Big)^2 = 1. \tag{10.2.8}$$

Equation (10.2.8) is called the eiconal equation. We shall show that if a function f *satisfies both equations (10.2.7) and (10.2.8), then it is a linear function. Or, equivalently, a harmonic solution of the eiconal equation is a linear function.*

From equation (10.2.8) there is a continuous function $\theta = \theta(x_1, x_2)$ *such that*

$$\frac{\partial f}{\partial x_1} = \cos\theta, \qquad \frac{\partial f}{\partial x_2} = \sin\theta. \tag{10.2.9}$$

The closeness condition implies

$$\frac{\partial(\cos\theta)}{\partial x_2} = \frac{\partial(\sin\theta)}{\partial x_1},$$

which is equivalent to

$$\cos\theta\frac{\partial\theta}{\partial x_1} + \sin\theta\frac{\partial\theta}{\partial x_2} = 0. \tag{10.2.10}$$

Differentiating in (10.2.9) with respect to x_1 *and* x_2 *yields*

$$\frac{\partial^2 f}{\partial x_1^2} = -\sin\theta\frac{\partial\theta}{\partial x_1}, \qquad \frac{\partial^2 f}{\partial x_2^2} = \cos\theta\frac{\partial\theta}{\partial x_2}.$$

Adding and using (10.2.7) we obtain

$$\cos\theta\frac{\partial\theta}{\partial x_2} - \sin\theta\frac{\partial\theta}{\partial x_1} = 0. \tag{10.2.11}$$

Relations (10.2.10) and (10.2.11) can be written as a system

$$\begin{pmatrix} \cos\theta & \sin\theta \\ -\sin\theta & \cos\theta \end{pmatrix} \begin{pmatrix} \dfrac{\partial\theta}{\partial x_1} \\ \dfrac{\partial\theta}{\partial x_2} \end{pmatrix} = \begin{pmatrix} 0 \\ 0 \end{pmatrix},$$

with the solution $\dfrac{\partial\theta}{\partial x_1} = 0$, $\dfrac{\partial\theta}{\partial x_2} = 0$. *Therefore,* θ *is a constant function. This implies that* f *is linear, i.e.* $f(x_1, x_2) = c_1x_1 + c_2x_2$, *with* $c_i \in \mathbb{R}$.

Exercise 10.2.6 *If W_t and $\tilde{W}_t$ are two independent Brownian motions and $\rho \in [-1, 1]$ is a constant, use Lévy's theorem to show that the process $X_t = \rho W_t + \sqrt{1-\rho^2}\tilde{W}_t$ is a Brownian motion (use that $dW_t \, d\tilde{W}_t = 0$).*

Exercise 10.2.7 *Consider the process $X_t = \int_0^t f(s)\, dW_s$, with f square integrable function.*

(a) Show that X_t is a continuous martingale with the quadratic variation $\langle X, X\rangle_t = \int_0^t f^2(s)\, ds$;

(b) Apply Lévy's theorem to find all functions $f(s)$ for which X_t is a Brownian motion.

Exercise 10.2.8 *Let $X_t = e^{-t}\int_0^t e^s \, dW_s$, with W_s Brownian motion.*

(a) Show that $\langle X, X\rangle_t = t$ for any $t \geq 0$;

(b) Is X_t a Brownian motion?

The next exercise states that a Brownian motion is preserved by an orthogonal transform.

Exercise 10.2.9 *Let $W_t = (W_t^1, W_t^2)$ be a Brownian motion in the plane (i.e. W_t^i one-dimensional independent Brownian motions) and define the process $B_t = (B_t^1, B_t^2)$ by*

$$\begin{aligned} B_t^1 &= \cos\theta \, W_t^1 + \sin\theta \, W_t^2 \\ B_t^2 &= -\sin\theta \, W_t^1 + \cos\theta \, W_t^2, \end{aligned}$$

for a fixed angle $\theta \in \mathbb{R}$.

(a) Show that B_t^1, B_t^2 are Brownian motions;

(b) Prove that B_t^1, B_t^2 are independent processes.

Exercise 10.2.10 *Consider the process $Y_t = tW_{\frac{1}{t}}$, if $t > 0$ and $Y_0 = 0$, where W_t is a Brownian motion. Prove that Y_t is a Brownian motion.*

10.3 Time Change for Martingales

A martingale satisfying certain properties can always be considered as a Brownian motion running at a modified time clock. The next result is provided without proof. The interested reader can consult Karatzas and Shreve [26].

Theorem 10.3.1 (Dambis, Dubins and Schwarz, 1965) *Let M_t be a continuous, square integrable $\mathcal{F}_t$-martingale satisfying $\lim_{t\to\infty}\langle M, M\rangle_t = \infty$, a.s. Then M_t can be written as a time-transformed Brownian motion as*

$$M_t = B_{\langle M,M\rangle_t},$$

where B_t is a one-dimensional Brownian motion. Moreover, if we define for each $s \geq 0$ the stopping time

$$T(s) = \inf\{t \geq 0; \langle M, M \rangle_t > s\},$$

then the time-changed process

$$B_s = M_{T(s)}$$

is a $\mathcal{G}_s$-Brownian motion, where $\mathcal{G}_s = \mathcal{F}_{T(s)}$, $0 \leq s$.

Example 10.3.2 (Scaled Brownian motion) *Let W_t be a Brownian motion and consider the process $X_t = cW_t$, which is a continuous martingale. Assume $c \neq 0$. Since $(dX_t)^2 = (c\,dW_t)^2 = c^2 dt$, the quadratic variation of X_t is*

$$\langle X, X \rangle_t = \int_0^t (dX_s)^2 = \int_0^t c^2\, ds = c^2 t \to \infty, \qquad t \to \infty.$$

Then there is a Brownian motion $\widetilde{W}_t$ such that $X_t = \widetilde{W}_{c^2 t}$. This can also be written as $\frac{1}{c}\widetilde{W}_{c^2 t} = W_t$. Substituting $s = c^2 t$, yields $\widetilde{W}_s = cW_{s/c^2}$. Therefore, if W_s is a Brownian motion, then the process cW_{s/c^2} is also a Brownian motion. In particular, if $c = -1$, then $-W_s$ is a Brownian motion.

Example 10.3.3 *This is an application of the previous example. Let $T > 0$. Using the scaling property of the Brownian motion and a change of variables, we have the following identities in law*

$$\begin{aligned} \int_0^T W_t^n\, dt &= \int_0^T \left(cW_{t/c^2}\right)^n dt = c^n \int_0^T W_{t/c^2}^n\, dt \\ &= c^n \int_0^{T/c^2} W_s^n\, c^2\, ds = c^{n+2} \int_0^{T/c^2} W_s^n\, ds, \end{aligned}$$

for any $c > 0$. Then set $c = \sqrt{T}$ and obtain the following identity

$$\int_0^T W_t^n\, dt \overset{\text{law}}{=} T^{1+n/2} \int_0^1 W_s^n\, ds.$$

This relation can be easily verified for $n = 1$, when both sides are normally distributed as $N(0, T^3/3)$, see section 3.3.

Example 10.3.4 *It is known that the process $X_t = W_t^2 - t$ is a continuous martingale with respect to $\mathcal{F} = \sigma\{W_s; s \leq t\}$. An application of Ito's formula yields*

$$dX_t = 2W_t\, dW_t,$$

so the quadratic variation is

$$\langle X, X\rangle_t = \int_0^t (dX_s)^2 = 4\int_0^t W_s^2\, ds := \alpha_t.$$

Therefore, there is a Brownian motion $\widetilde{W}_t$ *such that* $X_t = \widetilde{W}_{\alpha_t}$. *This is equivalent to stating that given a Brownian motion* W_t, *its square can be written as*

$$W_t^2 = t + \widetilde{W}_{\alpha_t}.$$

Example 10.3.5 (Brownian Bridge) *The Brownian bridge is provided by formula (8.5.13). The Ito integral* $M_t = \int_0^t \frac{1}{1-s}\, dW_s$ *can be written as a Brownian motion as in the following. Since* $(dM_t)^2 = \frac{1}{(1-t)^2}\, dt$, *the quadratic variation becomes*

$$\langle M, M\rangle_t = \int_0^t \frac{1}{(1-s)^2}\, ds = \frac{t}{1-t}.$$

Then there is a Brownian motion B_t *such that* $M_t = B_{\frac{t}{1-t}}$, *and hence*

$$(1-t)\int_0^t \frac{1}{1-s}\, dW_s = (1-t)B_{\frac{t}{1-t}} = \tilde{B}_{t(1-t)},$$

where $\tilde{B}$ *is also a Brownian motion. It follows that the Brownian bridge formula (8.5.13) can be written equivalently as*

$$X_t = a(1-t) + bt + \tilde{B}_{t(1-t)}, \qquad o \le t \le 1.$$

This process has the characteristics of a Brownian motion, while satisfying the boundary conditions $X_0 = a$ *and* $X_1 = b$.

Example 10.3.6 (Lamperti's property) *Let* B_t *be a Brownian motion and consider the integrated geometric Brownian motion*

$$A_t = \int_0^t e^{2B_s}\, ds, \qquad t \ge 0.$$

We note that the process A_t *is continuous and strictly increasing in* t, *with* $A_0 = 0$ *and* $\lim_{t\to\infty} A_t = \infty$. *Therefore, there is an inverse process* T_u, *i.e.*

$$A_{T_u} = u, \qquad u \ge 0.$$

Applying the chain rule yields the derivative

$$\frac{dT_u}{du} = e^{-2B_{T_u}}. \tag{10.3.12}$$

Ito's formula provides

$$de^{B_t} = e^{B_t} dB_t + \frac{1}{2} e^{B_t}\, dt,$$

which can be written equivalently as

$$e^{B_t} = 1 + \frac{1}{2}\int_0^t e^{B_s}\, ds + M_t, \tag{10.3.13}$$

where $M_t = \int_0^t e^{B_s}\, dB_s$. *Since* M_t *is a continuous martingale, with*

$$\langle M, M\rangle_t = \int_0^t (dM_s)^2 = \int_0^t e^{2B_s}\, ds = A_t,$$

by Theorem 10.3.1 there is a Brownian motion W_t *such that* $M_t = W_{A_t}$, *or equivalently,* $M_{T_u} = W_u$, $u \geq 0$. *Then replacing* t *by* T_u *in equation (10.3.13) yields*

$$e^{B_{T_u}} = 1 + \frac{1}{2}\int_0^{T_u} e^{B_s}\, ds + W_u, \tag{10.3.14}$$

which can be written, after applying the chain rule, in differential notation as

$$de^{B_{T_u}} = \frac{1}{2} e^{B_{T_u}} \frac{dT_u}{du}\, du + dW_u.$$

Substituting the derivative of T_u *from (10.3.12), the foregoing relation becomes*

$$de^{B_{T_u}} = \frac{1}{2e^{B_{T_u}}} du + dW_u.$$

Denoting $R_u = e^{B_{T_u}}$, *then*

$$dR_u = \frac{1}{2R_u} du + dW_u,$$

which is equation (10.2.5) for $n = 2$, *i.e.* R_u *is a Bessel process. Substituting* $u = A_t$ *and* $t = T_u$ *in* $R_u = e^{B_{T_u}}$ *yields*

$$\boxed{e^{B_t} = R_{A_t}.}$$

Hence, the geometric Brownian motion e^{B_t} *is a time-transformed Bessel process in the plane,* R_{A_t}. *For a generalized Lamperti property in the case of a Brownian motion with drift, see Yor [48]. See also Lamperti [31].*

Lemma 10.3.7 *Let $\varphi(t)$ be a continuous differentiable, with $\varphi'(t) > 0$, $\varphi(0) = 0$ and $\lim_{t\to\infty} \varphi(t) = \infty$. Then given the one-dimensional Brownian motion B_t, there is another one-dimensional Brownian motion W_t such that*

$$W_{\varphi(t)} = \int_0^t \sqrt{\varphi'(s)}\, dB_s, \qquad \forall t \geq 0. \tag{10.3.15}$$

Proof: The process $X_t = \int_0^t \sqrt{\varphi'(s)}\, dB_s$ is a continuous martingale, with the quadratic variation

$$\langle X, X \rangle_t = \int_0^t (dX_s)^2 = \int_0^t (\sqrt{\varphi'(s)}\, dB_s)^2 = \int_0^t \varphi'(s)\, ds = \varphi(t).$$

Then by Theorem 10.3.1 there is a Brownian motion W_t such that $W_{\varphi(t)} = X_t$. ∎

Example 10.3.8 *If $\varphi(t) = c^2 t$, $c \neq 0$, then Lemma 10.3.7 yields $W_{c^2 t} = cB_t$. This is equivalent to stating that for a given Brownian motion B_t, the process*

$$W_s = cB_{s/c^2}, \qquad s \geq 0$$

is also a Brownian motion.

Remark 10.3.9 The right side of (10.3.15) is a Wiener integral. Hence, under certain conditions, a Wiener integral becomes a time-scaled Brownian motion.

Exercise 10.3.10 *Let B_t be a Brownian motion. Prove that there is another Brownian motion W_t such that*

$$W_{e^{a^2 t} - 1} = a \int_0^t e^{\frac{a^2 s}{2}}\, dB_s.$$

Exercise 10.3.11 (Ornstein-Uhlenbeck process) *Consider the equation*

$$dX_t = -qX_t dt + \alpha dW_t, \qquad X_0 = 0,$$

with q and α constants, with $q > 0$.

(a) *Show that the solution is given by*

$$X_t = \alpha e^{-qt} \int_0^t e^{qu}\, dW_u;$$

(b) *Show that there is a Brownian motion B_t such that*

$$X_t = \alpha e^{-qt} B_{(e^{2qt}-1)/(2q)}.$$

Exercise 10.3.12 *Consider the equation*

$$dX_t = -\frac{\alpha}{2}X_t dt + \alpha dW_t, \qquad X_0 = 0,$$

with α constant. Show that there is a Brownian motion $\tilde{B}_t$ such that

$$X_t = e^{-\alpha t/2}\tilde{B}_{\alpha e^{\alpha t}}.$$

Theorem 10.3.13 (Time change formula for Ito integrals) *Let φ be a function as in Lemma 10.3.7, and F a continuous function. Then given a Brownian motion B_t, there is another Brownian motion W_t such that*

$$\int_\alpha^\beta F(u)\,dW_u = \int_{\varphi^{-1}(\alpha)}^{\varphi^{-1}(\beta)} F(\varphi(t))\sqrt{\varphi'(t)}\,dB_t, \qquad a.s. \tag{10.3.16}$$

Proof: First, we will prove formula (10.3.16) informally, and then we will check that the identity holds in law. Formula (10.3.15) can be written in the equivalent differential form as

$$dW_{\varphi(t)} = \sqrt{\varphi'(t)}\,dB_t.$$

Then for a continuous function g we have

$$\int_a^b g(t)dW_{\varphi(t)} = \int_a^b g(t)\sqrt{\varphi'(t)}\,dB_t. \tag{10.3.17}$$

Using a change of variable, the left side integral becomes

$$\int_a^b g(t)dW_{\varphi(t)} = \int_{\varphi(a)}^{\varphi(b)} g(\varphi^{-1}(u))\,dW_u. \tag{10.3.18}$$

Relations (10.3.17) and (10.3.18) imply

$$\int_{\varphi(a)}^{\varphi(b)} g(\varphi^{-1}(u))\,dW_u = \int_a^b g(t)\sqrt{\varphi'(t)}\,dB_t.$$

Substituting $F = g\circ\varphi^{-1}$, $\alpha = \varphi(a)$, and $\beta = \varphi(b)$ yields

$$\int_\alpha^\beta F(u)\,dW_u = \int_{\varphi^{-1}(\alpha)}^{\varphi^{-1}(\beta)} F(\varphi(t))\sqrt{\varphi'(t)}\,dB_t. \tag{10.3.19}$$

Each of the sides of formula (10.3.16)

$$X = \int_\alpha^\beta F(u)\,dW_u, \quad Y = \int_{\varphi^{-1}(\alpha)}^{\varphi^{-1}(\beta)} F(\varphi(t))\sqrt{\varphi'(t)}\,dB_t$$

is a random variable, which is given as a Wiener integral, so they are both normally distributed. Therefore, in order to show that they are equal in law, it suffices to show that they have the same first two moments.

From the properties of Wiener integrals we have that $E[X_t] = E[Y_t] = 0$ and

$$\begin{aligned} Var(X) &= \int_\alpha^\beta F^2(u)\,du \\ Var(Y) &= \int_{\varphi^{-1}(\alpha)}^{\varphi^{-1}(\beta)} \Big(F(\varphi(t))\sqrt{\varphi'(t)}\Big)^2\,dt \\ &= \int_{\varphi^{-1}(\alpha)}^{\varphi^{-1}(\beta)} F(\varphi(t))^2\varphi'(t)\,dt = \int_\alpha^\beta F^2(u)\,du, \end{aligned}$$

so $Var(X) = Var(Y)$. Hence $X = Y$ in law. ■

Example 10.3.14 *For $\varphi(t) = \tan t$, formula (10.3.16) becomes*

$$\int_\alpha^\beta F(u)\,dW_u = \int_{\tan^{-1}\alpha}^{\tan^{-1}\beta} F(\tan t)\sec t\,dB_t.$$

If $F(u) = \dfrac{1}{1+u^2}$ and $\alpha = 0$, then we obtain

$$\int_0^\beta \frac{1}{1+u^2}\,dW_u = \int_0^{\tan^{-1}\beta} \frac{1}{\sec t}\,dB_t,$$

which, after substituting $v = \tan^{-1}\beta$, implies

$$\int_0^{\tan v} \frac{1}{1+u^2}\,dW_u = \int_0^v \cos t\,dB_t.$$

Making $v \nearrow \frac{\pi}{2}$ yields

$$\int_0^\infty \frac{1}{1+u^2}\,dW_u = \int_0^{\pi/2} \cos t\,dB_t.$$

For the sake of completeness, we include next a stochastic variant of Fubini's theorem. For a more general variant of this theorem, the reader is referred to Ikeda and Watanabe [23].

Theorem 10.3.15 (Stochastic Fubini) *If $f : \mathbb{R}_+ \times \mathbb{R}_+ \to \mathbb{R}$ is a bounded measurable function, then*

$$\int_0^t \int_0^T f(s,r)\,dr\,dW_s = \int_0^T \int_0^t f(s,t)\,dW_s\,dr.$$

10.4 Girsanov's Theorem

In this section we shall present and prove a version of Girsanov's theorem, which will suffice for the purpose of proposed applications. The main usage of Girsanov's theorem is the reduction of drift. Consequently, Girsanov's theorem applies in finance where it shows that in the study of security markets the differences between the mean rates of return can be removed. For a gentle introduction into this subject the interested reader can consult Baxter and Rennie [5], or Neftici [35].

We shall recall first a few basic notions. Let $(\Omega, \mathcal{F}, P)$ be a probability space. When dealing with an $\mathcal{F}_t$-martingale on the aforementioned probability space, the filtration $\mathcal{F}_t$ is considered to be the σ-algebra generated by the given Brownian motion W_t, i.e. $\mathcal{F}_t = \sigma\{W_u; 0 \leq u \leq s\}$. By default, a martingale is considered with respect to the probability measure P, in the sense that the expectations involve an integration with respect to P

$$\mathbb{E}^P[X] = \int_\Omega X(\omega) dP(\omega).$$

We have not used the upper script until now since there was no doubt which probability measure was used. In this section we shall also use another probability measure given by

$$dQ = M_T dP,$$

where M_T is an exponential process. This means that $Q : \mathcal{F} \to \mathbb{R}$ is given by

$$Q(A) = \int_A dQ = \int_A M_T dP, \qquad \forall A \in \mathcal{F}.$$

Since $M_T > 0$, $M_0 = 1$, using the martingale property of M_t yields

$$\begin{aligned} Q(A) &> 0, \qquad A \neq \emptyset; \\ Q(\Omega) &= \int_\Omega M_T dP = \mathbb{E}^P[M_T] = \mathbb{E}^P[M_T|\mathcal{F}_0] = M_0 = 1, \end{aligned}$$

which shows that Q is a probability on $\mathcal{F}$, and hence $(\Omega, \mathcal{F}, Q)$ becomes a probability space. Furthermore, if X is a random variable, then

$$\begin{aligned} \mathbb{E}^Q[X] &= \int_\Omega X(\omega)\, dQ(\omega) = \int_\Omega X(\omega) M_T(\omega)\, dP(\omega) \\ &= \mathbb{E}^P[X M_T]. \end{aligned}$$

The following result will play a central role in proving Girsanov's theorem:

Lemma 10.4.1 *Let X_t be the Ito process*

$$dX_t = u(t)dt + dW_t, \qquad X_0 = 0,\ 0 \le t \le T,$$

with $u(s)$ a bounded function. Consider the exponential process

$$M_t = e^{-\int_0^t u(s)\,dW_s - \frac{1}{2}\int_0^t u^2(s)\,ds}.$$

Then X_t is an $\mathcal{F}_t$-martingale with respect to the measure

$$dQ(\omega) = M_T(\omega)dP(\omega).$$

Proof: We need to prove that X_t is an $\mathcal{F}_t$-martingale with respect to Q, so it suffices to show the following three properties:

1. Integrability of X_t. This part usually follows from standard manipulations of norms estimations. We shall do it here in detail. Integrating in the equation of X_t between 0 and t provides

$$X_t = \int_0^t u(s)\,ds + W_t. \tag{10.4.20}$$

We start with an estimation of the expectation with respect to P

$$\begin{aligned} \mathbb{E}^P[X_t^2] &= \mathbb{E}^P\Big[\Big(\int_0^t u(s)\,ds\Big)^2 + 2\int_0^t u(s)\,ds\,W_t + W_t^2\Big] \\ &= \Big(\int_0^t u(s)\,ds\Big)^2 + 2\int_0^t u(s)\,ds\,\mathbb{E}^P[W_t] + \mathbb{E}^P[W_t^2] \\ &= \Big(\int_0^t u(s)\,ds\Big)^2 + t < \infty, \qquad \forall 0 \le t \le T, \end{aligned}$$

where the last inequality follows from the norm estimation

$$\begin{aligned} \int_0^t u(s)\,ds &\le \int_0^t |u(s)|\,ds \le \Big[t\int_0^t |u(s)|^2\,ds\Big]^{1/2} \\ &\le \Big[t\int_0^T |u(s)|^2\,ds\Big]^{1/2} = T^{1/2}\|u\|_{L^2[0,T]}. \end{aligned}$$

Next we obtain an estimation with respect to Q

$$\begin{aligned} \mathbb{E}^Q[|X_t|]^2 &= \Big(\int_\Omega |X_t| M_T\,dP\Big)^2 \le \int_\Omega |X_t|^2\,dP \int_\Omega M_T^2\,dP \\ &= \mathbb{E}^P[X_t^2]\,\mathbb{E}^P[M_T^2] < \infty, \end{aligned}$$

since $\mathbb{E}^P[X_t^2] < \infty$ and $\mathbb{E}^P[M_T^2] = e^{\int_0^T u(s)^2\,ds} = e^{\|u\|^2_{L^2[0,T]}}$, see Exercise 10.1.9.

2. $\mathcal{F}_t$-mesurability of X_t. This follows from equation (10.4.20) and the fact that W_t is $\mathcal{F}_t$-measurable.

3. Conditional expectation of X_t. From Examples 10.1.3 and 10.1.6 recall that for any $0 \leq t \leq T$:

(*i*) M_t is an $\mathcal{F}_t$-martingale with respect to probability measure P;

(*ii*) $X_t M_t$ is an $\mathcal{F}_t$-martingale with respect to probability measure P.

We need to verify that

$$\mathbb{E}^Q[X_t|\mathcal{F}_s] = X_s, \qquad \forall s \leq t,$$

which can be written as

$$\int_A X_t \, dQ = \int_A X_s \, dQ, \qquad \forall A \in \mathcal{F}_s.$$

Since $dQ = M_T dP$, the previous relation becomes

$$\int_A X_t M_T \, dP = \int_A X_s M_T \, dP, \qquad \forall A \in \mathcal{F}_s.$$

This can be written in terms of conditional expectation as

$$\mathbb{E}^P[X_t M_T|\mathcal{F}_s] = \mathbb{E}^P[X_s M_T|\mathcal{F}_s]. \qquad (10.4.21)$$

We shall prove this identity by showing that both terms are equal to $X_s M_s$. Since X_s is $\mathcal{F}_s$-predictable and M_t is a martingale, the right side term becomes

$$\mathbb{E}^P[X_s M_T|\mathcal{F}_s] = X_s \mathbb{E}^P[M_T|\mathcal{F}_s] = X_s M_s, \qquad \forall s \leq T.$$

Let $s < t$. Using the tower property (see Proposition 2.12.6, part 3), the left side term becomes

$$\begin{aligned} \mathbb{E}^P[X_t M_T|\mathcal{F}_s] &= \mathbb{E}^P\big[\mathbb{E}^P[X_t M_T|\mathcal{F}_t]|\mathcal{F}_s\big] = \mathbb{E}^P\big[X_t \mathbb{E}^P[M_T|\mathcal{F}_t]|\mathcal{F}_s\big] \\ &= \mathbb{E}^P\big[X_t M_t|\mathcal{F}_s\big] = X_s M_s, \end{aligned}$$

where we used that M_t and $X_t M_t$ are martingales and X_t is $\mathcal{F}_t$-measurable. Hence (10.4.21) holds and X_t is an $\mathcal{F}_t$-martingale with respect to the probability measure Q. ■

Proposition 10.4.2 *Consider the process*

$$X_t = \int_0^t u(s) \, ds + W_t, \qquad 0 \leq t \leq T,$$

with $u \in L^2[0, T]$ a deterministic function, and let $dQ = M_T dP$. Then

$$\mathbb{E}^Q[X_t^2] = t.$$

Proof: Denote $U(t) = \int_0^t u(s)\,ds$. Then

$$\begin{aligned}\mathbb{E}^Q[X_t^2] &= \mathbb{E}^P[X_t^2 M_T] = \mathbb{E}^P[U^2(t)M_T + 2U(t)W_tM_T + W_t^2 M_T] \\ &= U^2(t)\mathbb{E}^P[M_T] + 2U(t)\mathbb{E}^P[W_tM_T] + \mathbb{E}^P[W_t^2 M_T]. \quad (10.4.22)\end{aligned}$$

From Exercise 10.1.9 (a) we have $\mathbb{E}^P[M_T] = 1$. In order to compute $\mathbb{E}^P[W_tM_T]$ we use the tower property and the martingale property of M_t

$$\begin{aligned}\mathbb{E}^P[W_tM_T] &= \mathbb{E}^P[\mathbb{E}^P[W_tM_T|\mathcal{F}_t]] = \mathbb{E}^P[W_t\mathbb{E}^P[M_T|\mathcal{F}_t]] \\ &= \mathbb{E}^P[W_tM_t]. \quad (10.4.23)\end{aligned}$$

Using the product rule

$$\begin{aligned}d(W_tM_t) &= M_t dW_t + W_t dM_t + dW_t dM_t \\ &= \big(M_t - u(t)M_tW_t\big)dW_t - u(t)M_t dt,\end{aligned}$$

where we used $dM_t = -u(t)M_t dW_t$. Integrating between 0 and t yields

$$W_tM_t = \int_0^t \big(M_s - u(s)M_sW_s\big)dW_s - \int_0^t u(s)M_s\,ds.$$

Taking the expectation and using the property of Ito integrals we have

$$E[W_tM_t] = -\int_0^t u(s)E[M_s]\,ds = -\int_0^t u(s)\,ds = -U(t). \quad (10.4.24)$$

Substituting into (10.4.23) yields

$$\mathbb{E}^P[W_tM_T] = -U(t). \quad (10.4.25)$$

For computing $\mathbb{E}^P[W_t^2 M_T]$ we proceed in a similar way

$$\begin{aligned}\mathbb{E}^P[W_t^2M_T] &= \mathbb{E}^P[\mathbb{E}^P[W_t^2M_T|\mathcal{F}_t]] = \mathbb{E}^P[W_t^2\mathbb{E}^P[M_T|\mathcal{F}_t]] \\ &= \mathbb{E}^P[W_t^2M_t]. \quad (10.4.26)\end{aligned}$$

Using the product rule yields

$$\begin{aligned}d(W_t^2M_t) &= M_t d(W_t^2) + W_t^2 dM_t + d(W_t^2)dM_t \\ &= M_t(2W_t dW_t + dt) - W_t^2\big(u(t)M_t dW_t\big) \\ &\qquad - (2W_t dW_t + dt)\big(u(t)M_t dW_t\big) \\ &= M_tW_t\big(2 - u(t)W_t\big)dW_t + \big(M_t - 2u(t)W_tM_t\big)dt.\end{aligned}$$

Integrate between 0 and t

$$W_t^2M_t = \int_0^t [M_sW_s(2 - u(s)W_s)]\,dW_s + \int_0^t \big(M_s - 2u(s)W_sM_s\big)\,ds,$$

and take the expected value to get

$$\begin{aligned}\mathbb{E}^P[W_t^2 M_t] &= \int_0^t \big(E[M_s] - 2u(s)E[W_s M_s]\big)\, ds \\ &= \int_0^t \big(1 + 2u(s)U(s)\big)\, ds \\ &= t + U^2(t),\end{aligned}$$

where we used (10.4.24). Substituting into (10.4.26) yields

$$\mathbb{E}^P[W_t^2 M_T] = t + U^2(t). \tag{10.4.27}$$

Substituting (10.4.25) and (10.4.27) into relation (10.4.22) yields

$$\mathbb{E}^Q[X_t^2] = U^2(t) - 2U(t)^2 + t + U^2(t) = t, \tag{10.4.28}$$

which ends the proof of the proposition. ■

Now we are prepared to prove one of the most important results of Stochastic Calculus.

Theorem 10.4.3 (Girsanov's Theorem) *Let $u \in L^2[0,T]$ be a deterministic function. Then the process*

$$X_t = \int_0^t u(s)\, ds + W_t, \qquad 0 \le t \le T$$

is a Brownian motion with respect to the probability measure Q given by

$$dQ = e^{-\int_0^T u(s)\, dW_s - \frac{1}{2}\int_0^T u(s)^2\, ds} dP.$$

Proof: In order to prove that X_t is a Brownian motion on the probability space $(\Omega, \mathcal{F}, Q)$ we shall apply Lévy's characterization theorem, see Theorem 10.2.1. Lemma 10.4.1 implies that the process X_t satisfies the following properties:

1. $X_0 = 0$;

2. X_t is continuous in t;

3. X_t is a square integrable $\mathcal{F}_t$-martingale on the space $(\Omega, \mathcal{F}, Q)$. Using Proposition 10.4.2, the martingale property of W_t, and the additivity and the tower property of expectations yields

$$\begin{aligned}\mathbb{E}^Q[(X_t - X_s)^2] &= \mathbb{E}^Q[X_t^2] - 2\mathbb{E}^Q[X_t X_s] + \mathbb{E}^Q[X_s^2] \\ &= t - 2\mathbb{E}^Q[X_t X_s] + s \\ &= t - 2\mathbb{E}^Q[\mathbb{E}^Q[X_t X_s|\mathcal{F}_s]] + s \\ &= t - 2\mathbb{E}^Q[X_s \mathbb{E}^Q[X_t|\mathcal{F}_s]] + s \\ &= t - 2\mathbb{E}^Q[X_s^2] + s \\ &= t - 2s + s = t - s.\end{aligned}$$

4. The quadratic variation of X_t is

$$\langle X, X\rangle_t = \int_0^t (dX_s)^2 = \int_0^t ds = t.$$

■

Choosing $u(s) = \lambda$, constant, we obtain the following consequence.

Corollary 10.4.4 *Let W_t be a Brownian motion on the probability space $(\Omega, \mathcal{F}, P)$. Then the process*

$$X_t = \lambda t + W_t, \qquad 0 \leq t \leq T$$

is a Brownian motion on the probability space $(\Omega, \mathcal{F}, Q)$, where

$$dQ = e^{-\frac{1}{2}\lambda^2 T - \lambda W_T}\, dP.$$

This result states that a Brownian motion with drift can be viewed as a regular Brownian motion under a certain change of the probability measure.

Exercise 10.4.5 *Show that for any random variable X on $(\Omega, \mathcal{F})$ we have*

$$\mathbb{E}^P[X] = \mathbb{E}^Q[XM_T^{-1}].$$

Exercise 10.4.6 *Show that:*

(a) $\mathbb{E}^P[f(\lambda t + W_t)] = \mathbb{E}^Q[f(B_t)\, M_T^{-1}]$ for any continuous function f;

(b) $\mathbb{E}^P[\int_0^T e^{\lambda t + W_t}\, dt] = \mathbb{E}^Q[\int_0^T e^{B_t}\, dt\, M_T^{-1}]$, where B_t is a Q-Brownian motion.

Proposition 10.4.7 (Reduction of drift formulas) *Let W_t be a Brownian motion and f a measurable function. Then*

(i) $\mathbb{E}[f(\lambda t + W_t)] = e^{-\frac{\lambda^2 t}{2}}\mathbb{E}[f(W_t)e^{\lambda W_t}]$;

(ii) $\mathbb{E}[f(W_t)] = e^{-\frac{\lambda^2 t}{2}}\mathbb{E}[f(\lambda t + W_t)e^{-\lambda W_t}]$.

Proof: (*i*) Let W_t be a Brownian motion on the space $(\Omega, \mathcal{F}, P)$. By Girsanov's theorem, the process $X_t = \lambda t + W_t$ can be considered as a Brownian motion on $(\Omega, \mathcal{F}, Q)$ and we have

$$\begin{aligned}
\mathbb{E}[f(\lambda t + W_t)] &= \mathbb{E}^P[f(\lambda t + W_t)] = \mathbb{E}^Q[f(X_t)M_T^{-1}] = \mathbb{E}^Q[f(X_t)e^{\frac{\lambda^2 T}{2}+\lambda W_T}] \\
&= \mathbb{E}^Q[f(X_t)e^{\frac{\lambda^2 T}{2}+\lambda(X_T-\lambda T)}] \\
&= e^{-\frac{\lambda^2 T}{2}}\mathbb{E}^Q[f(X_t)e^{\lambda X_t}e^{\lambda(X_T-X_t)}] \\
&= e^{-\frac{\lambda^2 T}{2}}\mathbb{E}^Q[f(X_t)e^{\lambda X_t}]\,\mathbb{E}^Q[e^{\lambda(X_T-X_t)}] \\
&= e^{-\frac{\lambda^2 T}{2}}\mathbb{E}^Q[f(X_t)e^{\lambda X_t}]e^{\frac{\lambda^2}{2}(T-t)} \\
&= e^{-\frac{\lambda^2 t}{2}}\mathbb{E}^Q[f(X_t)e^{\lambda X_t}] \\
&= e^{-\frac{\lambda^2 t}{2}}E[f(W_t)e^{\lambda W_t}].
\end{aligned}$$

(ii) We apply Girsanov's theorem for the Q-Brownian motion $X_t = \lambda t + W_t$ and obtain

$$\begin{aligned}
\mathbb{E}^Q[f(X_t)] &= \mathbb{E}^P[f(\lambda t + W_t)M_T] = \mathbb{E}^P[f(\lambda t + W_t)e^{-\frac{\lambda^2 T}{2}-\lambda W_T}] \\
&= \mathbb{E}^P[f(\lambda t + W_t)e^{-\lambda W_t}e^{-\frac{\lambda^2 T}{2}-\lambda(W_T-W_t)}] \\
&= e^{-\frac{\lambda^2 T}{2}}\mathbb{E}^P[f(\lambda t + W_t)e^{-\lambda W_t}]\,\mathbb{E}^P[e^{-\lambda(W_T-W_t)}] \\
&= e^{-\frac{\lambda^2 t}{2}}\mathbb{E}^P[f(\lambda t + W_t)e^{-\lambda W_t}].
\end{aligned}$$

Replacing X_t by W_t in the first term yields the desired formula. ∎

Exercise 10.4.8 *Use the reduction of drift formulas and Example 8.10.7 to show*

(a) $\mathbb{E}[(\lambda t + W_t)^{2k}e^{-\lambda W_t}] = \dfrac{(2k)!}{2^k k!}t^k e^{\frac{\lambda^2 t}{2}}$;

(b) $\mathbb{E}[(\lambda t + W_t)^{2k+1}e^{-\lambda W_t}] = 0$.

Exercise 10.4.9 *Use the reduction of drift formula to show*

(a) $\mathbb{E}[\sin(t + \sigma W_t)] = e^{-\sigma^2 t/2}\sin t$;

(b) $\mathbb{E}[\cos(t + \sigma W_t)] = e^{-\sigma^2 t/2}\cos t$.

Exercise 10.4.10 *Use the reduction of drift formulas to find*

(a) $\mathbb{E}[\cos(\lambda t + W_t)e^{-\lambda W_t}]$;

(b) $\mathbb{E}[\sin(\lambda t + W_t)e^{-\lambda W_t}]$.

Exercise 10.4.11 *Let W_t be a Brownian motion on the space $(\Omega, \mathcal{F}, P)$, and $dQ = M_T dP$. Show that*

$$\mathbb{E}^Q[e^{\lambda t + W_t}] = e^{\frac{t}{2}}$$

(a) by a direct computation;

(b) using Girsanov's theorem.

Exercise 10.4.12 *Let $X_t = \lambda t + W_t$, with W_t a P-Brownian motion.*

(a) *Find $\mathbb{E}^P[X_t]$ and $\mathbb{E}^Q[X_t]$;*

(b) *Find $\mathbb{E}^P[X_t^2]$ and $\mathbb{E}^Q[X_t^2]$.*

Exercise 10.4.13 *Use Jensen's inequality to show that for any convex, measurable function f we have*

$$\mathbb{E}[f(\lambda t + W_t)e^{-\lambda W_t}] \geq f(0)e^{\frac{\lambda^2 t}{2}}.$$

Exercise 10.4.14 *Use the reduction of drift formulas to show*

(a) $\mathbb{E}[W_t e^{-\lambda W_t}] = -\lambda t e^{\frac{\lambda^2 t}{2}}$;

(b) $\mathbb{E}[W_t^2 e^{-\lambda W_t}] = (t + \lambda^2 t^2)e^{\frac{\lambda^2 t}{2}}$.

Exercise 10.4.15 *Consider the stochastic process $X_t = \frac{\lambda t^2}{2} + W_t$, where W_t is a Brownian motion on $(\Omega, \mathcal{F}, P)$.*

(a) *Find the probability measure dQ such that X_t becomes a Q-Brownian motion;*

(b) *Compute explicitly $\mathbb{E}^P[X_t M_T]$;*

(c) *Use $\mathbb{E}^P[X_t M_T] = \mathbb{E}^Q[X_t] = 0$ to find a formula for*

$$\mathbb{E}^P[W_t e^{-\lambda \int_0^t s\, dW_s}].$$

Remark 10.4.16 Girsanov's theorem can be used to compute, at least in theory, expectations of the form

$$\mathbb{E}[f(W_t)e^{\int_0^t g(s)\, dW_s}],$$

with f and g continuous functions.

Exercise 10.4.17 *Use the drift reduction formula to express $Var[f(\lambda t + W_t)]$.*

Chapter 11

Some Applications of Stochastic Calculus

In this chapter we shall present a few applications of stochastic calculus to a few applied domains of mathematics. The main idea is that some parameters, which in the case of deterministic Calculus are kept constant during the evolution of the process, in this case are influenced by the exterior white noise, which is modeled by the informal derivative of a Brownian motion, $\frac{dW_t}{dt}$. This way, the ordinary differential equations become stochastic differential equations, and their solutions are stochastic processes. For more applications of the white noise in chemistry and electricity one can consult the book of Gardiner [20]. For further applications to queueing theory the reader is referred to Ross [43]. For financial economics applications, see Sondermann [45], and for stochastical modeling of oil prices, see Postali and Picchetti [39]. For an application to car pricing in a stochastic environment see Alshamary and Calin [1] and [2].

11.1 White Noise

The *white noise* is used in applications as an idealization of a random noise that is independent at different times and has a very large fluctuation at any time. It can be successfully applied to problems involving an outside noise influence, such as trajectory of small particles which diffuse in a liquid due to the molecular bombardments, or signal processing, where it models the completely unpredictable "static" influence. The fact that the noise is not biased towards any specific "frequency", gives it its name "white noise". In this chapter we shall study a few applications of the white noise in kinematics, population growth, radioactive decay and filtering problems.

The white noise will be denoted by $\mathcal{N}_t$ and considered as a stochastic process. The effect of the noise during the time interval dt is normally distributed with mean zero and variance dt, and is given as an infinitesimal jump of a Brownian motion, $\mathcal{N}_t dt = dB_t$. Thus, it is convenient sometimes to represent the white noise informally as a derivative of a Brownian motion,

$$\mathcal{N}_t = \frac{dB_t}{dt}.$$

Since B_t is nowhere differentiable, the aforementioned derivative does not make sense classically. However, it makes sense in the following "generalized sense":

$$\int_{\mathbb{R}} \mathcal{N}_t f(t)\, dt = -\int_{\mathbb{R}} B_t f'(t)\, dt,$$

for any compact supported, smooth function f. Hence, from this point of view, the white noise $\mathcal{N}_t$ is a *generalized function* or a *distribution*. In the following we shall state its relation with the Dirac distribution δ_0, which is defined in the generalized sense as

$$\int_{\mathbb{R}} \delta_0(t) f(t)\, dt = f(0),$$

for any compact supported, smooth function f.

In order to study the white noise $\mathcal{N}_t$, we should investigate first the process

$$X_t^{(\epsilon)} = \frac{1}{\epsilon}\Big(B(t+\epsilon) - B(t)\Big), \quad t \geq 0,$$

which models the rate of change of a Brownian motion $B(t)$. Since we have

$$\begin{aligned} \mathbb{E}[X_t^{(\epsilon)}] &= \frac{1}{\epsilon}\Big(\mathbb{E}[B(t+\epsilon)] - \mathbb{E}[B(t)]\Big) = 0 \\ Var(X_t^{(\epsilon)}) &= \frac{1}{\epsilon^2}(t+\epsilon-t) = \frac{1}{\epsilon}, \end{aligned}$$

the limiting process $\mathcal{N}_t = \lim\limits_{\epsilon \searrow 0} X_t^{(\epsilon)}$ will have zero mean and infinite variance.

Consider $s < t$ and choose $\epsilon > 0$ small enough, such that $(s, s+\epsilon) \cap (t, t+\epsilon) = \emptyset$. Using the properties of Brownian motions, the differences $B(s+\epsilon) - B(s)$ and $B(t+\epsilon) - B(t)$ are independent. Hence, the random variables $\mathcal{N}_s$ and $\mathcal{N}_t$ are independent for $s < t$.

In the following we shall compute the covariance of the process $\mathcal{N}_t$. For reasons which will be clear later we shall extend the parameter t to take values in the entire real line $\mathbb{R}$. This can be done by defining the Brownian motion $B(t)$ as

$$B(t) = \begin{cases} W_1(t), & \text{if } t \geq 0 \\ W_2(-t), & \text{if } t < 0, \end{cases}$$

where $W_1(t)$ and $W_2(t)$ are two independent Brownian motions. First, we compute the covariance of the process $X_t^{(\epsilon)}$. Assume $s < t$. Then using the formula $\mathbb{E}[B(u)B(v)] = \min\{u, v\}$, we have

$$\begin{aligned}
Cov(X_s^{(\epsilon)}, X_t^{(\epsilon)}) &= \mathbb{E}[X_s^{(\epsilon)} X_t^{(\epsilon)}] \\
&= \frac{1}{\epsilon^2}\Big(\mathbb{E}\big[\big(B(s+\epsilon) - B(s)\big)\big(B(t+\epsilon) - B(t)\big)\big]\Big) \\
&= \frac{1}{\epsilon^2}\Big(\mathbb{E}\big[\big(B(s+\epsilon)B(t+\epsilon)\big] - \mathbb{E}\big[B(s+\epsilon)B(t)\big] \\
&\quad -\mathbb{E}\big[B(s)B(t+\epsilon)\big] + \mathbb{E}\big[B(s)B(t)\big]\big]\Big) \\
&= \frac{1}{\epsilon^2}\Big(s + \epsilon - \min\{s+\epsilon, t\} - s + s\Big) \\
&= \frac{1}{\epsilon^2}\Big(s + \epsilon - \Big(s + \min\{\epsilon, t - s\}\Big)\Big) \\
&= \frac{1}{\epsilon}\Big(1 - \min\Big\{1, \frac{t-s}{\epsilon}\Big\}\Big) = \frac{1}{\epsilon}\max\Big\{1 - \frac{t-s}{\epsilon}, 0\Big\}.
\end{aligned}$$

For any s, t we can derive the more general formula

$$Cov(X_s^{(\epsilon)}, X_t^{(\epsilon)}) = \frac{1}{\epsilon}\max\Big\{1 - \frac{|t-s|}{\epsilon}, 0\Big\}. \tag{11.1.1}$$

Consider the test function

$$\varphi_\epsilon(\tau) = \frac{1}{\epsilon}\max\Big\{1 - \frac{\tau}{\epsilon}, 0\Big\} = \begin{cases} \frac{1}{\epsilon}\Big(1 - \frac{\tau}{\epsilon}\Big), & \text{if } |\tau| \leq \epsilon \\ 0, & \text{if } |\tau| > \epsilon, \end{cases}$$

which verifies $\varphi_\epsilon(\tau) \geq 0$, $\varphi_\epsilon(0) = \dfrac{1}{\epsilon}$ and

$$\int_{\mathbb{R}} \varphi_\epsilon(\tau)\, d\tau = \int_{-\epsilon}^{\epsilon} \frac{1}{\epsilon}\Big(1 - \frac{\tau}{\epsilon}\Big)\, d\tau = 1.$$

Therefore, we have

$$\lim_{\epsilon \searrow 0} \varphi_\epsilon(\tau) = \delta_0(\tau),$$

where δ_0 is the Dirac distribution centered at 0. In fact, the above limit has the following meaning

$$\lim_{\epsilon \searrow 0} \int_{\mathbb{R}} \varphi_\epsilon(\tau) f(\tau)\, d\tau = f(0),$$

for any test function f.

Since the covariance formula (11.1.1) can also be expressed as

$$Cov(X_s^{(\epsilon)}, X_t^{(\epsilon)}) = \varphi_\epsilon(t-s),$$

then

$$Cov(\mathcal{N}_s, \mathcal{N}_t) = \lim_{\epsilon \searrow 0} Cov(X_s^{(\epsilon)}, X_t^{(\epsilon)}) = \lim_{\epsilon \searrow 0} \varphi_\epsilon(t-s) = \delta_0(t-s).$$

We arrive at the following definition of the white noise.

Definition 11.1.1 *A white noise is a generalized stochastic process $\mathcal{N}_t$, which is stationary and Gaussian, with mean and covariance given by*

$$\begin{aligned} \mathbb{E}[\mathcal{N}_t] &= 0 \\ Cov(\mathcal{N}_s, \mathcal{N}_t) &= \delta_0(t-s). \end{aligned}$$

11.2 Stochastic Kinematics

During a race, a cyclist has average speed m. However, his speed varies in time. Sometimes the cyclist exceeds the speed m, but he gets tired after a while and slows down. If the cyclist's speed decreases under the mean m, then he recuperates the muscle power and is able to speed up again. The cyclist's instantaneous velocity v_t satisfies a mean reverting process described by the equation

$$dv_t = a(m - v_t)dt + \sigma dW_t,$$

where σ and a are two positive constants that correspond to the volatility and rate at which the velocity is pulled towards the mean m. The solution is given by

$$v_t = m + (v_0 - m)e^{-at} + \sigma e^{-at} \int_0^t e^{as}\, dW_s. \tag{11.2.2}$$

Since the last term is a Wiener integral, the speed v_t is normally distributed with mean and variance

$$\begin{aligned} \mathbb{E}[v_t] &= m + (v_0 - m)e^{-at} \\ Var(v_t) &= \frac{\sigma^2}{2a}(1 - e^{-2at}). \end{aligned}$$

The expectation of the speed as of time u is given by the conditional expectation given the information $\mathcal{F}_u$ available at time u

$$\mathbb{E}[v_t|\mathcal{F}_u] = m + (v_0 - m)e^{-at} + \sigma e^{-at} \int_0^u e^{as}\, dW_s.$$

The cyclist's stochastic coordinate at time t is obtained integrating the velocity

$$\begin{aligned} x_t &= x_0 + \int_0^t v_s\,ds \\ &= x_0 + \int_0^t [m + (v_0 - m)e^{-as} + \sigma e^{-as}\int_0^s e^{au}\,dW_u]\,ds \\ &= x_0 + mt + (v_0 - m)\frac{1 - e^{-at}}{a} + \sigma\int_0^t\int_0^s e^{a(u-s)}\,dW_u\,ds. \end{aligned}$$

The expected coordinate is

$$\begin{aligned} \mathbb{E}[x_t] &= x_0 + \int_0^t \mathbb{E}[v_s]\,ds \\ &= x_0 + \int_0^t [m + (v_0 - m)e^{-at}]\,ds \\ &= x_0 + mt + (v_0 - m)\frac{1 - e^{-at}}{a} \\ &= x_{unif}(t) + (v_0 - m)\frac{1 - e^{-at}}{a}. \end{aligned}$$

The term $x_{unif}(t) = x_0 + mt$ denotes the coordinate the cyclist would have if moving at the constant velocity m. The difference $v_0 - m$ provides the following upper and lower bounds

$$(1 - at/2)|v_0 - m|t \le |\mathbb{E}[x_t] - x_{unif}(t)| \le |v_0 - m|t.$$

This shows that the error between the expected coordinate and the coordinate of a uniform move is at most linear in time and is controlled by the difference $v_0 - m$. Therefore, the expected coordinate is the classical coordinate, $\mathbb{E}[x_t] = x_{unif}(t)$, if and only if $v_0 = m$.

The acceleration a_t is obtained as the derivative of velocity v_t with respect to time

$$\begin{aligned} a_t &= \frac{dv_t}{dt} \\ &= -a(v_0 - m)e^{-at} - a\sigma e^{-at}\int_0^t e^{as}\,dW_s + \sigma\frac{dW_t}{dt} \\ &= a_0 e^{-at} - a\sigma e^{-at}\int_0^t e^{as}\,dW_s + \sigma\frac{dW_t}{dt}, \end{aligned} \tag{11.2.3}$$

where a_0 is the initial acceleration. The first term is a deterministic function, the second term is a normally distributed random variable of zero mean, while $\sigma\frac{dW_t}{dt}$ is the white noise term. Since $\mathbb{E}\Big[\frac{dW_t}{dt}\Big] = 0$, the long run limit of the expectation becomes $\mathbb{E}[a_t] = a_0 e^{-at} \to 0$, as $t \to \infty$.

Let M denote the mass of the cyclist and $F_t = Ma_t$ be the muscle force developed at time t. The work done by the cyclist between instances 0 and t is given by

$$\mathbb{W} = \int_{x_0}^{x_t} F_s\,dx_s = M\int_{x_0}^{x_t} a_s\,dx_s = M\int_0^t a_s v_s\,ds,$$

where the velocity and the acceleration are given by (11.2.2) and (11.2.3). Computing the exact expression of $\mathbb{W}$ is tedious. However, using the properties of Ito integrals one can compute $\mathbb{E}[\mathbb{W}]$, see Exercise 11.2.1.

Since the square of velocity is given by

$$\begin{aligned} v_t^2 &= \Big(m + (v_0 - m)e^{-at}\Big)^2 + 2\sigma e^{-at}\Big(m + (v_0 - m)e^{-at}\Big)\int_0^t e^{as}\,dW_s \\ &\quad + \sigma^2 e^{-2at}\Big(\int_0^t e^{as}\,dW_s\Big)^2, \end{aligned}$$

the expectation of the kinetic energy becomes

$$\begin{aligned} \mathbb{E}\Big[\frac{Mv_t^2}{2}\Big] &= \frac{M}{2}\Big[\Big(m + (v_0 - m)e^{-at}\Big)^2 + 0 + \sigma^2 e^{-2at}\mathbb{E}\Big[\Big(\int_0^t e^{as}\,dW_s\Big)^2\Big]\Big] \\ &= \frac{M}{2}\Big[\Big(m + (v_0 - m)e^{-at}\Big)^2 + \frac{\sigma^2}{2a}(1 - e^{-2at})\Big]. \end{aligned}$$

Exercise 11.2.1 *Show that*

$$\mathbb{E}[\mathbb{W}] = M\Big\{a_0 m\frac{1 - e^{-at}}{a} + \Big[a_0(v_0 - m) + \frac{\sigma^2}{2}\Big]\frac{1 - e^{-2at}}{2a} - \frac{\sigma^2}{2}t\Big\}.$$

Exercise 11.2.2 *A snowflake in falling motion is described by the equation*

$$dv_t = g\,dt + \sigma v_t\,dW_t, \qquad v_0 = 0,$$

where g and σ are positive constants.

(*a*) *Find* $\mathbb{E}[v_t]$;

(*b*) *Compute* $\mathbb{E}[a_t]$;

(*c*) *Solve the equation to find a formula for the velocity* v_t;

(*d*) *Find a formula for the acceleration* a_t.

Exercise 11.2.3 *A particle with the initial velocity $v_0 = 1$ m/sec decelerates in a noisy way according to the equation*

$$dv_t = -2v_t dt + 0.3\,dW_t.$$

(*a*) *What is the probability that at the time* $t = 3$ *sec the velocity of the particle is less than* 0.1 *m/sec?*

(*b*) *Find the work done by the environment on the particle in order to decelerate it from* $v_0 = 1$ *m/sec to* $v = 0.5$ *m/sec.*

Exercise 11.2.4 *A snowball rolls downhill with the velocity given by the equation*

$$dv_t = 0.6\, v_t\, dt + 0.2\, dW_t, \qquad v_0 = 0.$$

(*a*) *Find the velocity* v_t;

(*b*) *What is the probability that the velocity is greater than* 10 *m/sec at* $t = 20$ *sec?*

11.3 Radioactive Decay

Consider a radioactive atom which contains $N(t)$ nuclei at time t. Assume the number of nuclei which decay during the time interval Δt is Poisson distributed

$$P\Big(N(t) - N(t + \Delta t) = n\Big) = \frac{\lambda^n (\Delta t)^n}{n!} e^{-\lambda \Delta t},$$

where the constant λ stands for the decay rate. For a small time interval Δt the previous formula becomes

$$P\Big(N(t) - N(t + \Delta t) = 1\Big) = \lambda \Delta t,$$

i.e. the probability of the occurrence of one decay in a small time interval is proportional with the time interval. The probability of the complementary event is

$$P\Big(N(t) - N(t + \Delta t) = 0\Big) = 1 - \lambda \Delta t. \tag{11.3.4}$$

Divide the interval $[0, t]$ into n equidistant subintervals

$$0 = t_0 < t_1 < t_2 < \cdots < t_{n-1} < t_n = t,$$

with $\Delta t = t_{k+1} - t_k = t/n$. The event of not having any decays during the interval $[0, t]$ can be expressed as

$$\{N(0) - N(t) = 0\} = \bigcap_{k=0}^{n-1} \{N(t_k) - N(t_{k+1}) = 0\}.$$

Since the increments of a Poisson process are independent, we have

$$P\Big(N(0) - N(t) = 0\Big) = \bigcap_{k=0}^{n-1} P\Big(N(t_k) - N(t_{k+1}) = 0\Big) = (1 - \lambda \Delta t)^n = \Big(1 - \frac{\lambda t}{n}\Big)^n.$$

Let $n \to \infty$ and obtain

$$P\Big(N(0) - N(t) = 0\Big) = e^{-\lambda t}.$$

For $N(t)$ large enough, this probability represents the fraction of nuclei that survived the decay during the time interval $[0, t]$. Since the percentage of nuclei that are still "alive" after time t is represented by the quotient $N(t)/N(0)$, we have

$$\frac{N(t)}{N(0)} = e^{-\lambda t}.$$

This relation, written as $N(t) = N(0)e^{-\lambda t}$, is the *law of radioactive decay.*

Now we shall develop a differential equation for $N(t)$. Relation (11.3.4) states that the fraction of nuclei that resist the decay during the time interval $[t, t + \Delta t]$ is

$$\frac{N(t + \Delta t)}{N(t)} = 1 - \lambda \Delta t.$$

Cross multiplying and subtracting $N(t)$ yields

$$N(t) - N(t + \Delta t) = -\lambda N(t)\Delta t. \tag{11.3.5}$$

Assuming that the period of observation is infinitely fine, $\Delta t \to dt$, the equation becomes

$$dN(t) = -\lambda N(t)\, dt.$$

This describes the kinetics of the radioactive decay, stating that the change in the number of nuclei $dN(t)$ during the time interval dt is proportional with the number of nuclei $N(t)$. Solving the aforementioned equation, we obtain again the law of radioactive decay

$$N(t) = N(0)e^{-kt}.$$

Noisy radioactive decay In real life relation (11.3.5) does not hold exactly, and some errors of measurement or counting are involved. These will be added as a noisy term

$$N(t) - N(t + \Delta t) = -\lambda N(t)\Delta t + \text{"noise"}.$$

For Δt small, this becomes a stochastic differential equation

$$dN(t) = -\lambda N(t)\, dt + \sigma dW_t,$$

with σ positive constant and W_t Brownian motion. The obtained equation is called *Langevin's equation.* We shall solve it as a linear stochastic differential equation. Multiplying by the integrating factor $e^{\lambda t}$ yields

$$d(e^{\lambda t} N(t)) = \sigma e^{\lambda t} dW_t.$$

Integrating yields

$$e^{\lambda t} N(t) = N(0) + \sigma \int_0^t e^{\lambda s}\, dW_s.$$

Hence the solution is

$$N(t) = N(0)e^{-\lambda t} + \sigma e^{-\lambda t} \int_0^t e^{\lambda s}\, dW_s. \tag{11.3.6}$$

This is the *Ornstein-Uhlenbeck process*. Since the last term is a Wiener integral, by Proposition 8.2.1 we have that $N(t)$ is Gaussian with the mean

$$\mathbb{E}[N(t)] = N(0)e^{-\lambda t} + E\Big[\sigma \int_0^t e^{\lambda(s-t)}\, dW_s\Big] = N(0)e^{-\lambda t}$$

and variance

$$Var[N(t)] = Var\Big[\sigma \int_0^t e^{\lambda(s-t)}\, dW_s\Big] = \frac{\sigma^2}{2\lambda}(1 - e^{-2\lambda t}).$$

Using Exercise 10.3.11 we can write the Gaussian term as a Brownian motion under a time change as

$$\int_0^t e^{\lambda s}\, dW_s = B_{(e^{2\lambda t}-1)/(2\lambda)},$$

with B_t Brownian motion. Hence the solution can also be written as

$$N(t) = N(0)e^{-\lambda t} + \sigma e^{-\lambda t} B_{(e^{2\lambda t}-1)/(2\lambda)}.$$

Using the expansion

$$e^{2\lambda t} = 1 + 2\lambda t + o(t^2), \qquad t \to 0,$$

then $(e^{2\lambda t} - 1)/(2\lambda) = t + o(t^2)$, and hence the following approximation holds for t small

$$N(t) = N(0)e^{-\lambda t} + \sigma e^{-\lambda t} B_t.$$

Exercise 11.3.1 *Let $N(t)$ be a noisy radioactive decay. Define the half time h as*

$$h = \inf\{t > 0; N(t) \leq \frac{1}{2} N(0)\}.$$

(*a*) *Prove that* $\mathbb{E}[e^{\lambda h}] = 2$;

(*b*) *Use Jensen's inequality to show that* $\mathbb{E}[h] \leq \frac{\ln 2}{\lambda}$.

Exercise 11.3.2 *Consider a machine which consists initially of N distinct parts. Assume the number of parts which get defective during the time interval Δt is Poisson distributed*

$$P\Big(N(t) - N(t + \Delta t) = n\Big) = \frac{\lambda^n(\Delta t)^n}{n!}e^{-\lambda\Delta t}.$$

(a) What is the probability that all parts still function perfectly at time t?

(b) Find the time t such that the 90% of the machine functions perfectly at time t.

Exercise 11.3.3 *A living organism has initially N_0 cells. Assume the number of cells which die during the time interval Δt is Poisson distributed*

$$P\Big(N(t) - N(t + \Delta t) = n\Big) = \frac{\lambda^n(\Delta t)^n}{n!}e^{-\lambda\Delta t}.$$

The organism dies when at least 30% *of its cells are dead. Find an approximation of the death time of the organism.*

11.4 Noisy Pendulum

The small oscillations of a free simple pendulum can be described by the linear equation

$$\ddot{\theta}(t) = -k^2\theta(t), \tag{11.4.7}$$

where $\theta(t)$ is the angle between the string and the vertical direction. If the exterior perturbations are modeled by a white noise process, $\mathcal{N}_t$, then the pendulum equation under small deviations writes as

$$\ddot{\theta}(t) + k^2\theta(t) = \sigma\mathcal{N}_t, \tag{11.4.8}$$

where k and σ are constants and the noise is given informally as $\mathcal{N}_t = \frac{dB_t}{dt}$. The general solution of equation (11.4.8) can be expressed as the sum

$$\theta(t) = \theta_p(t) + \theta_0(t), \tag{11.4.9}$$

where $\theta_p(t)$ is a particular solution of (11.4.8) and $\theta_0(t)$ is the solution of the associated homogeneous equation (11.4.7).

Standard ODE methods provide

$$\theta_0(t) = c_1\cos(kt) + c_2\sin(kt),$$

with $c_1, c_2 \in \mathbb{R}$. The particular solution $\theta_p(t)$ can be obtained by the method of variable coefficients. We are looking for a particular solution

$$\theta_p(t) = u_1(t)\cos(kt) + u_2(t)\sin(kt), \tag{11.4.10}$$

where $u_1(t)$ and $u_2(t)$ are two differentiable functions, which will be determined later. Assuming

$$u_1'(t)\cos(kt) + u_2'(t)\sin(kt) = 0, \tag{11.4.11}$$

differentiating yields

$$\begin{aligned}\theta_p'(t) &= u_1'(t)\cos(kt) + u_2'(t)\sin(kt) - ku_1(t)\sin(kt) + ku_2(t)\cos(kt)\\ &= -ku_1(t)\sin(kt) + ku_2(t)\cos(kt).\end{aligned}$$

Then the second derivative is

$$\begin{aligned}\theta_p''(t) &= -ku_1'(t)\sin(kt) + ku_2'(t)\cos(kt)\\ &\quad -k^2u_1(t)\cos(kt) - k^2u_2(t)\sin(kt).\end{aligned} \tag{11.4.12}$$

Substituting (11.4.12) and (11.4.10) into (11.4.8) yields

$$-ku_1'(t)\sin(kt) + ku_2'(t)\cos(kt) = \sigma\frac{dB_t}{dt}, \tag{11.4.13}$$

where we used the informal notation for the white noise $\mathcal{N}_t = \dfrac{dB_t}{dt}$. Equations (11.4.11) and (11.4.13) yield the ODEs system in u_1 and u_2

$$u_1'(t)\cos(kt) + u_2'(t)\sin(kt) = 0 \tag{11.4.14}$$

$$-u_1'(t)\sin(kt) + u_2'(t)\cos(kt) = \frac{\sigma}{k}\frac{dB_t}{dt}. \tag{11.4.15}$$

The reduction method and integration produces the following solutions

$$u_1(t) = -\frac{\sigma}{k}\int_0^t \sin(ks)\,dB_s \tag{11.4.16}$$

$$u_2(t) = \frac{\sigma}{k}\int_0^t \cos(ks)\,dB_s. \tag{11.4.17}$$

These represent the effect of the white noise dB_s along the solutions trajectories $\sin(ks)$ and $\cos(ks)$. From the properties of Wiener integrals, it follows that $u_1(t)$ and $u_2(t)$ have normal distributions with the mean, variances and covariance given by

$$\begin{aligned}
\mathbb{E}[u_1(t)] &= \mathbb{E}[u_2(t)] = 0 \\
Var[u_1(t)] &= \frac{\sigma^2}{k^2}\int_0^t \sin^2(ks)\,ds = \frac{\sigma^2}{k^2}\Big(\frac{t}{2} - \frac{\sin(2kt)}{4k}\Big) \\
Var[u_2(t)] &= \frac{\sigma^2}{k^2}\int_0^t \cos^2(ks)\,ds = \frac{\sigma^2}{k^2}\Big(\frac{t}{2} + \frac{\sin(2kt)}{4k}\Big)
\end{aligned}$$

$$\begin{aligned}
Cov[u_1(t), u_2(t)] &= \mathbb{E}[u_1(t)u_2(t)] \\
&= E\Big[-\frac{\sigma}{k}\int_0^t \sin(ks)\,dB_s, \frac{\sigma}{k}\int_0^t \cos(ks)\,dB_s\Big] \\
&= -\frac{\sigma^2}{k^2}\int_0^t \sin(ks)\cos(ks)\,ds = -\frac{\sigma^2}{k^3}\frac{\sin^2(kt)}{2}.
\end{aligned}$$

The particular solution (11.4.10) becomes

$$\theta_p(t) = \Big(-\frac{\sigma}{k}\int_0^t \sin(ks)\,dB_s\Big)\cos(kt) + \Big(\frac{\sigma}{k}\int_0^t \cos(ks)\,dB_s\Big)\sin(kt). \tag{11.4.18}$$

Hence the general solution for the pendulum equation given by (11.4.9) is

$$\theta(t) = \Big(c_1 - \frac{\sigma}{k}\int_0^t \sin(ks)\,dB_s\Big)\cos(kt) + \Big(c_2 + \frac{\sigma}{k}\int_0^t \cos(ks)\,dB_s\Big)\sin(kt),$$

where the constants c_1 and c_2 depend on the initial data as

$$c_1 = \theta(0), \qquad c_2 = \frac{\theta'(0)}{k}.$$

It is worth noting that $\theta(t)$ is not normally distributed. This follows from the fact that $Cov[u_1(t), u_2(t)] \neq 0$ implies that $u_1(t)$ and $u_2(t)$ are not independent. However, we are able to compute the mean and variance as in the following

$$\begin{aligned}
\mathbb{E}[\theta(t)] &= \theta_0(t) = c_1\cos(kt) + c_2\sin(kt) \\
Var[\theta(t)] &= Var[\theta_p(t)] = \cos^2(kt)Var(u_1(t)) + \sin^2(kt)Var(u_2(t)) \\
&\quad +2\sin(kt)\cos(kt)Cov(u_1(t), u_2(t)) \\
&= \frac{\sigma^2 t}{2k^2} - \frac{\sigma^2}{4k^3}\sin(2kt).
\end{aligned}$$

We shall present in the following another method for finding X_t, involving an integrating factor. Considering $X_1(t) = \theta(t)$, $X_2(t) = \dot{\theta}(t)$, the pendulum equation

$$\ddot{\theta}(t) = -k^2\theta(t) + \sigma\frac{dB_t}{dt}$$

becomes a first order system of stochastic differential equations

$$\begin{aligned} dX_1(t) &= X_2(t)\,dt \\ dX_2(t) &= -k^2 X_1(t)\,dt + \sigma\,dB_t. \end{aligned}$$

Denoting

$$X_t = \begin{pmatrix} X_1 \\ X_2 \end{pmatrix}, \qquad A = \begin{pmatrix} 0 & 1 \\ -k^2 & 0 \end{pmatrix}, \quad K = \begin{pmatrix} 0 \\ \sigma \end{pmatrix},$$

the aforementioned system becomes a linear matrix stochastic differential equation

$$dX_t = AX_t\,dt + K\,dB_t.$$

Multiplying by the integrating factor e^{-At} yields the exact equation

$$d(e^{-At}X_t) = e^{-At}K\,dB_t.$$

Integrating we obtain the solution

$$X_t = e^{At}X_0 + \int_0^t e^{A(t-s)}K\,dB_s. \tag{11.4.19}$$

Since $A^2 = -k^2\mathbb{I}_2$, a computation of the exponential of At involving a power series provides

$$\begin{aligned} e^{At} &= \sum_{n\geq 0}\frac{A^n t^n}{n!} = \sum_{n\geq 0}\frac{A^{2n}t^{2n}}{(2n)!} + \sum_{n\geq 0}\frac{A^{2n+1}t^{2n+1}}{(2n+1)!} \\ &= \sum_{n\geq 0}\frac{(-1)^n k^{2n}t^{2n}}{(2n)!}\mathbb{I}_2 + \frac{1}{k}\sum_{n\geq 0}\frac{(-1)^n k^{2n+1}t^{2n+1}}{(2n+1)!}A \\ &= \cos(kt)\mathbb{I}_2 + \frac{1}{k}\sin(kt)A \\ &= \begin{pmatrix} \cos(kt) & 0 \\ 0 & \cos(kt) \end{pmatrix} + \begin{pmatrix} 0 & \frac{1}{k}\sin(kt) \\ -k\sin(kt) & 0 \end{pmatrix} \\ &= \begin{pmatrix} \cos(kt) & \frac{1}{k}\sin(kt) \\ -k\sin(kt) & \cos(kt) \end{pmatrix}. \end{aligned}$$

The expectation of X_t is given by

$$\begin{aligned} \mathbb{E}[X_t] = e^{At}X_0 &= \begin{pmatrix} \cos(kt) & \frac{1}{k}\sin(kt) \\ -k\sin(kt) & \cos(kt) \end{pmatrix}\begin{pmatrix} \theta(0) \\ \dot\theta(0) \end{pmatrix} \\ &= \begin{pmatrix} \cos(kt)\theta(0) + \frac{1}{k}\sin(kt)\dot\theta(0) \\ -k\sin(kt)\theta(0) + \cos(kt)\dot\theta(0) \end{pmatrix}. \end{aligned}$$

Considering each component separately, implies

$$\begin{aligned}\mathbb{E}[\theta(t)] &= \cos(kt)\theta(0) + \frac{1}{k}\sin(kt)\dot{\theta}(0)\\ \mathbb{E}[\dot{\theta}(t)] &= -k\sin(kt)\theta(0) + \cos(kt)\dot{\theta}(0).\end{aligned}$$

Since

$$e^{A(t-s)}K = \begin{pmatrix} \frac{\sigma}{k}\sin(k(t-s)) \\ \sigma\cos(k(t-s)) \end{pmatrix},$$

the integral term of (11.4.19) can be computed as

$$\int_0^t e^{A(t-s)}K\,dB_s = \begin{pmatrix} \dfrac{\sigma}{k}\displaystyle\int_0^t \sin(k(t-s))\,dB_s \\ \\ \sigma\displaystyle\int_0^t \cos(k(t-s))\,dB_s \end{pmatrix}.$$

Since any electronic circuit is mathematically equivalent to a pendulum equation, a similar method of study can be applied to it. It is worth noting that the analysis of noise in electronic circuits was developed as early as 1920s by Rice [42] and Schottky [44].

Exercise 11.4.1 *Consider an electric circuit, in which the charge Q_t at time t satisfies the equation*

$$\frac{d^2Q_t}{dt^2} + 3\frac{dQ_t}{dt} + 2Q_t = \mathcal{N}_t,$$

where the external force is just the influence of the white noise $\mathcal{N}_t = \dfrac{dW_t}{dt}$.

(a) Find the solution of the homogeneous equation;

(b) Show that a particular solution is given by

$$Q_t^p = \int_0^t \left(e^{-(t-s)} - e^{-2(t-s)}\right) dW_s.$$

(c) Find the general solution, Q_t, and show that $\mathbb{E}[Q_t]$ satisfies the homogeneous equation.

Exercise 11.4.2 *(a) Solve the following stochastic differential equation*

$$\begin{aligned}dX_t &= Y_t\,dt + \alpha\,dW_t^1\\ dY_t &= -X_t\,dt + \beta\,dW_t^2,\end{aligned}$$

where (W_t^1, W_t^2) is a 2-dimensional Brownian motion and α and β are constants.

(b) Use part (a) to find a solution for the following stochastic pendulum equation

$$\ddot{\theta}_t + \theta_t = \beta\dot{W}_t^2 + \alpha\ddot{W}_t^1.$$

11.5 Stochastic Population Growth

This section presents a few population growth models driven by noisy growth rates. This implies that the population size is stochastic.

Exponential growth model The population at time t, denoted by P_t, satisfies the growth equation

$$dP_t = r_t\, P_t\, dt, \tag{11.5.20}$$

where r_t is the stochastic growth rate. Assume that r_t oscillates irregularly around some deterministic average function $a(t)$

$$r_t = a(t) + \text{"noise"}.$$

If the size of the white noise is β, then

$$\text{"noise"} = \beta \mathcal{N}_t = \beta \frac{dB_t}{dt}.$$

Substituting in (11.5.20) yields the following SDE

$$dP_t = a(t)P_t dt + \beta P_t dB_t, \tag{11.5.21}$$

where $\beta > 0$ is a constant and B_t is a Brownian motion. The equation (11.5.21) can be reduced to an exact equation multiplying by the integrating factor $\rho_t = e^{\frac{1}{2}\beta^2 t - \beta B_t}$. Ito's formula provides

$$d\rho_t = \rho_t(\beta^2 dt - \beta dt),$$

and hence $d\rho_t\, dP_t = -\beta^2 Y_t\, dt$. Denoting $Y_t = \rho_t P_t$ and applying the product rule yields

$$\begin{aligned} dY_t &= d(\rho_t P_t) = d\rho_t\, P_t + \rho_t dP_t + d\rho_t\, dP_t \\ &= (\beta^2 Y_t + a(t)Y_t - \beta^2 Y_t)dt + (\beta Y_t - \beta Y_t)dB_t \\ &= a(t)Y_t dt. \end{aligned}$$

Hence the process Y_t satisfies the deterministic equation

$$dY_t = a(t)Y_t\, dt$$

with the initial condition $Y_0 = \rho_0 P_0 = P_0$. Integrating yields

$$Y_t = P_0 e^{\int_0^t a(s)\, ds}.$$

Solving for P_t we obtain

$$\boxed{P_t = P_0 e^{\int_0^t a(s)\, ds - \beta^2 t/2 + \beta B_t}.} \tag{11.5.22}$$

Using $\mathbb{E}[e^{\beta B_t}] = e^{\beta^2 t/2}$, we obtain

$$\mathbb{E}[P_t] = P_0 e^{\int_0^t a(s)\,ds - \beta^2 t/2} \mathbb{E}[e^{\beta B_t}] = P_0 e^{\int_0^t a(s)\,ds}.$$

It is worth noting that the function $P(t) = \mathbb{E}[P_t]$ satisfies the deterministic equation

$$dP(t) = a(t)\,P(t)dt.$$

The population P_t provided by formula (11.5.22) is log-normally distributed. In fact, $\ln \dfrac{P_t}{P_0}$ is normally distributed

$$\ln \frac{P_t}{P_0} \sim N(m(t), \beta^2 t),$$

with the mean given by

$$m(t) = \int_0^t a(s)ds - \frac{\beta^2 t}{2}.$$

Exercise 11.5.1 *Consider the population given by the formula* (11.5.22).

(*a*) *Find the probability distribution function of* P_t.

(*b*) *Find the probability density function of* P_t.

Exercise 11.5.2 *A bacteria population has an intrinsic growth rate of* $r = 0.08$ *and noise size* $\beta = 0.01$ *per day. If the population starts with* 10,000 *bacteria, find the probability that there are more than* 11,000 *bacteria after 2 days.*

Exercise 11.5.3 *A population has a noisy growth rate given by* $r_t = t^2 + \frac{dW_t}{dt}$. *Find the doubling time* T, *which satisfies*

$$\mathbb{E}[P_T] = 2P_0.$$

Population growth in a stochastic and crowded environment In the previous exponential growth model the population can increase indefinitely. A more realistic model was obtained by P.F. Verhust in 1832 (and rediscovered by R. Pearl in the twentieth century), who assumed that due to competition the population also tends to decrease at a rate proportional with the number of encounters between the population members, which is proportional with the square of the population size

$$dP(t) = rP(t)dt - kP(t)^2 dt. \qquad (11.5.23)$$

The constant r is the intrinsic growth rate, i.e. the relative rate at which the population would increase if there were no restrictions on the population. The positive constant k reflects the damping effect on the population growth caused by competition for resources between the members of the same population. The solution of the equation (11.5.23) is given by the *logistic function*

$$P(t) = \frac{P_0 K}{P_0 + (K - P_0)e^{-rt}}, \tag{11.5.24}$$

where $K = r/k$ is the saturation level, or carrying capacity of the environment. This represents the equilibrium level to which the population, regardless of its initial size, will tend in the long run

$$K = \lim_{t\to\infty} P(t).$$

One of the stochastic models for the population growth in a stochastic and competitive environment is obtained keeping in equation (11.5.23) the rate k constant, while considering a noisy intrinsic rate of growth

$$dP_t = (r + \beta N_t)P_t\, dt - kP_t^2 dt. \tag{11.5.25}$$

This equation can be written equivalently as

$$dP_t = rP_t\, dt - kP_t^2 dt + \beta P_t\, dB_t, \tag{11.5.26}$$

where the positive constant β measures the size of the noise of the system. Rewriting the equation as

$$dP_t = kP_t(K - P_t)dt + \beta P_t\, dB_t,$$

and multiplying by the integrating factor $\rho_t = e^{\frac{1}{2}\beta^2 t - \beta B_t}$ leads to the exact equation

$$d(\rho_t P_t) = k\rho_t P_t(K - P_t)dt.$$

Substituting $Y_t = \rho_t P_t$ yields the equation

$$dY_t = kY_t(K - \rho_t^{-1} Y_t)dt. \tag{11.5.27}$$

In order to solve (11.5.27) we shall make the new substitution $Z_t = \dfrac{e^{rt}}{Y_t}$. Since $(dY_t)^2 = 0$, Ito's formula provides

$$\begin{aligned} dZ_t &= r\frac{e^{rt}}{Y_t}dt - \frac{1}{Y_t^2}e^{rt}dY_t \\ &= rZ_t dt - \frac{1}{Y_t^2}e^{rt}kY_t(K - \rho_t^{-1}Y_t)dt \\ &= rZ_t dt - kZ_t(K - \rho_t^{-1}Y_t)dt \\ &= kZ_t Y_t \rho_t^{-1} dt \\ &= ke^{rt}\rho_t^{-1} dt, \end{aligned}$$

where we used that $r = kK$ and $Z_t Y_t = e^{rt}$. The process Z_t satisfies the integrable equation

$$dZ_t = ke^{rt}\rho_t^{-1}dt$$

with the solution

$$Z_t = Z_0 + k\int_0^t e^{rs}\rho_s^{-1}\,ds.$$

Since $Z_0 = \frac{1}{Y_0} = \frac{1}{P_0}$, substituting back we obtain the following expression for the population

$$P_t = \rho_t^{-1}Y_t = \rho_t^{-1}\frac{e^{rt}}{Z_t} = \frac{e^{rt-\beta^2 t/2+\beta B_t}}{\frac{1}{P_0} + k\int_0^t e^{rs}\rho_s^{-1}\,ds}.$$

In the following we shall manipulate the previous expression in order to make it look as close as possible to the logistic equation (11.5.24).

$$\begin{aligned}
P_t &= \frac{e^{rt-\beta^2 t/2+\beta B_t}}{\frac{1}{P_0} + k\int_0^t e^{rs}\rho_s^{-1}\,ds} = \frac{P_0 e^{rt}e^{-\beta^2 t/2+\beta B_t}}{1 + kP_0\int_0^t e^{rs}\rho_s^{-1}\,ds} = \frac{P_0 K e^{rt}e^{-\beta^2 t/2+\beta B_t}}{K + rP_0\int_0^t e^{rs}\rho_s^{-1}\,ds} \\
&= \frac{P_0 K e^{-\beta^2 t/2+\beta B_t}}{Ke^{-rt} + rP_0 e^{-rt}\int_0^t e^{rs}\rho_s^{-1}\,ds} \\
&= \frac{P_0 K e^{-\beta^2 t/2+\beta B_t}}{(K - P_0)e^{-rt} + P_0\Big(1 + r\int_0^t e^{rs}\rho_s^{-1}\,ds\Big)e^{-rt}}.
\end{aligned}$$

If in the previous formula we let $\beta = 0$, and hence $\rho_s = 1$, we obtain exactly the expression (11.5.24).

A more sophisticated model is obtained if in equation (11.5.23) both rates r and k are noisy, with the size of the noise proportional with the rates as follows

$$r_t = r + \alpha r\frac{dB_t}{dt}, \qquad k_t = k + \alpha k\frac{dB_t}{dt}.$$

We note that both rates are driven by the same uncertainty source. Substituting in the equation yields the following SDE

$$dP_t = (rP_t - kP_t^2)dt + \alpha(rP_t - kP_t^2)dB_t.$$

It can be shown that this equation has a unique strong solution, but the discussion of this subject is beyond the level of this textbook.

Population growth in a stochastic catastrophic environment In the previous model the population tends to decrease due to competition and limited space. In the present model the population decreases suddenly due to some unexpected catastrophic events, such as earthquakes, wars, diseases, natural calamities, etc. The SDE satisfied by the population in this case is

$$dP_t = rP_t\,dt - \beta P_t\,dN_t, \tag{11.5.28}$$

where N_t is a Poisson process with rate λ. The positive constant β is a measure of the size of the drop in the instantaneous relative change $\dfrac{dP_t}{P_t}$ and takes values in $(0,1)$.

We shall construct a solution as in the following. Let S_k denote the kth jumping time for the Poisson process, i.e. $N_{S_k} = k$ and $N_{S_{k-}} = k-1$. Consider $t \in [0, S_1)$. Since there are no jumps in this interval the population satisfies the stochastic differential equation

$$dP_t = rP_t dt, \qquad 0 \le t < S_1,$$

with the solution given by the usual formula $P_t = P_0 e^{rt}$, for $t \in [0, S_1)$. In particular, when $t = S_{1-}$, we have

$$P_{S_{1-}} = P_0 e^{rS_1}. \tag{11.5.29}$$

Since at the jumping time S_1 we have

$$\frac{dP_{S_1}}{P_{S_{1-}}} = \frac{P_{S_1} - P_{S_{1}-}}{P_{S_1-}} = -\beta,$$

then $P_{S_1} = (1-\beta)P_{S_{1-}}$. Combining with formula (11.5.29) yields

$$P_{S_1} = (1-\beta)P_{S_{1-}} = P_0 e^{rS_1}(1-\beta). \tag{11.5.30}$$

Because there are no jumps in the interval $[S_1, S_2)$ the following differential equation holds

$$dP_t = rP_t dt, \qquad S_1 \le t < S_2,$$

with the solution

$$P_t = P_{S_1} e^{r(t-S_1)}.$$

Combining with (11.5.30) yields

$$P_t = P_{S_1} e^{r(t-S_1)} = P_0 e^{rS_1}(1-\beta)e^{r(t-S_1)} = P_0 e^{rt}(1-\beta), \quad S_1 \le t < S_2.$$

The effect of passing over a jump is to multiply the solution by the factor $(1-\beta)$. Using this observation we obtain inductively

$$P_t = P_0 e^{rt}(1-\beta)^n, \quad S_n \le t < S_{n+1}.$$

Replacing n by N_t we arrive at the following expression for the population in a stochastic catastrophic environment

$$\boxed{P_t = P_0 e^{rt}(1-\beta)^{N_t}.} \tag{11.5.31}$$

In the next paragraph we shall compute the mean of P_t using conditional expectations

$$\begin{aligned}
\mathbb{E}[P_t] &= P_0 e^{rt}\mathbb{E}[(1-\beta)^{N_t}] = P_0 e^{rt}\sum_{n\geq 0}\mathbb{E}[(1-\beta)^{N_t}|N_t = n]P(N_t = n) \\
&= P_0 e^{rt}\sum_{n\geq 0}(1-\beta)^n P(N_t = n) = P_0 e^{rt}\sum_{n\geq 0}(1-\beta)^n \frac{\lambda^n t^n}{n!}e^{-\lambda t} \\
&= P_0 e^{rt}e^{-\lambda t}e^{(1-\beta)\lambda t} = P_0 e^{(r-\beta\lambda)t}.
\end{aligned}$$

To conclude, if $r < \beta\lambda$, the population tends to decrease and then disappear in the long run. A population with $r = \beta\lambda$ has on average a constant size.

Next we shall evaluate the probability of the event $\{P_t \leq x\}$ for a given $x > 0$. We consider the following transformation of events

$$\begin{aligned}
\{P_t \leq x\} &= \{P_0 e^{(r-\beta\lambda)t} \leq x\} = \{P_0 e^{rt}(1-\beta)^{N_t} \leq x\} = \{(1-\beta)^{N_t} \leq \frac{x}{P_0}e^{-rt}\} \\
&= \Big\{N_t \geq \frac{\ln\left(\frac{x}{P_0}\right) - rt}{\ln(1-\beta)}\Big\},
\end{aligned}$$

where we used that $\ln(1-\beta) < 0$. Denoting $y_t = \dfrac{\ln\left(\frac{x}{P_0}\right) - rt}{\ln(1-\beta)}$, the probability can now be evaluated as in the following

$$P(P_t \leq x) = e^{-\lambda t}\sum_{k\geq y_t}\frac{\lambda^k t^k}{k!}.$$

Exercise 11.5.4 *An ant colony of* 1,000 *ants grows at the intrinsic rate* $r =$ 0.30 *per month. However, each rainfall kills* 2% *of the ant population, and it rains* 5 *times per year.*

(*a*) *Write the stochastic differential equation for the ant population size;*

(*b*) *What is the probability that there are at least* 2,000 *ants in the colony at the end of the first year?*

(*c*) *What is the expected size of the colony after* 2 *years?*

Population growth with stochastic harvesting Besides the effect of exponential growth at the intrinsic rate r, we shall also assume that the population is harvested at the stochastic rate $C_t = \rho + \sigma \frac{dW_t}{dt}$, where $\rho > 0$ is the mean harvesting rate and σ measures the size of the noise. The population at time t, denoted by P_t satisfies the following equation

$$dP_t = (rP_t - C_t)dt,$$

which can be written equivalently as

$$dP_t = rP_t dt - \rho dt - \sigma dW_t. \tag{11.5.32}$$

Multiplying by e^{-rt} and solving it as a linear stochastic differential equation yields the solution

$$P_t = e^{rt}\Big(P_0 - \frac{\rho}{r}\Big) + \sigma e^{rt}\int_0^t e^{-rs}\, dW_s. \tag{11.5.33}$$

This implies that P_t is normally distributed with the mean and variance given by

$$\mathbb{E}[P_t] = e^{rt}\Big(P_0 - \frac{\rho}{r}\Big) \tag{11.5.34}$$

$$Var[P_t] = \frac{\sigma^2}{2r}(e^{2rt} - 1). \tag{11.5.35}$$

Exercise 11.5.5 *Show that the stochastic process* (11.5.33) *is the solution of equation* (11.5.32).

Exercise 11.5.6 *Prove formulas* (11.5.34) *and* (11.5.35).

11.6 Pricing Zero-coupon Bonds

A bond is a financial instrument which pays back at the end of its lifetime, T, an amount equal to B. If the contract does not provide any payments until maturity, the bond is called a *zero-coupon bond* or a *discount bond.* The value of the bond at time t is denoted by $B(t)$. In the case when the yield r is constant, i.e., if

$$\frac{dB(t)}{B(t)} = r\, dt,$$

then the bond value is given by the familiar expression $B(t) = Be^{-\int_t^T r\, ds} = Be^{-r(T-t)}$. We shall assume next the case when the yield of the bond is affected by the noise in the market

$$r_t = r + \sigma\frac{dW_t}{dt}, \qquad r > 0,$$

where σ is a positive constant that controls the size of the noise. This leads to the following stochastic process

$$\begin{aligned} B_t &= Be^{-\int_t^T r_s\,ds} = Be^{-\int_t^T r\,ds}e^{-\int_t^T \sigma\,dW_s} \\ &= Be^{-r(T-t)}e^{-\sigma(W_T-W_t)}. \end{aligned}$$

The bond value is the conditional expectation of B_t given the information in the market until time t

$$\begin{aligned} \mathbb{E}[B_t|\mathcal{F}_t] &= Be^{-r(T-t)}\mathbb{E}[e^{-\sigma(W_T-W_t)}|\mathcal{F}_t] \\ &= Be^{-r(T-t)}\mathbb{E}[e^{-\sigma(W_T-W_t)}] \\ &= Be^{-r(T-t)}e^{\frac{1}{2}\sigma^2(T-t)} \\ &= Be^{-(r-\frac{1}{2}\sigma^2)(T-t)}. \end{aligned}$$

According to the previous model, we note that in the case when the market noise is small, $r > \frac{\sigma^2}{2}$, the bond value appreciates, having the maximum value, B, at $t = T$. If the noise is large, $\frac{\sigma^2}{2} > r$, the bond depreciates and has to be sold as soon as possible.

Exercise 11.6.1 *Use Ito's formula to show that the process B_t satisfies the equation*

$$\begin{aligned} dB(t) &= \Big(r + \frac{\sigma^2}{2}\Big)B(t)dt + \sigma B(t)dW_t \\ B(T) &= B. \end{aligned}$$

11.7 Finding the Cholesterol Level

The amount of cholesterol in the blood of a person at time t is described by a stochastic process denoted by C_t. The body cholesterol is either manufactured by the body or it is absorbed from the intaken food. If E denotes the daily rate at which cholesterol is eaten and C_0 stands for the normal level of cholesterol for a health person, then C_t satisfies the following stochastic differential equation

$$dC_t = a(C_0 - C_t)dt + bEdt + \sigma dW_t, \tag{11.7.36}$$

where a is a production parameter, b is the absorption parameter, and σ is the size of the noise in the measurement. Solving (11.7.36) as a linear equation we obtain the solution

$$C_t = C_0e^{-at} + (C_0 + \frac{b}{a}E)(1 - e^{-at}) + \sigma e^{-at}\int_0^t e^{as}\,dW_s.$$

From the properties of Wiener integrals it follows that C_t is normally distributed, with mean

$$\mathbb{E}[C_t] = C_0 e^{-at} + (C_0 + \frac{b}{a}E)(1 - e^{-at})$$

and variance

$$Var[C_t] = \frac{\sigma^2}{2a}(1 - e^{-2at}).$$

If the diet is kept for a long time, then the cholesterol level becomes the normal random variable

$$C_t \sim N\Big(C_0 + \frac{b}{a}E, \frac{\sigma^2}{2a}\Big), \qquad t \to \infty.$$

In order to evaluate the health of a particular person who is on a given diet (i.e. when E is kept constant for a long time) we have to find the probability that the cholesterol level is over a given acceptable level M. Using the long run normality of C_t, this probability can be evaluated as

$$P(C_t > M) = \sqrt{\frac{a}{\pi}\frac{1}{\sigma}} \int_M^\infty e^{-\frac{(x - C_0 - \frac{b}{a}E)^2}{\sigma^2}}\, dx. \qquad (11.7.37)$$

Exercise 11.7.1 *Assuming that the intaken food does not have a constant cholesterol level, the term E is replaced by $E +$ "noise". Write and solve the corresponding stochastic differential equation.*

Exercise 11.7.2 *The normal cholesterol level in the blood is $C_0 = 200$ (milligrams per deciliter), the production parameter is $a = 0.1$ and the absorption parameter for a particular person is $b = 0.15$. What is the maximum daily intake E of cholesterol such that the long run level of cholesterol is less than 220 with a probability of 95%?*

11.8 Photon's Escape Time from the Sun

This section deals with the computation of a rough approximation of the time taken by a photon of light to travel from the center of the sun to its surface. Fusion reactions occur in the core of the sun due to high heat and pressure. These reactions release high energy photons (gamma rays), which are absorbed in only a few millimeters by the solar plasma particles and then are emitted again in a random direction, see Fig. 11.1.

The mathematical model for such a photon emitted and randomly redirected by plasma particles is a Brownian motion process. If X_t denotes the

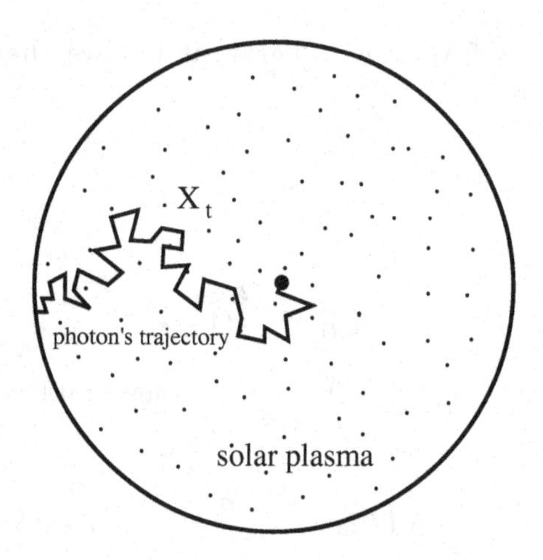

Figure 11.1: *The photon bounces back and forth in its effort to emerge to the sun's surface.*

coordinates vector in $\mathbb{R}^3$ of the photon at time t, where the center of the sun is assigned the zero coordinate, then

$$X_t = a + \sigma W_t, \tag{11.8.38}$$

where $a = X_0$ is the location where the photon was initially created, σ is the dispersion function and $W_t = (W_t^1, W_t^2, W_t^3)$ is a 3-dimensional Brownian process in $\mathbb{R}^3$. The dispersion is given by Einstein's formula

$$\sigma^2 = 2kT/(6\pi\eta\alpha),$$

where k is Boltzmann's constant, T denotes the absolute temperature, α is the diameter of the photon, and η is the viscosity of the solar plasma. For the sake of simplicity we shall assume in the following that σ is constant.

The time when the photon reaches the surface of the sun is a random variable denoted by τ. We shall use a similar method as the one described in section 9.6 to find the expected time $\mathbb{E}[\tau]$. If R_S denotes the radius of the sun, the time necessary for a photon to emerge to the sun's surface is the exit time

$$\tau = \inf\{t > 0; |X_t| \geq R_S\}. \tag{11.8.39}$$

Since the infinitesimal generator of the process (11.8.38) is the operator

$$\frac{\sigma^2}{2}\Delta = \frac{\sigma^2}{2}(\partial_{x_1}^2 + \partial_{x_2}^2 + \partial_{x^3}^2),$$

using the function $f(x) = |x|^2 = x_1^2 + x_2^2 + x_3^2$ in Dynkin's formula

$$\mathbb{E}[f(X_\tau)] = f(x) + \mathbb{E}\Big[\int_0^\tau \frac{\sigma^2}{2}\Delta f(X_s)\, ds\Big]$$

yields

$$R_S^2 = |a|^2 + \mathbb{E}\Big[\int_0^T 3\sigma^2\,ds\Big],$$

and hence

$$\mathbb{E}[\tau] = \frac{R_S^2 - |a|^2}{3\sigma^2}. \tag{11.8.40}$$

In particular, if the photon is emitted from the sun's center, the expected emerging time to the surface is

$$\mathbb{E}[\tau] = \frac{R_S^2}{3\sigma^2}. \tag{11.8.41}$$

Using the numerical values for the sun's radius and photon's diffusion given by $R_S = 6.955 \times 10^5$ km and $\sigma^2 = 0.0025\,\text{km}^2/\text{sec}$ (this corresponds to a 50 meters per second radial photon displacement), formula (11.8.41) yields the approximate value $\mathbb{E}[\tau] \approx 2$ million years. Some other sources compute slightly different values, but the idea is that it takes a really long time for a photon to leave the sun's interior. This is huge compared with the only 8 minutes spent by the photon on its way to earth.

It is worth noting that if a star has its radius 100 times the sun's radius, then the expected emerging time multiplies by a factor of 10^4. This means $\mathbb{E}[\tau] \approx 2 \times 10^{10}$ years (20 billion years), which is longer than the entire age of the universe ($\approx$ 14 billion years)! Hence, it is possible that a photon created in the center of the star has not found its way out to the surface yet. Since the life span of a star is usually around 10 billion years, the photon will probably not get out of the star during its life span.

11.9 Filtering Theory

In early 1960s Kalman and Bucy found a procedure for estimating the state of a signal that satisfies a noisy linear differential equation based on a series of noisy observations. This is known now under the name of Kalman-Bucy filter. This theory has useful applications in signal processing of aerospace tracking, GPS location systems, radars, MRI medical imaging, statistical quality control, and any other applications dealing with reducing or filtering noise out of an observed system. For more examples on this subject as well as the complete proofs, the reader can consult Øksendal [37].

Assume ζ_t is the *input process*, i.e., a process that describes the state of a stochastic system at time t, which needs to be observed. This process has some "noise" built in and its evolution satisfies a given linear stochastic differential equation

$$d\zeta_t = a(t)\zeta_t\,dt + b(t)dW_t,$$

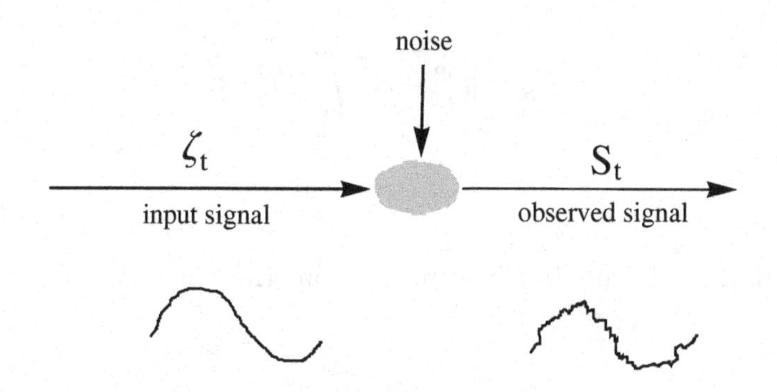

Figure 11.2: *The input signal ζ_t and the observed signal S_t.*

where $a(t)$ and $b(t)$ are given deterministic functions and W_t is a Brownian motion, independent of the initial value ζ_0. We assume that ζ_t is observed continuously with the actual observations $S_t = \zeta_t +$ "noise", see Fig. 11.2. If the white noise is given by "noise" $= \beta(t)\dfrac{dB_t}{dt}$, then

$$S_t\, dt = \zeta_t\, dt + \beta(t) dB_t. \tag{11.9.42}$$

Introducing the cumulative observation process

$$Q_t = \int_0^t S_u\, du$$

and using that the information (σ-algebras) induced by the processes Q_t and S_t are the same,

$$\mathcal{F}_t^Q = \mathcal{F}_t^S,$$

it follows that equation (11.9.42) can be replaced by the more mathematically useful formula

$$dQ_t = \zeta_t\, dt + \beta(t) dB_t, \tag{11.9.43}$$

with B_t independent of W_t. From now on, Q_t will be considered as the *observation process* instead of S_t. It is easier to work with the cummulative observation process Q_t rather than the actual observations S_t.

The filtering problem can now be stated as:

Given the observations Q_s, $0 \leq s \leq t$, satisfying equation (11.9.43), *find the best estimate $\widehat{\zeta_t}$ of the state ζ_t.*

One of the best estimators $\widehat{\zeta_t}$, which is mathematically tractable, is the one which minimizes the mean square error

$$R(t) = \mathbb{E}[(\zeta_t - \widehat{\zeta_t})^2].$$

This means that for any other square integrable random variable Y, which is measurable with respect to the field $\mathcal{F}_t^Q$, we have the inequality

$$\mathbb{E}[(\zeta_t - \widehat{\zeta}_t)^2] \le \mathbb{E}[(\zeta_t - Y)^2].$$

It turns out that the best estimator $\widehat{\zeta}_t$ coincides with the conditional expectation of ζ_t given the information induced by Q_s, $0 \le s \le t$, namely,

$$\widehat{\zeta}_t = \mathbb{E}[\zeta_t | \mathcal{F}_t^Q].$$

The computation of the best estimator $\widehat{\zeta}_t$ is provided by the following central result in filtering theory.

Theorem 11.9.1 (Kalman-Bucy filter) *Let the state of a system ζ_t satisfy the equation*

$$d\zeta_t = a(t)\zeta_t\,dt + b(t)dW_t, \tag{11.9.44}$$

where $a(t)$ and $b(t)$ are deterministic functions. Assume that the random variable ζ_0 and the Brownian motion W_t are independent, and let $\mathbb{E}[\zeta_0] = \mu$, $Var[\zeta_0] = \sigma^2$. Assume the observation Q_t satisfy the equation

$$dQ_t = \alpha(t)\zeta_t\,dt + \beta(t)dB_t, \qquad Q_0 = 0, \tag{11.9.45}$$

with deterministic functions $\alpha(t)$, $\beta(t)$ and Brownian motion B_t, independent of W_t and ζ_0.

Then the conditional expectation $\widehat{\zeta}_t = \mathbb{E}[\zeta_t|\mathcal{F}_t^Q]$ is the solution of the linear stochastic differential equation

$$d\widehat{\zeta}_t = U(t)\widehat{\zeta}_t dt + V(t)dQ_t, \qquad \widehat{\zeta}_0 = \mu, \tag{11.9.46}$$

with

$$V(t) = \frac{\alpha(t)}{\beta^2(t)}R(t), \qquad U(t) = a(t) - \frac{\alpha^2(t)}{\beta^2(t)}R(t),$$

and $R(t)$ satisfying the deterministic Riccati equation

$$\frac{dR(t)}{dt} = b^2(t) + 2a(t)R(t) - \frac{\alpha^2(t)}{\beta^2(t)}R^2(t), \quad R(0) = \sigma^2.$$

Moreover, the least mean square error is given by $R(t) = \mathbb{E}[(\zeta_t - \widehat{\zeta}_t)^2]$.

The process $\widehat{\zeta}_t$ is called *the Kalman-Bucy filter* of the filtering problem (11.9.44)-(11.9.45). Furthermore, if

$$\lim_{t\to\infty} R(t) = 0,$$

we say that $\widehat{\zeta}_t$ is an *exact asymptotic estimation* of ζ_t.

Explicit solutions In the following we shall deal with the closed form solution of equation (11.9.46). First, we shall solve (11.9.44) as a linear equation. Denote $A(t) = e^{\int_0^t a(s)\,ds}$ and multiply by $e^{-A(t)}$ to get

$$d(\zeta_t e^{-A(t)}) = b(t)e^{-A(t)}\,dW_t.$$

Integrating and solving for ζ_t we obtain

$$\zeta_t = \zeta_0 e^{A(t)} + \int_0^t e^{A(t)-A(s)} b(s)\,dW_s. \tag{11.9.47}$$

The mean and variance of the input process ζ_t are

$$\begin{aligned} \mathbb{E}[\zeta_t] &= \mathbb{E}[\zeta_0]e^{A(t)} = \mu e^{A(t)}; \\ Var[\zeta_t] &= \int_0^t e^{2[A(t)-A(s)]} b^2(s)\,ds, \end{aligned}$$

where for the second identity we used the properties of Wiener integrals.

Now, equation (11.9.46) can be solved as a linear equation. After multiplying by the factor

$$\rho(t)^{-1} = \exp\{-\int_0^t U(s)\,ds\}$$

the equation becomes

$$d(\rho(t)^{-1}\widehat{\zeta}_t) = \rho(t)^{-1}V(t)dQ_t.$$

Integrating, we obtain the filter solution

$$\begin{aligned} \widehat{\zeta}_t &= \mu\rho_t + \rho(t)\int_0^t \rho(s)^{-1}V(s)\,dQ_s \\ &= \mu\rho_t + \rho(t)\int_0^t \rho(s)^{-1}V(s)\alpha(s)\zeta_s\,ds + \rho_t\int_0^t \rho_s(s)^{-1}V(s)\beta(s)\,dB_s, \end{aligned}$$

with ζ_t given by (11.9.47).

It is worth noting that the expectation of the filter is given by

$$\begin{aligned} \mathbb{E}[\widehat{\zeta}_t] &= \mathbb{E}[\mathbb{E}[\zeta_t|\mathcal{F}_t^Q]] = \mathbb{E}[\zeta_t] \\ &= \mu e^{A(t)}. \end{aligned}$$

Example 11.9.2 (Noisy observations of a random variable) *Assume ζ is a random variable which needs to be measured. Its known mean and variance are given by $\mathbb{E}[\zeta] = \mu$ and $Var[\zeta] = \sigma^2$. For instance, we may consider ζ to be the heart rate per minute, or the cholesterol blood level, or the systolic* *blood pressure for a particular person. Observing any of the aforementioned variables involve measurement errors that are described by the noise factor. The actual observations are given by a process* $S_t = \zeta +$ "noise", *which can be written in terms of the cumulative observations as*

$$dQ_t = \zeta dt + \beta dB_t,$$

with $\beta > 0$ *constant. This is equivalent to* $Q_t = \zeta t + \beta B_t$, *where the noise factor represented by the Brownian motion* B_t *is assumed independent of* ζ. *The filtering problem in this case means we have to find a stochastic process* $\widehat{\zeta}_t$ *which is the best approximation of* ζ *up to time* t, *given the measurements*

$$Q_s = \zeta s + \beta B_s, \qquad 0 \leq s \leq t.$$

In order to find the filter $\widehat{\zeta}_t$ *consider the constant process* $\zeta_t = \zeta$ *and associate the filtering problem*

$$\begin{aligned} d\zeta_t &= 0 \\ dQ_t &= \zeta dt + \beta dB_t. \end{aligned}$$

Since in this case $\alpha(t) = 1$, $\beta(t) = \beta$, $a(t) = 0$, *and* $b(t) = 0$, *the Riccati equation becomes*

$$\frac{dR(t)}{dt} = -\frac{1}{\beta^2} R^2(t),$$

with the solution $R(t) = \dfrac{\beta^2}{t - C\beta^2}$ *depending on the parameter* C. *Using the initial condition* $R(0) = \sigma^2$, *we obtain*

$$R(t) = \frac{\sigma^2\beta^2}{\sigma^2 t + \beta^2}.$$

Then the coefficients $V(t)$ *and* $U(t)$ *take the following form*

$$\begin{aligned} V(t) &= \frac{1}{\beta^2} R(t) = \frac{\sigma^2}{\sigma^2 t + \beta^2} \\ U(t) &= -\frac{1}{\beta^2} R(t) = -\frac{\sigma^2}{\sigma^2 t + \beta^2}. \end{aligned}$$

Equation (11.9.46) becomes

$$d\widehat{\zeta}_t + \frac{\sigma^2}{\sigma^2 t + \beta^2}\widehat{\zeta}_t dt = \frac{\sigma^2}{\sigma^2 t + \beta^2} dQ_t.$$

Multiplying by the integrating factor

$$e^{\int_0^t \frac{\sigma^2}{\sigma^2 s + \beta^2}\, ds} = \frac{\sigma^2 t + \beta^2}{\beta^2}$$

we obtain the exact differential equation

$$d\Big(\frac{\sigma^2 t+\beta^2}{\beta^2}\widehat{\zeta}_t\Big) = \frac{\sigma^2}{\beta^2}\,dQ_t,$$

which after integration provides the following formula for the best estimate

$$\begin{aligned}\widehat{\zeta}_t &= \frac{\beta^2}{\sigma^2 t+\beta^2}\widehat{\zeta}_0 + \frac{\sigma^2}{\sigma^2 t+\beta^2}Q_t \\ &= \frac{1}{\sigma^2 t+\beta^2}(\beta^2\mu+\sigma^2 Q_t).\end{aligned}$$

It is worth noting that the foregoing formula implies that the best estimate of the random variable ζ, given the continuous observations Q_s, $0\le s\le t$, depends only on the last observation Q_t. In the case of discrete observations, the best estimate will depend on each observation, see the next example.

Example 11.9.3 *Consider a random variable ζ, with $\mathbb{E}[\zeta]=\mu$ and $Var(\zeta)=\sigma^2$, which is observed n times with the results*

$$\begin{aligned}S_1 &= \zeta+\epsilon_1\\ S_2 &= \zeta+\epsilon_2\\ \cdots & \quad \cdots\cdots\\ S_n &= \zeta+\epsilon_n,\end{aligned}$$

where ϵ_j is an error random variable, independent of ζ, with $\mathbb{E}[\epsilon_j]=0$, and $Var(\epsilon_j)=m^2$. We shall assume in the beginning that the errors ϵ_j are independent. The goal is to find the best estimate $\widehat{\zeta}$ of ζ given the observations S_j, $1\le j\le n$,

$$\widehat{\zeta}=\mathbb{E}[\zeta|S_1,\cdots,S_n].$$

Consider the affine space generated by the observations S_j

$$\mathcal{L}(S)=\{\alpha_0+\sum_{j=1}^{n}c_jS_j\,;\,\alpha_0,c_j\in\mathbb{R}\}.$$

It makes sense to look for the best estimator as an element $\widehat{\zeta}\in\mathcal{L}(S)$ such that the distance $\mathbb{E}[(\widehat{\zeta}-\zeta)^2]$ is minimized. This occurs when the estimator $\widehat{\zeta}$ is the orthogonal projection of ζ on the space $\mathcal{L}(S)$. This means we have to determine the constants α_0, c_j such that the following $n+1$ conditions are satisfied

$$\begin{aligned}\mathbb{E}[\widehat{\zeta}] &= \mathbb{E}[\zeta]\\ \mathbb{E}[(\widehat{\zeta}-\zeta)S_j] &= 0,\quad j=1,\cdots,n.\end{aligned}$$

Let $\widehat{\zeta} = \alpha_0 + c_1 S_1 + \cdots + c_n S_n$. *Since*

$$\mu = \mathbb{E}[\widehat{\zeta}] = \alpha_0 + \mu(c_1 + \cdots + c_n)$$

it follows that $\alpha_0 = c_0\mu$, *with*

$$c_0 + c_1 + c_2 + \cdots c_n = 1. \tag{11.9.48}$$

Therefore, $\widehat{\zeta}$ *belongs to the convex hull of* $\{\mu, S_1, \cdots, S_n\}$, *i.e.*

$$\widehat{\zeta} = c_0\mu + c_1 S_1 + \cdots + c_n S_n.$$

A computation provides

$$\begin{aligned}
\mathbb{E}[(\widehat{\zeta} - \zeta)S_j] &= \mathbb{E}[\widehat{\zeta} S_j] - \mathbb{E}[\zeta S_j] \\
&= c_0\mu\mathbb{E}[S_j] + \sum_{k=1}^{n} c_k \mathbb{E}[S_k S_j] - \mathbb{E}[\zeta(\zeta + \epsilon_j)] \\
&= c_0\mu^2 + \sum_{k=1}^{n} c_k \mathbb{E}[(\zeta + \epsilon_k)(\zeta + \epsilon_j)] - \mathbb{E}[\zeta^2] - \mathbb{E}[\zeta]\mathbb{E}[\epsilon_j] \\
&= c_0\mu^2 + \mathbb{E}[\zeta^2]\sum_{k=1}^{n} c_k + m^2 \sum_{k=1}^{n} c_k \delta_{kj} - \mathbb{E}[\zeta^2] \\
&= c_0\mu^2 + \mathbb{E}[\zeta^2](1 - c_0) + m^2 c_j - \mathbb{E}[\zeta^2] \\
&= c_0(\mu^2 - \mathbb{E}[\zeta^2]) + m^2 c_j \\
&= -c_0\sigma^2 + m^2 c_j.
\end{aligned}$$

The orthogonality condition $\mathbb{E}[(\widehat{\zeta} - \zeta)S_j] = 0$ *implies*

$$c_j = \frac{c_0\sigma^2}{m^2}, \quad j = 1, \cdots, n. \tag{11.9.49}$$

Substituting in (11.9.48) provides an equation for c_0, *which has the solution*

$$c_0 = \frac{m^2}{m^2 + n\sigma^2},$$

and hence (11.9.49) becomes

$$c_j = \frac{\sigma^2}{m^2 + n\sigma^2}.$$

In conclusion, the best estimator of ζ, *given* n *observations* $S_1, \cdots, S_n$, *is given by*

$$\begin{aligned}
\widehat{\zeta} &= c_0\mu + \sum_{k=1}^{n} c_k S_k \\
&= \frac{m^2\mu}{m^2 + n\sigma^2} + \frac{\sigma^2}{m^2 + n\sigma^2}\sum_{k=1}^{n} S_k.
\end{aligned}$$

It is worth noting that each observation is equally weighted. If the number of observations is large, $n \to \infty$, then $\widehat{\zeta} \approx \frac{1}{n}\sum_{k=1}^{n} S_k$. This means that the best approximation is given by the average of the observations.

All previous formulas are valid for the case of independent observations ϵ_j. Assume now that the covariance matrix

$$\rho_{ij} = Cov(\epsilon_k, \epsilon_j) = \mathbb{E}[\epsilon_k \epsilon_j]$$

is not necessarily diagonal. A similar computation leads in this case to the formula

$$\mathbb{E}[(\widehat{\zeta} - \zeta)S_j] = -c_0\sigma^2 + \sum_{k=1}^{n} c_k \rho_{kj}.$$

The orthogonality formula becomes

$$\sum_{k=1}^{n} c_k \rho_{kj} = c_0\sigma^2,$$

which is a linear system in c_k. Under the assumption that the matrix (ρ_{kj}) is non-singular, let (ρ^{kj}) be its inverse matrix, so the solution of the aforementioned linear system is

$$c_k = c_0\sigma^2 \sum_{j=1}^{n} \rho^{kj}, \qquad k = 1, \cdots, n. \tag{11.9.50}$$

Substituting into (11.9.48) and solving for the coefficient c_0 yields

$$c_0 = \frac{1}{1 + \sigma^2 \sum_{k,j=1}^{n} \rho^{kj}}.$$

Then formula (11.9.50) provides the other coefficients

$$c_k = \frac{\sum_{j=1}^{n} \rho^{kj}}{\sigma^{-2} + \sum_{k,j=1}^{n} \rho^{kj}}, \qquad k = 1, \cdots, n.$$

Hence, the best estimator of ζ in terms of the covariance matrix is given by

$$\widehat{\zeta} = \frac{\mu}{1 + \sigma^2 \sum_{k,j=1}^{n} \rho^{kj}} + \sum_{k=1}^{n} \frac{\sum_{j=1}^{n} \rho^{kj}}{\sigma^{-2} + \sum_{k,j=1}^{n} \rho^{kj}} S_k.$$

Example 11.9.4 *Consider that the observed state is the blood cholesterol level, C_t, which is given by the stochastic equation (11.7.36)*

$$dC_t = a(C_0 - C_t)dt + bEdt + \sigma dW_t, \tag{11.9.51}$$

where a is a production parameter, b is the absorption parameter, and σ is the size of the noise in the measurement. The observation process is given by

$$dQ_t = C_t dt + \beta\, dB_t, \qquad Q_0 = 0.$$

We need to find the best estimate $\widehat{C_t}$ of the cholesterol level C_t, given the noisy observations Q_s, $s \le t$. However, Kalman-Bucy filter cannot be applied directly to (11.9.51). We need to perform first the substitution

$$\zeta_t = C_t - C_0 - \frac{b}{a}E,$$

which transforms (11.9.51) into the linear stochastic differential equation

$$d\zeta_t = -a\zeta_t dt + \sigma dW_t.$$

Let $Z_t = Q_t - \Big(C_0 + \dfrac{b}{a}E\Big)t$. We note that the σ-algebras induced by Z_t and Q_t are the same, $\mathcal{F}_t^Z = \mathcal{F}_t^Q$, and hence we can consider Z_t as an observation process instead of Q_t, satisfying

$$dZ_t = \zeta_t dt + \beta dB_t, \qquad Z_0 = 0.$$

The Riccati equation associated with the filtering problem having the input process ζ_t and observations Z_t is given by

$$\frac{dR(t)}{dt} = -2aR(t) - \frac{1}{\beta^2}R^2(t) + \sigma^2, \quad R(0) = 0,$$

where we used that $\mathbb{E}[\zeta_0] = -\frac{b}{a}E$ and $Var(\zeta_0) = 0$. The solution is given by

$$R(t) = \frac{\alpha_1(1 - e^{2Kt})}{1 - \frac{\alpha_1}{\alpha_2}e^{2Kt}},$$

where

$$\begin{aligned}
\alpha_1 &= -a\beta^2 - \beta\sqrt{a^2\beta^2 + \sigma^2} \\
\alpha_2 &= -a\beta^2 + \beta\sqrt{a^2\beta^2 + \sigma^2} \\
K &= \sqrt{a^2 + (\sigma/\beta)^2}.
\end{aligned}$$

Solving for the estimation process $\widehat{\zeta_t}$, we obtain the formula

$$\widehat{\zeta_t} = -\frac{b}{a}Ee^{-\int_0^t H(s)\,ds} + \frac{1}{\beta^2}\int_0^t e^{-\int_s^t H(u)\,du}R(s)\,dZ_s,$$

where

$$H(s) = a + \frac{1}{\beta^2}R(s).$$

Using that

$$\lim_{t\to\infty} R(t) = \alpha_2,$$

we obtain the following long run behavior for the estimation

$$\lim_{t\to\infty} \widehat{\zeta_t} = -\frac{b}{a}Ee^{-Kt} + \frac{\alpha_2}{\beta^2}e^{-Kt}\int_0^t e^{Ks}\,dZ_s.$$

This can be transformed in terms of $\widehat{C_t}$ and Q_t as follows

$$\lim_{t\to\infty} \widehat{C_t} = C_0 + (1-e^{-Kt})\Big(\frac{b}{a}E - \frac{\alpha_2}{K\beta^2}(C_0 + \frac{b}{a}E)\Big) + \frac{\alpha_2}{\beta^2}e^{-Kt}\int_0^t e^{Ks}\,dQ_s.$$

Example 11.9.5 *A modern application of filtering theory is in GPS location. The input signal is the true position, while the observation process is the GPS position. The position is measured by an inertial navigation system (INS). This has transducers that measure acceleration, which leads to the vehicle position by double integration and specifying its initial position. Since there are always errors in the acceleration measurement, this leads to noise in the position location.*

The INS position can be checked often by observing the GPS receiver. Hence, the INS position estimate can be corrected by using periodic GPS observations, which are also susceptible to error of measurement. Finding the optimal way to use the GPS measurements in order to correct the INS estimates is a stochastic filtering problem, see Bain and Crisan [4].

Exercise 11.9.6 *Consider the filtering problem*

$$\begin{aligned} d\zeta_t &= -\frac{1+t}{2}dt + dW_t, \quad \mu = 0, \sigma = 1 \\ dQ_t &= \frac{1}{1+t}\zeta_t dt + \frac{1}{1+t}dB_t, \qquad Q_0 = 0. \end{aligned}$$

(a) Show that the associated Riccati equation is given by

$$\frac{dR(t)}{dt} = 1 - (1+t)R_t - R_t^2, \quad R(0) = 1.$$

(b) Verify that the solution of the Riccati equation is

$$R(t) = \frac{1}{1+t}.$$

(b) Compute the best estimator for ζ_t and show that

$$\widehat{\zeta_t} = \frac{1}{1+t}e^{-\frac{1}{2}(t+t^2/2)}\int_0^t (1+s)e^{\frac{1}{2}(s+s^2/2)}\,dQ_s.$$

Exercise 11.9.7 *In a Kalman-Bucy filter the state ζ and the observations Q_t are given by $d\zeta_t = dW_t$, with ζ_0 being normally distributed with mean 1 and variance 2, and $dQ_t = \zeta_t dt + 2dB_t$, with $Q_0 = 0$. Find the best estimate $\widehat{\zeta_t}$ for ζ_t.*

Chapter 12

Hints and Solutions

Here we give the hints and solutions to selected exercises. The reader should be able to derive the solutions to the rest based on what he has learnt from the examples in the chapters.

Chapter 2

Exercise 2.9.7 (a) For any random variables A,B, and variable $\lambda \in \mathbb{R}$, integrating in the inequality

$$\Big(A(\omega)\lambda + B(\omega)\Big)^2 \geq 0, \qquad \forall \omega \in \Omega$$

implies

$$\int_\Omega \Big(A(\omega)\lambda + B(\omega)\Big)^2 dP(\omega) \geq 0.$$

After expanding and collecting the powers of λ, this can be written as

$$\Big(\int_\Omega A^2(\omega)\, dP(\omega)\Big)\lambda^2 + 2\Big(\int_\Omega A(\omega)B(\omega)\, dP(\omega)\Big)\lambda + \int_\Omega B^2(\omega)\, dP(\omega) \geq 0.$$

Substituting

$$a = \int_\Omega A^2(\omega)\, dP(\omega),\ b = 2\Big(\int_\Omega A(\omega)B(\omega)\, dP(\omega)\Big),\ c = \int_\Omega B^2(\Omega)\, dP(\omega),$$

the previous inequality becomes

$$a\lambda^2 + b\lambda + c \geq 0, \qquad \forall \lambda \in \mathbb{R}.$$

This occurs when $b^2 \leq 4ac$, which in this case becomes

$$\mathbb{E}[AB]^2 \leq \mathbb{E}[A^2]\mathbb{E}[B^2].$$

(*b*) Substitute $A = X - \mu_X$ and $B = Y - \mu_B$ and obtain

$$\begin{aligned}\mathbb{E}[(X-\mu_X)(Y-\mu_Y)] &\le \sqrt{\mathbb{E}[(X-\mu_X)^2]}\sqrt{\mathbb{E}[(Y-\mu_Y)^2]} \Longleftrightarrow \\ Cov(X,Y) &\le \sigma_X\sigma_Y.\end{aligned}$$

(*c*) They are proportional.

Exercise 2.10.1 Let $X \sim N(\mu, \sigma^2)$. Then the distribution function of Y is

$$\begin{aligned}F_Y(y) &= P(Y<y) = P(\alpha X+\beta<y) = P\Big(X<\frac{y-\beta}{\alpha}\Big) \\ &= \frac{1}{\sqrt{2\pi}\sigma}\int_{-\infty}^{\frac{y-\beta}{\alpha}} e^{-\frac{(x-\mu)^2}{2\sigma^2}}\,dx = \frac{1}{\sqrt{2\pi}\alpha\sigma}\int_{-\infty}^{y} e^{-\frac{\left(z-(\alpha\mu+\beta)\right)^2}{2\alpha^2\sigma^2}}\,dz \\ &= \frac{1}{\sqrt{2\pi}\sigma'}\int_{-\infty}^{y} e^{-\frac{(z-\mu')^2}{2(\sigma')^2}}\,dz,\end{aligned}$$

with $\mu' = \alpha\mu+\beta$, $\sigma' = \alpha\sigma$.

Exercise 2.10.5 (*a*) Making $t = n$ yields $\mathbb{E}[Y^n] = \mathbb{E}[e^{nX}] = e^{\mu n + n^2\sigma^2/2}$.
(*b*) Let $n = 1$ and $n = 2$ in (*a*) to get the first two moments and then use the formula of variance.

Exercise 2.12.7 The tower property

$$\mathbb{E}\big[\mathbb{E}[X|\mathcal{G}]|\mathcal{H}\big] = \mathbb{E}[X|\mathcal{H}], \qquad \mathcal{H}\subset\mathcal{G}$$

is equivalent to

$$\int_A \mathbb{E}[X|\mathcal{G}]\,dP = \int_A X\,dP, \qquad \forall A\in\mathcal{H}.$$

Since $A \in \mathcal{G}$, the previous relation holds by the definition of $\mathbb{E}[X|\mathcal{G}]$.

Exercise 2.12.8 (*a*) $|\Omega| = 2^4 = 16$;
(*b*) $P(A) = 3/8$, $P(B) = 1/2$, $P(C) = 1/4$;
(*c*) $P(A\cap B) = \frac{3}{16}$, $P(B\cap C) = \frac{3}{16}$;
(*d*) $P(A)P(B) = \frac{3}{16} = P(A\cap B)$, so A, B independent;
(*e*) $P(B)P(C) = \frac{2}{16} \neq \frac{3}{16} = P(B\cap C)$, so B, C independent; $P(B|C) = \frac{3}{4}$;
(*f*) we know the outcomes of the first two tosses but we do not know the order of the last two tosses;
(*g*) $A \in \mathcal{G}$ (true), $B \in \mathcal{F}$ (true), $C \in \mathcal{G}$ (true);
(*i*) $\mathbb{E}[X] = 0$, $\mathbb{E}[Y] = \frac{15}{16}$, $\mathbb{E}[X|\mathcal{G}] = X$, since X is $\mathcal{G}$-measurable.

Exercise 2.12.9 (*a*) Direct application of the definition.
(*b*) $P(A) = \int_A dP = \int_\Omega \chi_A(\omega)\,dP(\omega) = \mathbb{E}[\chi_A]$.

(*d*) $\mathbb{E}[\chi_A X] = \mathbb{E}[\chi_A]\mathbb{E}[X] = P(A)\mathbb{E}[X]$.
(*e*) We have the sequence of equivalencies

$$\begin{aligned}\mathbb{E}[X|\mathcal{G}] &= \mathbb{E}[X] \Leftrightarrow \int_A \mathbb{E}[X]\,dP = \int_A X\,dP, \forall A \in \mathcal{G} \Leftrightarrow \\ \mathbb{E}[X]P(A) &= \int_A X\,dP \Leftrightarrow \mathbb{E}[X]P(A) = \mathbb{E}[\chi_A],\end{aligned}$$

which follows from (*d*).

Exercise 2.13.5 If $\mu = \mathbb{E}[X]$ then

$$\begin{aligned}\mathbb{E}[(X - \mathbb{E}[X])^2] &= \mathbb{E}[X^2 - 2\mu X + \mu^2] = \mathbb{E}[X^2] - 2\mu^2 + \mu^2 \\ &= \mathbb{E}[X^2] - \mathbb{E}[X]^2 = Var[X].\end{aligned}$$

Exercise 2.13.6 From Exercise 2.13.5 we have $Var(X) = 0 \Leftrightarrow X = \mathbb{E}[X]$, i.e. X is a constant.

Exercise 2.13.7 The same proof as the one in Jensen's inequality.

Exercise 2.13.8 It follows from Jensen's inequality or using properties of integrals.

Exercise 2.13.13 (*a*) $m(t) = \mathbb{E}[e^{tX}] = \sum_k e^{tk} \frac{\lambda^k e^{-\lambda}}{k!} = e^{\lambda(e^t - 1)}$;
(*b*) It follows from the first Chernoff bound.

Exercise 2.13.16 Choose $f(x) = x^{2k+1}$ and $g(x) = x^{2n+1}$.

Exercise 2.14.2 By direct computation we have

$$\begin{aligned}\mathbb{E}[(X - Y)^2] &= \mathbb{E}[X^2] + \mathbb{E}[Y^2] - 2\mathbb{E}[XY] \\ &= Var(X) + \mathbb{E}[X]^2 + Var[Y] + \mathbb{E}[Y]^2 - 2\mathbb{E}[X]\mathbb{E}[Y] \\ &\quad + 2\mathbb{E}[X]\mathbb{E}[Y] - 2\mathbb{E}[XY] \\ &= Var(X) + Var[Y] + (\mathbb{E}[X] - \mathbb{E}[Y])^2 - 2Cov(X, Y).\end{aligned}$$

Exercise 2.14.3 (*a*) Since

$$\mathbb{E}[(X - X_n)^2] \geq \mathbb{E}[X - X_n]^2 = (\mathbb{E}[X_n] - \mathbb{E}[X])^2 \geq 0,$$

the Squeeze Theorem yields $\lim_{n\to\infty}(\mathbb{E}[X_n] - \mathbb{E}[X]) = 0$.
(*b*) Writing

$$X_n^2 - X^2 = (X_n - X)^2 - 2X(X - X_n),$$

and taking the expectation we get

$$\mathbb{E}[X_n^2] - \mathbb{E}[X^2] = \mathbb{E}[(X_n - X)^2] - 2\mathbb{E}[X(X - X_n)].$$

The right side tends to zero since

$$\begin{aligned}
\mathbb{E}[(X_n - X)^2] &\to 0 \\
|\mathbb{E}[X(X - X_n)]| &\leq \int_\Omega |X(X - X_n)|\, dP \\
&\leq \Big(\int_\Omega X^2\, dP\Big)^{1/2}\Big(\int_\Omega (X - X_n)^2\, dP\Big)^{1/2} \\
&= \sqrt{\mathbb{E}[X^2]\mathbb{E}[(X - X_n)^2]} \to 0.
\end{aligned}$$

(c) It follows from part (b).

(d) Apply Exercise 2.14.2.

Exercise 2.14.4 Using Jensen's inequality we have

$$\begin{aligned}
\mathbb{E}\big[\big(\mathbb{E}[X_n|\mathcal{H}] - \mathbb{E}[X|\mathcal{H}]\big)^2\big] &= \mathbb{E}\big[\big(\mathbb{E}[X_n - X|\mathcal{H}]\big)^2\big] \\
&\leq \mathbb{E}\big[\mathbb{E}[(X_n - X)^2|\mathcal{H}]\big] \\
&= \mathbb{E}[(X_n - X)^2] \to 0,
\end{aligned}$$

as $n \to 0$.

Exercise 2.16.7 The integrability of X_t follows from

$$\mathbb{E}[|X_t|] = \mathbb{E}\big[|\mathbb{E}[X|\mathcal{F}_t]|\big] \leq \mathbb{E}\big[\mathbb{E}[|X|\,|\mathcal{F}_t]\big] = \mathbb{E}[|X|] < \infty.$$

X_t is $\mathcal{F}_t$-predictable by the definition of the conditional expectation. Using the tower property yields

$$\mathbb{E}[X_t|\mathcal{F}_s] = \mathbb{E}\big[\mathbb{E}[X|\mathcal{F}_t]|\mathcal{F}_s\big] = \mathbb{E}[X|\mathcal{F}_s] = X_s, \qquad s < t.$$

Exercise 2.16.8 Since

$$\mathbb{E}[|Z_t|] = \mathbb{E}[|aX_t + bY_t + c|] \leq |a|\mathbb{E}[|X_t|] + |b|\mathbb{E}[|Y_t|] + |c| < \infty$$

then Z_t is integrable. For $s < t$, using the martingale property of X_t and Y_t we have

$$\mathbb{E}[Z_t|\mathcal{F}_s] = a\mathbb{E}[X_t|\mathcal{F}_s] + b\mathbb{E}[Y_t|\mathcal{F}_s] + c = aX_s + bY_s + c = Z_s.$$

Exercise 2.16.9 In general the answer is no for both (a) and (b). For instance, if $X_t = Y_t$ the process X_t^2 is not a martingale, since the Jensen's inequality

$$\mathbb{E}[X_t^2|\mathcal{F}_s] \geq \Big(\mathbb{E}[X_t|\mathcal{F}_s]\Big)^2 = X_s^2$$

is not necessarily an identity. For instance B_t^2 is not a martingale, with B_t the Brownian motion process.

Exercise 2.16.10 It follows from the identity

$$\mathbb{E}[(X_t - X_s)(Y_t - Y_s)|\mathcal{F}_s] = \mathbb{E}[X_tY_t - X_sY_s|\mathcal{F}_s].$$

Exercise 2.16.11 (*a*) Let $Y_n = S_n - \mathbb{E}[S_n]$. We have

$$\begin{aligned} Y_{n+k} &= S_{n+k} - \mathbb{E}[S_{n+k}] \\ &= Y_n + \sum_{j=1}^{k} X_{n+j} - \sum_{j=1}^{k} \mathbb{E}[X_{n+j}]. \end{aligned}$$

Using the properties of expectation we have

$$\begin{aligned} \mathbb{E}[Y_{n+k}|\mathcal{F}_n] &= Y_n + \sum_{j=1}^{k} \mathbb{E}[X_{n+j}|\mathcal{F}_n] - \sum_{j=1}^{k} \mathbb{E}[\mathbb{E}[X_{n+j}]|\mathcal{F}_n] \\ &= Y_n + \sum_{j=1}^{k} \mathbb{E}[X_{n+j}] - \sum_{j=1}^{k} \mathbb{E}[X_{n+j}] \\ &= Y_n. \end{aligned}$$

(*b*) Let $Z_n = S_n^2 - Var(S_n)$. The process Z_n is an $\mathcal{F}_n$-martingale iff

$$\mathbb{E}[Z_{n+k} - Z_n|\mathcal{F}_n] = 0.$$

Let $U = S_{n+k} - S_n$. Using the independence we have

$$\begin{aligned} Z_{n+k} - Z_n &= (S_{n+k}^2 - S_n^2) - \big(Var(S_{n+k} - Var(S_n)\big) \\ &= (S_n + U)^2 - S_n^2 - \big(Var(S_{n+k} - Var(S_n)\big) \\ &= U^2 + 2US_n - Var(U), \end{aligned}$$

so

$$\begin{aligned} \mathbb{E}[Z_{n+k} - Z_n|\mathcal{F}_n] &= \mathbb{E}[U^2] + 2S_n\mathbb{E}[U] - Var(U) \\ &= \mathbb{E}[U^2] - (\mathbb{E}[U^2] - \mathbb{E}[U]^2) \\ &= 0, \end{aligned}$$

since $\mathbb{E}[U] = 0$.

Exercise 2.16.12 Let $\mathcal{F}_n = \sigma(X_k; k \le n)$. Using the independence

$$\mathbb{E}[|P_n|] = \mathbb{E}[|X_0|] \cdots \mathbb{E}[|X_n|] < \infty,$$

so $|P_n|$ integrable. Taking out the predictable part we have

$$\begin{aligned}\mathbb{E}[P_{n+k}|\mathcal{F}_n] &= \mathbb{E}[P_n X_{n+1}\cdots X_{n+k}|\mathcal{F}_n] = P_n\mathbb{E}[X_{n+1}\cdots X_{n+k}|\mathcal{F}_n] \\ &= P_n\mathbb{E}[X_{n+1}]\cdots\mathbb{E}[X_{n+k}] = P_n.\end{aligned}$$

Exercise 2.16.13 (*a*) Since the random variable $Y = \theta X$ is normally distributed with mean $\theta\mu$ and variance $\theta^2\sigma^2$, then

$$\mathbb{E}[e^{\theta X}] = e^{\theta\mu + \frac{1}{2}\theta^2\sigma^2}.$$

Hence $\mathbb{E}[e^{\theta X}] = 1$ iff $\theta\mu + \frac{1}{2}\theta^2\sigma^2 = 0$ which has the nonzero solution $\theta = -2\mu/\sigma^2$.

(*b*) Since $e^{\theta X_i}$ are independent, integrable and satisfy $\mathbb{E}[e^{\theta X_i}] = 1$, by Exercise 2.16.12 we get that the product $Z_n = e^{\theta S_n} = e^{\theta X_1}\cdots e^{\theta X_n}$ is a martingale.

Chapter 3

Exercise 3.1.4 B_t starts at 0 and is continuous in t. By Proposition 3.1.2 B_t is a martingale with $\mathbb{E}[B_t^2] = t < \infty$. Since $B_t - B_s \sim N(0, |t-s|)$, then $\mathbb{E}[(B_t - B_s)^2] = |t-s|$.

Exercise 3.1.9 It is obvious that $X_t = W_{t+t_0} - W_{t_0}$ satisfies $X_0 = 0$ and that X_t is continuous in t. The increments are normal distributed $X_t - X_s = W_{t+t_0} - W_{s+t_0} \sim N(0, |t-s|)$. If $0 < t_1 < \cdots < t_n$, then $0 < t_0 < t_1 + t_0 < \cdots < t_n + t_0$. The increments $X_{t_{k+1}} - X_{t_k} = W_{t_{k+1}+t_0} - W_{t_k+t_0}$ are obviously independent and stationary.

Exercise 3.1.10 Let $s < t$. Then we have

$$X_t - X_s = \frac{1}{\sqrt{\lambda}}(W_{\lambda t} - W_{\lambda s}) \sim \frac{1}{\sqrt{\lambda}}N\big(0, \lambda(t-s)\big) = N(0, t-s).$$

The other properties are obvious.

Exercise 3.1.11 Apply Property 3.1.7.

Exercise 3.1.12 Using the moment generating function, we get $\mathbb{E}[W_t^3] = 0$, $\mathbb{E}[W_t^4] = 3t^2$.

Exercise 3.1.13 (*a*) Let $s < t$. Then

$$\begin{aligned}\mathbb{E}[(W_t^2 - t)(W_s^2 - s)] &= \mathbb{E}\Big[\mathbb{E}[(W_t^2 - t)(W_s^2 - s)]|\mathcal{F}_s\Big] \\ &= \mathbb{E}\Big[(W_s^2 - s)\mathbb{E}[(W_t^2 - t)]|\mathcal{F}_s\Big] \\ &= \mathbb{E}\Big[(W_s^2 - s)^2\Big] = \mathbb{E}\Big[W_s^4 - 2sW_s^2 + s^2\Big] \\ &= \mathbb{E}[W_s^4] - 2s\mathbb{E}[W_s^2] + s^2 = 3s^2 - 2s^2 + s^2 = 2s^2.\end{aligned}$$

(*b*) Using part (*a*) we have

$$\begin{aligned} 2s^2 &= \mathbb{E}[(W_t^2 - t)(W_s^2 - s)] \\ &= \mathbb{E}[W_s^2 W_t^2] - s\mathbb{E}[W_t^2] - t\mathbb{E}[W_s^2] + ts \\ &= \mathbb{E}[W_s^2 W_t^2] - st. \end{aligned}$$

Therefore $\mathbb{E}[W_s^2 W_t^2] = ts + 2s^2$.
(*c*) $Cov(W_t^2, W_s^2) = \mathbb{E}[W_s^2 W_t^2] - \mathbb{E}[W_s^2]\mathbb{E}[W_t^2] = ts + 2s^2 - ts = 2s^2$.
(*d*) $Corr(W_t^2, W_s^2) = \frac{2s^2}{2ts} = \frac{s}{t}$, where we used

$$Var(W_t^2) = \mathbb{E}[W_t^4] - \mathbb{E}[W_t^2]^2 = 3t^2 - t^2 = 2t^2.$$

Exercise 3.1.14 (*a*) The distribution function of Y_t is given by

$$\begin{aligned} F(x) &= P(Y_t \le x) = P(tW_{1/t} \le x) = P(W_{1/t} \le x/t) \\ &= \int_0^{x/t} \phi_{1/t}(y)\,dy = \int_0^{x/t} \sqrt{t/(2\pi)} e^{ty^2/2}\,dy \\ &= \int_0^{x/\sqrt{t}} \frac{1}{\sqrt{2\pi}} e^{-u^2/2}\,du. \end{aligned}$$

(*b*) The probability density of Y_t is obtained by differentiating $F(x)$

$$p(x) = F'(x) = \frac{d}{dx}\int_0^{x/\sqrt{t}} \frac{1}{\sqrt{2\pi}} e^{-u^2/2}\,du = \frac{1}{\sqrt{2\pi}} e^{-x^2/2},$$

and hence $Y_t \sim N(0, t)$.
(*c*) Using that Y_t has independent increments we have

$$\begin{aligned} Cov(Y_s, Y_t) &= \mathbb{E}[Y_s Y_t] - \mathbb{E}[Y_s]\mathbb{E}[Y_t] = \mathbb{E}[Y_s Y_t] \\ &= \mathbb{E}\Big[Y_s(Y_t - Y_s) + Y_s^2\Big] = \mathbb{E}[Y_s]\mathbb{E}[Y_t - Y_s] + \mathbb{E}[Y_s^2] \\ &= 0 + s = s. \end{aligned}$$

(*d*) Since

$$Y_t - Y_s = (t-s)(W_{1/t} - W_0) - s(W_{1/s} - W_{1/t})$$

$$\mathbb{E}[Y_t - Y_s] = (t-s)\mathbb{E}[W_{1/t}] - s\mathbb{E}[W_{1/s} - W_{1/t}] = 0,$$

and

$$\begin{aligned} Var(Y_t - Y_s) &= \mathbb{E}[(Y_t - Y_s)^2] = (t-s)^2\frac{1}{t} + s^2\left(\frac{1}{s} - \frac{1}{t}\right) \\ &= \frac{(t-s)^2 + s(t-s)}{t} = t - s. \end{aligned}$$

Exercise 3.1.15 (*a*) Applying the definition of expectation we have

$$\begin{aligned}\mathbb{E}[|W_t|] &= \int_{-\infty}^{\infty} |x| \frac{1}{\sqrt{2\pi t}} e^{-\frac{x^2}{2t}}\, dx = \int_0^{\infty} 2x \frac{1}{\sqrt{2\pi t}} e^{-\frac{x^2}{2t}}\, dx \\ &= \frac{1}{\sqrt{2\pi t}} \int_0^{\infty} e^{-\frac{y}{2t}}\, dy = \sqrt{2t/\pi}.\end{aligned}$$

(*b*) Since $\mathbb{E}[|W_t|^2] = \mathbb{E}[W_t^2] = t$, we have

$$Var(|W_t|) = \mathbb{E}[|W_t|^2] - \mathbb{E}[|W_t|]^2 = t - \frac{2t}{\pi} = t\Big(1 - \frac{2}{\pi}\Big).$$

Exercise 3.1.16 By the martingale property of $W_t^2 - t$ we have

$$\mathbb{E}[W_t^2|\mathcal{F}_s] = \mathbb{E}[W_t^2 - t|\mathcal{F}_s] + t = W_s^2 + t - s.$$

Exercise 3.1.17 (*a*) Expanding

$$(W_t - W_s)^3 = W_t^3 - 3W_t^2 W_s + 3W_t W_s^2 - W_s^3$$

and taking the expectation

$$\begin{aligned}\mathbb{E}[(W_t - W_s)^3|\mathcal{F}_s] &= \mathbb{E}[W_t^3|\mathcal{F}_s] - 3W_s\mathbb{E}[W_t^2] + 3W_s^2\mathbb{E}[W_t|\mathcal{F}_s] - W_s^3 \\ &= \mathbb{E}[W_t^3|\mathcal{F}_s] - 3(t-s)W_s - W_s^3,\end{aligned}$$

so

$$\mathbb{E}[W_t^3|\mathcal{F}_s] = 3(t-s)W_s + W_s^3,$$

since

$$\mathbb{E}[(W_t - W_s)^3|\mathcal{F}_s] = \mathbb{E}[(W_t - W_s)^3] = \mathbb{E}[W_{t-s}^3] = 0.$$

(*b*) Hint: Start from the expansion of $(W_t - W_s)^4$.

Exercise 3.2.3 Using that $e^{W_t - W_s}$ is stationary, we have

$$\mathbb{E}[e^{W_t - W_s}] = \mathbb{E}[e^{W_{t-s}}] = e^{\frac{1}{2}(t-s)}.$$

Exercise 3.2.4 (*a*)

$$\begin{aligned}\mathbb{E}[X_t|\mathcal{F}_s] &= \mathbb{E}[e^{W_t}|\mathcal{F}_s] = \mathbb{E}[e^{W_t - W_s} e^{W_s}|\mathcal{F}_s] \\ &= e^{W_s}\mathbb{E}[e^{W_t - W_s}|\mathcal{F}_s] = e^{W_s}\mathbb{E}[e^{W_t - W_s}] \\ &= e^{W_s} e^{t/2} e^{-s/2}.\end{aligned}$$

(*b*) This can also be written as

$$\mathbb{E}[e^{-t/2} e^{W_t}|\mathcal{F}_s] = e^{-s/2} e^{W_s},$$

which shows that $e^{-t/2}e^{W_t}$ is a martingale.
(*c*) From the stationarity we have

$$\mathbb{E}[e^{cW_t - cW_s}] = \mathbb{E}[e^{c(W_t - W_s)}] = \mathbb{E}[e^{cW_{t-s}}] = e^{\frac{1}{2}c^2(t-s)}.$$

Then for any $s < t$ we have

$$\begin{aligned}\mathbb{E}[e^{cW_t}|\mathcal{F}_s] &= \mathbb{E}[e^{c(W_t-W_s)}e^{cW_s}|\mathcal{F}_s] = e^{cW_s}\mathbb{E}[e^{c(W_t-W_s)}|\mathcal{F}_s] \\ &= e^{cW_s}\mathbb{E}[e^{c(W_t-W_s)}] = e^{cW_s}e^{\frac{1}{2}c^2(t-s)} = Y_s e^{\frac{1}{2}c^2 t}.\end{aligned}$$

Multiplying by $e^{-\frac{1}{2}c^2 t}$ yields the desired result.

Exercise 3.2.5 (*a*) Using Exercise 3.2.3 we have

$$\begin{aligned}Cov(X_s, X_t) &= \mathbb{E}[X_s X_t] - \mathbb{E}[X_s]\mathbb{E}[X_t] = \mathbb{E}[X_s X_t] - e^{t/2}e^{s/2} \\ &= \mathbb{E}[e^{W_s + W_t}] - e^{t/2}e^{s/2} = \mathbb{E}[e^{W_t - W_s}e^{2(W_s - W_0)}] - e^{t/2}e^{s/2} \\ &= \mathbb{E}[e^{W_t - W_s}]\mathbb{E}[e^{2(W_s - W_0)}] - e^{t/2}e^{s/2} = e^{\frac{t-s}{2}}e^{2s} - e^{t/2}e^{s/2} \\ &= e^{\frac{t+3s}{2}} - e^{(t+s)/2}.\end{aligned}$$

(*b*) Using Exercise 3.2.4 (*b*), we have

$$\begin{aligned}\mathbb{E}[X_s X_t] &= \mathbb{E}\Big[\mathbb{E}[X_s X_t|\mathcal{F}_s]\Big] = \mathbb{E}\Big[X_s \mathbb{E}[X_t|\mathcal{F}_s]\Big] \\ &= e^{t/2}\mathbb{E}\Big[X_s \mathbb{E}[e^{-t/2}X_t|\mathcal{F}_s]\Big] = e^{t/2}\mathbb{E}\Big[X_s\, e^{-s/2}X_s\Big] \\ &= e^{(t-s)/2}\mathbb{E}[X_s^2] = e^{(t-s)/2}\mathbb{E}[e^{2W_s}] \\ &= e^{(t-s)/2}e^{2s} = e^{\frac{t+3s}{2}},\end{aligned}$$

and continue like in part (*a*).

Exercise 3.2.6 Using the definition of the expectation we have

$$\begin{aligned}\mathbb{E}[e^{2W_t^2}] &= \int e^{2x^2}\phi_t(x)\,dx = \frac{1}{\sqrt{2\pi t}}\int e^{2x^2}e^{-\frac{x^2}{2t}}\,dx \\ &= \frac{1}{\sqrt{2\pi t}}\int e^{-\frac{1-4t}{2t}x^2}\,dx = \frac{1}{\sqrt{1-4t}},\end{aligned}$$

if $1 - 4t > 0$. Otherwise, the integral is infinite. We used the standard integral $\int e^{-ax^2} = \sqrt{\pi/a}$, $a > 0$.

Exercise 3.3.4 (*a*) It follows from the fact that Z_t is normally distributed; (*b*) Differentiate the moment generating function and evaluate it at $u = 0$.

Exercise 3.3.5 Using the definition of covariance we have

$$\begin{aligned} Cov\Big(Z_s, Z_t\Big) &= \mathbb{E}[Z_s Z_t] - \mathbb{E}[Z_s]\mathbb{E}[Z_t] = \mathbb{E}[Z_s Z_t] \\ &= \mathbb{E}\Big[\int_0^s W_u\,du \cdot \int_0^t W_v\,dv\Big] = \mathbb{E}\Big[\int_0^s \int_0^t W_u W_v\,dudv\Big] \\ &= \int_0^s \int_0^t \mathbb{E}[W_u W_v]\,dudv = \int_0^s \int_0^t \min\{u,v\}\,dudv \\ &= s^2\Big(\frac{t}{2} - \frac{s}{6}\Big), \qquad s < t. \end{aligned}$$

Exercise 3.3.6 (*a*) Using Exercise 3.3.5

$$\begin{aligned} Cov(Z_t, Z_t - Z_{t-h}) &= Cov(Z_t, Z_t) - Cov(Z_t, Z_{t-h}) \\ &= \frac{t^3}{3} - (t-h)^2\Big(\frac{t}{2} - \frac{t-h}{6}\Big) \\ &= \frac{1}{2}t^2 h + o(h). \end{aligned}$$

(*b*) Using $Z_t - Z_{t-h} = \int_{t-h}^t W_u\,du = hW_t + o(h)$,

$$\begin{aligned} Cov(Z_t, W_t) &= \frac{1}{h} Cov(Z_t, Z_t - Z_{t-h}) \\ &= \frac{1}{h}\Big(\frac{1}{2}t^2 h + o(h)\Big) = \frac{1}{2}t^2. \end{aligned}$$

Exercise 3.3.7 Let $s < u$. Since W_t has independent increments, taking the expectation in

$$e^{W_s + W_t} = e^{W_t - W_s} e^{2(W_s - W_0)}$$

we obtain

$$\begin{aligned} \mathbb{E}[e^{W_s + W_t}] &= \mathbb{E}[e^{W_t - W_s}]\mathbb{E}[e^{2(W_s - W_0)}] = e^{\frac{u-s}{2}} e^{2s} \\ &= e^{\frac{u+s}{2}} e^{s} = e^{\frac{u+s}{2}} e^{\min\{s,t\}}. \end{aligned}$$

Exercise 3.3.8 (*a*) $\mathbb{E}[X_t] = \int_0^t \mathbb{E}[e^{W_s}]\,ds = \int_0^t \mathbb{E}[e^{s/2}]\,ds = 2(e^{t/2} - 1)$
(*b*) Since $Var(X_t) = \mathbb{E}[X_t^2] - \mathbb{E}[X_t]^2$, it suffices to compute $\mathbb{E}[X_t^2]$. Using Exercise 3.3.7 we have

$$\begin{aligned} \mathbb{E}[X_t^2] &= \mathbb{E}\Big[\int_0^t e^{W_t}\,ds \cdot \int_0^t e^{W_u}\,du\Big] = \mathbb{E}\Big[\int_0^t \int_0^t e^{W_s} e^{W_u}\,dsdu\Big] \\ &= \int_0^t \int_0^t \mathbb{E}[e^{W_s + W_u}]\,dsdu = \int_0^t \int_0^t e^{\frac{u+s}{2}} e^{\min\{s,t\}}\,dsdu \\ &= \iint_{D_1} e^{\frac{u+s}{2}} e^{s} duds + \iint_{D_2} e^{\frac{u+s}{2}} e^{u} duds \\ &= 2\iint_{D_2} e^{\frac{u+s}{2}} e^{u} duds = \frac{4}{3}\Big(\frac{1}{2}e^{2t} - 2e^{t/2} + \frac{3}{2}\Big), \end{aligned}$$

where $D_1\{0 \le s < u \le t\}$ and $D_2\{0 \le u < s \le t\}$. In the last identity we applied Fubini's theorem. For finding the variance, use the formula $Var(X_t) = \mathbb{E}[X_t^2] - \mathbb{E}[X_t]^2$.

Exercise 3.3.9 (a) Splitting the integral at t and taking out the measurable part, we have

$$\begin{aligned}
\mathbb{E}[Z_T|\mathcal{F}_t] &= \mathbb{E}[\int_0^T W_u\,du|\mathcal{F}_t] = \mathbb{E}[\int_0^t W_u\,du|\mathcal{F}_t] + \mathbb{E}[\int_t^T W_u\,du|\mathcal{F}_t] \\
&= Z_t + \mathbb{E}[\int_t^T W_u\,du|\mathcal{F}_t] \\
&= Z_t + \mathbb{E}[\int_t^T (W_u - W_t + W_t)\,du|\mathcal{F}_t] \\
&= Z_t + \mathbb{E}[\int_t^T (W_u - W_t)\,du|\mathcal{F}_t] + W_t(T-t) \\
&= Z_t + \mathbb{E}[\int_t^T (W_u - W_t)\,du] + W_t(T-t) \\
&= Z_t + \int_t^T \mathbb{E}[W_u - W_t]\,du + W_t(T-t) \\
&= Z_t + W_t(T-t),
\end{aligned}$$

since $\mathbb{E}[W_u - W_t] = 0$.

(b) Let $0 < t < T$. Using (a) we have

$$\begin{aligned}
\mathbb{E}[Z_T - TW_T|\mathcal{F}_t] &= \mathbb{E}[Z_T|\mathcal{F}_t] - T\mathbb{E}[W_T|\mathcal{F}_t] \\
&= Z_t + W_t(T-t) - TW_t \\
&= Z_t - tW_t.
\end{aligned}$$

Exercise 3.3.10 (a)

$$\begin{aligned}
\mathbb{E}\Big[\int_t^T W_s^2\,ds|\mathcal{F}_t\Big] &= \mathbb{E}\Big[\int_t^T (W_s - W_t + W_t)^2\,ds|\mathcal{F}_t\Big] \\
&= \mathbb{E}\Big[\int_t^T (W_s - W_t)^2\,ds|\mathcal{F}_t\Big] \\
&\quad +2\mathbb{E}\Big[\int_t^T W_t(W_s - W_t)\,ds|\mathcal{F}_t\Big] + \mathbb{E}\Big[\int_t^T W_t^2\,ds|\mathcal{F}_t\Big] \\
&= \mathbb{E}\Big[\int_t^T (W_s - W_t)^2\,ds\Big] \\
&\quad +2W_t\mathbb{E}\Big[\int_t^T (W_s - W_t)\,ds\Big] + \mathbb{E}\Big[W_t^2(T-t)|\mathcal{F}_t\Big]
\end{aligned}$$

$$
\begin{aligned}
&= \mathbb{E}\Big[\int_0^{T-t} W_u^2\,du\Big] + 2W_t\mathbb{E}\Big[\int_0^{T-t} W_u\,du\Big] + W_t^2(T-t) \\
&= \frac{1}{2}(T-t)^2 + W_t^2(T-t).
\end{aligned}
$$

(*b*) Using part (*a*) we have

$$
\begin{aligned}
\mathbb{E}\Big[Y_T|\mathcal{F}_t\Big] &= \mathbb{E}\Big[\int_t^T W_s^2\,ds - TW_T^2 + \frac{1}{2}T^2|\mathcal{F}_t\Big] \\
&= \mathbb{E}\Big[\int_t^T W_s^2\,ds|\mathcal{F}_t\Big] - T\,\mathbb{E}\Big[W_T^2|\mathcal{F}_t\Big] + \frac{1}{2}T^2 \\
&= \int_0^t W_s^2\,ds + \frac{1}{2}(T-t)^2 + W_t^2(T-t) - TW_t^2 - T(T-t) + \frac{1}{2}T^2 \\
&= \int_0^t W_s^2\,ds + -tW_t^2 + \frac{1}{2}t^2 = Y_s.
\end{aligned}
$$

Exercise 3.4.1

$$
\begin{aligned}
\mathbb{E}[V_T|\mathcal{F}_t] &= \mathbb{E}\Big[e^{\int_0^t W_u\,du + \int_t^T W_u\,du}|\mathcal{F}_t\Big] \\
&= e^{\int_0^t W_u\,du}\mathbb{E}\Big[e^{\int_t^T W_u\,du}|\mathcal{F}_t\Big] \\
&= e^{\int_0^t W_u\,du}\mathbb{E}\Big[e^{\int_t^T (W_u - W_t)\,du + (T-t)W_t}|\mathcal{F}_t\Big] \\
&= V_t e^{(T-t)W_t}\mathbb{E}\Big[e^{\int_t^T (W_u - W_t)\,du}|\mathcal{F}_t\Big] \\
&= V_t e^{(T-t)W_t}\mathbb{E}\Big[e^{\int_t^T (W_u - W_t)\,du}\Big] \\
&= V_t e^{(T-t)W_t}\mathbb{E}\Big[e^{\int_0^{T-t} W_\tau\,d\tau}\Big] \\
&= V_t e^{(T-t)W_t} e^{\frac{1}{2}\frac{(T-t)^3}{3}}.
\end{aligned}
$$

Exercise 3.6.1

$$
\begin{aligned}
F(x) &= P(Y_t \le x) = P(\mu t + W_t \le x) = P(W_t \le x - \mu t) \\
&= \int_0^{x-\mu t} \frac{1}{\sqrt{2\pi t}} e^{-\frac{u^2}{2t}}\,du; \\
f(x) &= F'(x) = \frac{1}{\sqrt{2\pi t}} e^{-\frac{(x-\mu t)^2}{2t}}.
\end{aligned}
$$

Exercise 3.7.2 Since

$$
P(R_t \le \rho) = \int_0^\rho \frac{1}{t} x e^{-\frac{x^2}{2t}}\,dx,
$$

use the inequality

$$1 - \frac{x^2}{2t} < e^{-\frac{x^2}{2t}} < 1$$

to get the desired result.

Exercise 3.7.3 (*a*) The mean can be evaluated as

$$\begin{aligned}
\mathbb{E}[R_t] &= \int_0^\infty x p_t(x)\,dx = \int_0^\infty \frac{1}{t} x^2 e^{-\frac{x^2}{2t}}\,dx \\
&= \frac{1}{2t}\int_0^\infty y^{1/2} e^{-\frac{y}{2t}}\,dy = \sqrt{2t}\int_0^\infty z^{\frac{3}{2}-1} e^{-z}\,dz \\
&= \sqrt{2t}\,\Gamma\big(\frac{3}{2}\big) = \frac{\sqrt{2\pi t}}{2}.
\end{aligned}$$

(*b*) Since $\mathbb{E}[R_t^2] = \mathbb{E}[W_1(t)^2 + W_2(t)^2] = 2t$, then

$$Var(R_t) = 2t - \frac{2\pi t}{4} = 2t\Big(1 - \frac{\pi}{4}\Big).$$

Exercise 3.7.4

$$\begin{aligned}
\mathbb{E}[X_t] &= \frac{\mathbb{E}[X_t]}{t} = \frac{\sqrt{2\pi t}}{2t} = \sqrt{\frac{\pi}{2t}} \to 0, \qquad t \to \infty; \\
Var(X_t) &= \frac{1}{t^2} Var(R_t) = \frac{2}{t}(1 - \frac{\pi}{4}) \to 0, \qquad t \to \infty.
\end{aligned}$$

By Proposition 2.14.1 we get $X_t \to 0$, $t \to \infty$ in mean square.

Exercise 3.8.2

$$\begin{aligned}
P(N_t - N_s = 1) &= \lambda(t-s)e^{-\lambda(t-s)} \\
&= \lambda(t-s)\big(1 - \lambda(t-s) + o(t-s)\big) \\
&= \lambda(t-s) + o(t-s).
\end{aligned}$$

$$\begin{aligned}
P(N_t - N_s \geq 1) &= 1 - P(N_t - N_s = 0) + P(N_t - N_s = 1) \\
&= 1 - e^{-\lambda(t-s)} - \lambda(t-s)e^{-\lambda(t-s)} \\
&= 1 - \Big(1 - \lambda(t-s) + o(t-s)\Big) \\
&\quad -\lambda(t-s)\Big(1 - \lambda(t-s) + o(t-s)\Big) \\
&= \lambda^2(t-s)^2 = o(t-s).
\end{aligned}$$

Exercise 3.8.6 Write first as

$$\begin{aligned}
N_t^2 &= N_t(N_t - N_s) + N_t N_s \\
&= (N_t - N_s)^2 + N_s(N_t - N_s) + (N_t - \lambda t)N_s + \lambda t N_s,
\end{aligned}$$

then

$$\begin{aligned}
\mathbb{E}[N_t^2|\mathcal{F}_s] &= \mathbb{E}[(N_t - N_s)^2|\mathcal{F}_s] + N_s\mathbb{E}[N_t - N_s|\mathcal{F}_s] + \mathbb{E}[N_t - \lambda t|\mathcal{F}_s]N_s + \lambda t N_s \\
&= \mathbb{E}[(N_t - N_s)^2] + N_s\mathbb{E}[N_t - N_s] + \mathbb{E}[N_t - \lambda t]N_s + \lambda t N_s \\
&= \lambda(t-s) + \lambda^2(t-s)^2 + \lambda t N_s + N_s^2 - \lambda s N_s + \lambda t N_s \\
&= \lambda(t-s) + \lambda^2(t-s)^2 + 2\lambda(t-s)N_s + N_s^2 \\
&= \lambda(t-s) + [N_s + \lambda(t-s)]^2.
\end{aligned}$$

Hence $\mathbb{E}[N_t^2|\mathcal{F}_s] \neq N_s^2$ and hence the process N_t^2 is not an $\mathcal{F}_s$-martingale.

Exercise 3.8.7 (*a*)

$$\begin{aligned}
m_{N_t}(x) &= \mathbb{E}[e^{xN_t}] = \sum_{k\geq 0} e^{xk} P(N_t = k) \\
&= \sum_{k\geq 0} e^{xk} e^{-\lambda t} \frac{\lambda^k t^k}{k!} \\
&= e^{-\lambda t} e^{\lambda t e^x} = e^{\lambda t(e^x - 1)}.
\end{aligned}$$

(*b*) $\mathbb{E}[N_t^2] = m''_{N_t}(0) = \lambda^2 t^2 + \lambda t$. Similarly for the other relations.

Exercise 3.8.8 $\mathbb{E}[X_t] = \mathbb{E}[e^{N_t}] = m_{N_t}(1) = e^{\lambda t(e-1)}$.

Exercise 3.8.9 (*a*) Since $e^{xM_t} = e^{x(N_t - \lambda t)} = e^{-\lambda t x} e^{xN_t}$, the moment generating function is

$$\begin{aligned}
m_{M_t}(x) &= \mathbb{E}[e^{xM_t}] = e^{-\lambda t x}\mathbb{E}[e^{xN_t}] \\
&= e^{-\lambda t x} e^{\lambda t(e^x - 1)} = e^{\lambda t(e^x - x - 1)}.
\end{aligned}$$

(*b*) For instance

$$\mathbb{E}[M_t^3] = m'''_{M_t}(0) = \lambda t.$$

Since M_t is a stationary process, $\mathbb{E}[(M_t - M_s)^3] = \lambda(t-s)$.

Exercise 3.8.10

$$\begin{aligned}
Var[(M_t - M_s)^2] &= \mathbb{E}[(M_t - M_s)^4] - \mathbb{E}[(M_t - M_s)^2]^2 \\
&= \lambda(t-s) + 3\lambda^2(t-s)^2 - \lambda^2(t-s)^2 \\
&= \lambda(t-s) + 2\lambda^2(t-s)^2.
\end{aligned}$$

Exercise 3.11.3 (*a*) $\mathbb{E}[U_t] = \mathbb{E}\Big[\int_0^t N_u\,du\Big] = \int_0^t \mathbb{E}[N_u]\,du = \int_0^t \lambda u\,du = \frac{\lambda t^2}{2}$.

$$(b)\quad \mathbb{E}\Big[\sum_{k=1}^{N_t} S_k\Big] = \mathbb{E}[tN_t - U_t] = t\lambda t - \frac{\lambda t^2}{2} = \frac{\lambda t^2}{2}.$$

Exercise 3.11.4 (*a*) Use associativity of addition.
(*b*) Use that $T_1, \cdots, T_n$ are independent, exponentialy distributed, with $\mathbb{E}[T_k] = 1/\lambda$.

Exercise 3.11.6 The proof cannot go through because a product between a constant and a Poisson process is not a Poisson process.

Exercise 3.11.7 Let $p_X(x)$ be the probability density of X. If $p(x, y)$ is the joint probability density of X and Y, then $p_X(x) = \sum_y p(x, y)$. We have

$$\begin{aligned}
\sum_{y\geq 0} \mathbb{E}[X|Y = y]P(Y = y) &= \sum_{y\geq 0} \int x p_{X|Y=y}(x|y)P(Y = y)\,dx \\
&= \sum_{y\geq 0} \int \frac{xp(x, y)}{P(Y = y)} P(Y = y)\,dx \\
&= \int x \sum_{y\geq 0} p(x, y)\,dx \\
&= \int x p_X(x)\,dx = \mathbb{E}[X].
\end{aligned}$$

Exercise 3.11.9 (*a*) Since T_k has an exponential distribution with parameter λ

$$\mathbb{E}[e^{-\sigma T_k}] = \int_0^\infty e^{-\sigma x} \lambda e^{-\lambda x}\,dx = \frac{\lambda}{\lambda + \sigma}.$$

(*b*) We have

$$\begin{aligned}
U_t &= T_2 + 2T_3 + 3T_4 + \cdots + (n-2)T_{n-1} + (n-1)T_n + (t - S_n)n \\
&= T_2 + 2T_3 + 3T_4 + \cdots + (n-2)T_{n-1} + (n-1)T_n + nt \\
&\quad -n(T_1 + T_2 + \cdots + T_n) \\
&= nt - [nT_1 + (n-1)T_2 + \cdots + T_n].
\end{aligned}$$

(*c*) Using that the arrival times S_k, $k = 1, 2, \ldots n$, have the same distribution as the order statistics $U_{(k)}$ corresponding to n independent random variables uniformly distributed on the interval $(0, t)$, we get

$$\begin{aligned}
\mathbb{E}\Big[e^{-\sigma U_t}\Big| N_t = n\Big] &= \mathbb{E}[e^{-\sigma(tN_t - \sum_{k=1}^{N_t} S_k)}|N_t = n] \\
&= e^{-n\sigma t}\mathbb{E}[e^{\sigma \sum_{i=1}^n U_{(i)}}] = e^{-n\sigma t}\mathbb{E}[e^{\sigma \sum_{i=1}^n U_i}] \\
&= e^{-n\sigma t}\mathbb{E}[e^{\sigma U_1}] \cdots \mathbb{E}[e^{\sigma U_n}] \\
&= e^{-n\sigma t}\frac{1}{t}\int_0^t e^{\sigma x_1}\,dx_1 \cdots \frac{1}{t}\int_0^t e^{\sigma x_n}\,dx_n \\
&= \frac{(1 - e^{-\sigma t})^n}{\sigma^n t^n}.
\end{aligned}$$

(*d*) Using Exercise 3.11.7 we have

$$\begin{aligned} \mathbb{E}\Big[e^{-\sigma U_t}\Big] &= \sum_{n\geq 0} P(N_t=n)\mathbb{E}\Big[e^{-\sigma U_t}\Big|N_t=n\Big] \\ &= \sum_{n\geq 0} \frac{e^{-\lambda t^n}\lambda^n}{n!}\frac{(1-e^{-\sigma t})^n}{\sigma^n t^n} \\ &= e^{\lambda(1-e^{-\sigma t})/\sigma-\lambda}. \end{aligned}$$

Exercise 3.12.6 Use Doob's inequality for the submartingales W_t^2 and $|W_t|$, and use that $\mathbb{E}[W_t^2]=t$ and $\mathbb{E}[|W_t|]=\sqrt{2t/\pi}$, see Exercise 3.1.15 (*a*).

Exercise 3.12.7 Divide by t in the inequality from Exercise 3.12.6 part (*b*).

Exercise 3.12.10 Let $\sigma=n$ and $\tau=n+1$. Then

$$\mathbb{E}\Big[\sup_{n\leq t\leq n=1}\Big(\frac{N_t}{t}-\lambda\Big)^2\Big] \leq \frac{4\lambda(n+1)}{n^2}.$$

The result follows by taking $n\to\infty$ in the sequence of inequalities

$$0\leq \mathbb{E}\Big[\Big(\frac{N_t}{t}-\lambda\Big)^2\Big] \leq \mathbb{E}\Big[\sup_{n\leq t\leq n=1}\Big(\frac{N_t}{t}-\lambda\Big)^2\Big] \leq \frac{4\lambda(n+1)}{n^2}.$$

Chapter 4

Exercise 4.1.2 We have

$$\{\omega;\tau(\omega)\leq t\} = \begin{cases} \Omega, & \text{if } c\leq t \\ \emptyset, & \text{if } c>t \end{cases}$$

and use that $\emptyset,\Omega\in\mathcal{F}_t$.

Exercise 4.1.3 First we note that

$$\{\omega;\tau(\omega)\leq t\} = \bigcup_{0<s<t}\{\omega;|W_s(\omega)|\geq K\}. \tag{12.0.1}$$

This can be shown by double inclusion. Let $A_s=\{\omega;|W_s(\omega)|\geq K\}$.
" $\subset$ " Let $\omega\in\{\omega;\tau(\omega)\leq t\}$, so $\inf\{s>0;|W_s(\omega)|\geq K\}\leq t$. Then exists $\tau\geq u\geq t$ such that $|W_u(\omega)|\geq K$, and hence $\omega\in A_u$.
" $\supset$ " Let $\omega\in\bigcup_{0<s<t}\{\omega;|W_s(\omega)|\geq K\}$. Then there is $0<s<t$ such that $|W_s(\omega)|\geq K$. This implies $\tau(\omega)\leq s$ and since $s\leq t$ it follows that $\tau(\omega)\leq t$.
Since W_t is continuous, then (12.0.1) can also be written as

$$\{\omega;\tau(\omega)\leq t\} = \bigcup_{0<r<t,r\in\mathbb{Q}}\{\omega;|W_r(\omega)|\geq K\},$$

which implies $\{\omega; \tau(\omega) \leq t\} \in \mathcal{F}_t$ since $\{\omega; |W_r(\omega)| \geq K\} \in \mathcal{F}_t$, for $0 < r \leq t$. Hence τ is a stopping time. It is worth noting that $P(\tau < \infty) = 1$.

$$\begin{aligned} P(\{\omega; \tau(\omega) < \infty\}) &= P\Big(\bigcup_{0<s}\{\omega; |W_s(\omega)| > K\}\Big) > P(\{\omega; |W_s(\omega)| > K\}) \\ &= 1 - \int_{|x|<K} \frac{1}{\sqrt{2\pi s}} e^{-\frac{y^2}{2s}}\, dy > 1 - \frac{2K}{\sqrt{2\pi s}} \to 1, \quad s \to \infty. \end{aligned}$$

Exercise 4.1.4 Let $K_m = [a + \frac{1}{m}, b - \frac{1}{m}]$. We can write

$$\{\omega; \tau \leq t\} = \bigcap_{m\geq 1} \bigcup_{r<t, r\in\mathbb{Q}} \{\omega; X_r \notin K_m\} \in \mathcal{F}_t,$$

since $\{\omega; X_r \notin K_m\} = \{\omega; X_r \in \overline{K}_m\} \in \mathcal{F}_r \subset \mathcal{F}_t$.

Exercise 4.1.6 (a) No. The event $A = \{\omega; W_t(\omega)$ has a local maximum at time $t\}$ is not in $\sigma\{W_s; s \leq t\}$ but is in $\sigma\{W_s; s \leq t + \epsilon\}$.

Exercise 4.1.9 (a) We have $\{\omega; c\tau \leq t\} = \{\omega; \tau \leq t/c\} \in \mathcal{F}_{t/c} \subset \mathcal{F}_t$.

(b) $\{\omega; f(\tau) \leq t\} = \{\omega; \tau \leq f^{-1}(t)\} = \mathcal{F}_{f^{-1}(t)} \subset \mathcal{F}_t$, since $f^{-1}(t) \leq t$.

(c) Apply (b) with $f(x) = e^x$.

Exercise 4.1.11 If we let $G(n) = \{x; |x - a| < \frac{1}{n}\}$, then $\{a\} = \bigcap_{n\geq 1} G(n)$. Then $\tau_n = \inf\{t \geq 0; W_t \in G(n)\}$ are stopping times. Since $\sup_n \tau_n = \tau$, then τ is a stopping time.

Exercise 4.2.3 The relation is proved by verifying two cases:
(i) If $\omega \in \{\omega; \tau > t\}$ then $(\tau \wedge t)(\omega) = t$ and the relation becomes

$$M_\tau(\omega) = M_t(\omega) + M_\tau(\omega) - M_t(\omega).$$

(ii) If $\omega \in \{\omega; \tau \leq t\}$ then $(\tau \wedge t)(\omega) = \tau(\omega)$ and the relation is equivalent to the obvious relation

$$M_\tau = M_\tau.$$

Exercise 4.2.5 Taking the expectation in $\mathbb{E}[M_\tau|\mathcal{F}_\sigma] = M_\sigma$ yields $\mathbb{E}[M_\tau] = \mathbb{E}[M_\sigma]$, and then make $\sigma = 0$.

Exercise 4.3.9 Since $M_t = W_T^2 - t$ is a martingale with $\mathbb{E}[M_t] = 0$, by the Optional Stopping Theorem we get $\mathbb{E}[M_{\tau_a}] = \mathbb{E}[M_0] = 0$, so $\mathbb{E}[W_{\tau_a}^2 - \tau_a] = 0$, from where $\mathbb{E}[\tau_a] = \mathbb{E}[W_{\tau_a}^2] = a^2$, since $W_{\tau_a} = a$.

Exercise 4.3.10 (*a*) We have

$$\begin{aligned}
F(a) = P(X_t \le a) &= 1 - P(X_t > a) = 1 - P(\max_{0\le s\le t} W_s > a) \\
&= 1 - P(T_a \le t) = 1 - \frac{2}{\sqrt{2\pi}} \int_{|a|/\sqrt{t}}^{\infty} e^{-y^2/2}\, dy \\
&= \frac{2}{\sqrt{2\pi}} \int_0^{\infty} e^{-y^2/2}\, dy - \frac{2}{\sqrt{2\pi}} \int_{|a|/\sqrt{t}}^{\infty} e^{-y^2/2}\, dy \\
&= \frac{2}{\sqrt{2\pi}} \int_0^{|a|/\sqrt{t}} e^{-y^2/2}\, dy.
\end{aligned}$$

(*b*) The density function is $p(a) = F'(a) = \frac{2}{\sqrt{2\pi t}} e^{-a^2/(2t)}$, $a > 0$. Then

$$\begin{aligned}
\mathbb{E}[X_t] &= \int_0^{\infty} x p(x)\, dx = \frac{2}{\sqrt{2\pi t}} \int_0^{\infty} x e^{-x^2/(2t)}\, dy = \sqrt{\frac{2t}{\pi}} \\
\mathbb{E}[X_t^2] &= \frac{2}{\sqrt{2\pi t}} \int_0^{\infty} x^2 e^{-x^2/(2t)}\, dx = \frac{4t}{\sqrt{\pi}} \int_0^{\infty} y^2 e^{-y^2}\, dy \\
&= \frac{2}{\sqrt{2\pi t}} \frac{1}{2} \int_0^{\infty} u^{1/2} e^{-u}\, du = \frac{1}{\sqrt{2\pi t}} \Gamma(3/2) = t \\
Var(X_t) &= \mathbb{E}[X_t^2] - \mathbb{E}[X_t]^2 = t\Big(1 - \frac{2}{\pi}\Big).
\end{aligned}$$

Exercise 4.3.11 (*b*) $h(x) = \frac{2}{\sqrt{\pi t}} e^{-\frac{x^2}{4t}} \Big[2N\Big(\frac{x}{\sqrt{2t}}\Big) - 1\Big]$.

Exercise 4.3.16 It is recurrent since $P(\exists t > 0 : a < W_t < b) = 1$.

Exercise 4.4.4 Since

$$P(W_t > 0; t_1 \le t \le t_2) = \frac{1}{2} P(W_t \ne 0; t_1 \le t \le t_2) = \frac{1}{\pi} \arcsin\sqrt{\frac{t_1}{t_2}},$$

using the independence

$$P(W_t^1 > 0, W_t^2) = P(W_t^1 > 0) P(W_t^2 > 0) = \frac{1}{\pi^2}\Big(\arcsin\sqrt{\frac{t_1}{t_2}}\Big)^2.$$

The probability for $W_t = (W_t^1, W_t^2)$ to be in one of the quadrants is equal to $\frac{4}{\pi^2}\Big(\arcsin\sqrt{\frac{t_1}{t_2}}\Big)^2$.

Exercise 4.5.2 (*a*) We have

$$\begin{aligned}
P(X_t \text{ goes up to } \alpha) &= P(X_t \text{ goes up to } \alpha \text{ before down to } -\infty) \\
&= \lim_{\beta\to\infty} \frac{e^{2\mu\beta} - 1}{e^{2\mu\beta} - e^{-2\mu\alpha}} = 1.
\end{aligned}$$

Exercise 4.5.3

$$\begin{aligned} P(X_t \text{ never hits } -\beta) &= P(X_t \text{ goes up to } \infty \text{ before down to } -\beta) \\ &= \lim_{\alpha \to \infty} \frac{e^{2\mu\beta} - 1}{e^{2\mu\beta} - e^{-2\mu\alpha}}. \end{aligned}$$

Exercise 4.5.4 (a) Use that $\mathbb{E}[X_T] = \alpha p_\alpha - \beta(1 - p_\alpha)$; (b) $\mathbb{E}[X_T^2] = \alpha^2 p_\alpha + \beta^2(1 - p_\alpha)$, with $p_\alpha = \frac{e^{2\mu\beta}-1}{e^{2\mu\beta}-e^{-2\mu\alpha}}$; (c) Use $Var(T) = \mathbb{E}[T^2] - \mathbb{E}[T]^2$.

Exercise 4.5.7 Since $M_t = W_t^2 - t$ is a martingale, with $\mathbb{E}[M_t] = 0$, by the Optional Stopping Theorem we get $\mathbb{E}[W_T^2 - T] = 0$. Using $W_T = X_T - \mu T$ yields

$$\mathbb{E}[X_T^2 - 2\mu T X_T + \mu^2 T^2] = \mathbb{E}[T].$$

Then

$$\mathbb{E}[T^2] = \frac{\mathbb{E}[T](1 + 2\mu\mathbb{E}[X_T]) - \mathbb{E}[X_T^2]}{\mu^2}.$$

Substitute $\mathbb{E}[X_T]$ and $\mathbb{E}[X_T^2]$ from Exercise 4.5.4 and $\mathbb{E}[T]$ from Proposition 4.5.5.

Exercise 4.6.11 See the proof of Proposition 4.6.3.

Exercise 4.6.12

$$\mathbb{E}[T] = -\frac{d}{ds}\mathbb{E}[e^{-sT}]_{|s=0} = \frac{1}{\mu}(\alpha p_\alpha + \beta(1 - p_\beta)) = \frac{\mathbb{E}[X_T]}{\mu}.$$

Exercise 4.6.14 (b) Applying the Optional Stopping Theorem

$$\begin{aligned} \mathbb{E}[e^{cM_T - \lambda T(e^c - c - 1)}] = \mathbb{E}[X_0] &= 1 \\ \mathbb{E}[e^{ca - \lambda T f(c)}] &= 1 \\ \mathbb{E}[e^{-\lambda T f(c)}] &= e^{-ac}. \end{aligned}$$

Let $s = f(c)$, so $c = \varphi(s)$. Then $\mathbb{E}[e^{-\lambda s T}] = e^{-a\varphi(s)}$.

(c) Differentiating and taking $s = 0$ yields

$$\begin{aligned} -\lambda\mathbb{E}[T] &= -ae^{-a\varphi(0)}\varphi'(0) \\ &= -a\frac{1}{f'(0)} = -\infty, \end{aligned}$$

so $\mathbb{E}[T] = \infty$.

(d) The inverse Laplace transform $\mathcal{L}^{-1}\left(e^{-a\varphi(s)}\right)$ cannot be represented by elementary functions.

Exercise 4.9.9 Use $\mathbb{E}[(|X_t| - 0)^2] = \mathbb{E}[|X_t|^2] = \mathbb{E}[(X_t - 0)^2]$.

Exercise 4.10.3 (*a*) $L = 1$.
(*b*) A computation shows

$$\begin{aligned}\mathbb{E}[(X_t - 1)^2] &= \mathbb{E}[X_t^2 - 2X_t + 1] = \mathbb{E}[X_t^2] - 2\mathbb{E}[X_t] + 1 \\ &= Var(X_t) + (\mathbb{E}[X_t] - 1)^2.\end{aligned}$$

(*c*) Since $\mathbb{E}[X_t] = 1$, we have $\mathbb{E}[(X_t - 1)^2] = Var(X_t)$. Since

$$Var(X_t) = e^{-t}\mathbb{E}[e^{W_t}] = e^{-t}(e^{2t} - e^t) = e^t - 1,$$

then $\mathbb{E}[(X_t - 1)^2]$ does not tend to 0 as $t \to \infty$.

Exercise 4.11.4 Since M_t and N_t are martingales, then $M_t + N_t$ and $M_t - N_t$ are also martingales. Then $(M_t+N_t)^2 - \langle M+N, M+N\rangle_t$ and $(M_t-N_t)^2 - \langle M-N, M-N\rangle_t$ are martingales. Subtracting, yields $M_tN_t - \langle M, N\rangle_t$ martingale.

Exercise 4.11.8 (*a*) $\mathbb{E}[(dW_t)^2 - dt^2] = \mathbb{E}[(dW_t)^2] - dt^2 = 0$.

(*b*)
$$\begin{aligned}Var((dW_t)^2 - dt) &= \mathbb{E}[(dW_t^2 - dt)^2] = \mathbb{E}[(dW_t)^4 - 2dtdW_t + dt^2] \\ &= 3st^2 - 2dt \cdot 0 + dt^2 = 4dt^2.\end{aligned}$$

Exercise 4.12.2 (*a*) $dt\,dN_t = dt(dM_t + \lambda dt) = dt\,dM_t + \lambda dt^2 = 0$

(*b*) $dW_t\,dN_t = dW_t(dM_t + \lambda dt) = dW_t dM_t + \lambda dW_t dt = 0$.

Chapter 5

Exercise 5.2.3 (*a*) Use either the definition or the moment generation function to show that $\mathbb{E}[W_t^4] = 3t^2$. Using stationarity, $\mathbb{E}[(W_t - W_s)^4] = \mathbb{E}[W_{t-s}^4] = 3(t-s)^2$.

Exercise 5.4.1 (*a*) $\mathbb{E}[\int_0^T dW_t] = \mathbb{E}[W_T] = 0$.

(*b*) $\mathbb{E}[\int_0^T W_t\,dW_t] = \mathbb{E}[\frac{1}{2}W_T^2 - \frac{1}{2}T] = 0$.

(*c*) $Var(\int_0^T W_t\,dW_t) = \mathbb{E}[(\int_0^T W_t\,dW_t)^2] = \mathbb{E}[\frac{1}{4}W_T^2 + \frac{1}{4}T^2 - \frac{1}{2}TW_T^2] = \frac{T^2}{2}$.

Exercise 5.6.3 $X \sim N(0, \int_1^T \frac{1}{t}dt) = N(0, \ln T)$.

Exercise 5.6.4 $Y \sim N(0, \int_1^T t dt) = N\Big(0, \frac{1}{2}(T^2 - 1)\Big)$

Exercise 5.6.5 Normally distributed with zero mean and variance $\int_0^t e^{2(t-s)}\,ds = \frac{1}{2}(e^{2t}-1)$.

Exercise 5.6.6 Using the property of Wiener integrals, both integrals have zero mean and variance $\frac{7t^3}{3}$.

Exercise 5.6.7 The mean is zero and the variance is $t/3 \to 0$ as $t \to 0$.

Exercise 5.6.8 Since it is a Wiener integral, X_t is normally distributed with zero mean and variance

$$\int_0^t \Big(a + \frac{bu}{t}\Big)^2 du = (a^2 + \frac{b^2}{3} + ab)t.$$

Hence $a^2 + \frac{b^2}{3} + ab = 1$.

Exercise 5.6.9 Since both W_t and $\int_0^t f(s)\,dW_s$ have the mean equal to zero,

$$\begin{aligned} Cov\Big(W_t, \int_0^t f(s)\,dW_s\Big) &= \mathbb{E}[W_t, \int_0^t f(s)\,dW_s] = \mathbb{E}[\int_0^t dW_s \int_0^t f(s)\,dW_s] \\ &= \mathbb{E}[\int_0^t f(u)\,ds] = \int_0^t f(u)\,ds. \end{aligned}$$

The general result is

$$Cov\Big(W_t, \int_0^t f(s)\,dW_s\Big) = \int_0^t f(s)\,ds.$$

Choosing $f(u) = u^n$ yields the desired identity.

Exercise 5.8.6 Apply the expectation to

$$\Big(\sum_{k=1}^{N_t} f(S_k)\Big)^2 = \sum_{k=1}^{N_t} f^2(S_k) + 2\sum_{k\neq j}^{N_t} f(S_k)f(S_j).$$

Exercise 5.9.1 We have

$$\begin{aligned} \mathbb{E}\Big[\int_0^T e^{ks}\,dN_s\Big] &= \frac{\lambda}{k}(e^{kT}-1) \\ Var\Big(\int_0^T e^{ks}\,dN_s\Big) &= \frac{\lambda}{2k}(e^{2kT}-1). \end{aligned}$$

Chapter 6

Exercise 6.1.6 Let $X_t = \int_0^t e^{W_u}\,du$. Then

$$dG_t \;=\; d\Big(\frac{X_t}{t}\Big) = \frac{tdX_t - X_tdt}{t^2} = \frac{te^{W_t}dt - X_tdt}{t^2} = \frac{1}{t}\Big(e^{W_t} - G_t\Big)dt.$$

Exercise 6.2.4

(*a*) $e^{W_t}(1+\frac{1}{2}W_t)dt + e^{W_t}(1+W_t)dW_t$;

(*b*) $(6W_t + 10e^{5W_t})dW_t + (3 + 25e^{5W_t})dt$;

(*c*) $2e^{t+W_t^2}(1+W_t^2)dt + 2e^{t+W_t^2}W_tdW_t$;

(*d*) $n(t+W_t)^{n-2}\Big((t+W_t+\frac{n-1}{2})dt + (t+W_t)dW_t\Big)$;

(*e*) $\displaystyle \frac{1}{t}\left(W_t - \frac{1}{t}\int_0^t W_u\,du\right)dt$;

(*f*) $\displaystyle \frac{1}{t^\alpha}\left(e^{W_t} - \frac{\alpha}{t}\int_0^t e^{W_u}\,du\right)dt$.

Exercise 6.2.5

$$d(tW_t^2) = td(W_t^2) + W_t^2dt = t(2W_tdW_t + dt) + W_t^2dt = (t+W_t^2)dt + 2tW_tdW_t.$$

Exercise 6.2.6 (*a*) $tdW_t + W_tdt$;

(*b*) $e^t(W_tdt + dW_t)$;

(*c*) $(2-t/2)t\cos W_t\,dt - t^2\sin W_t\,dW_t$;

(*d*) $(\sin t + W_t^2\cos t)dt + 2\sin t\,W_t\,dW_t$;

Exercise 6.2.9 It follows from (6.2.11).

Exercise 6.2.10 Take the conditional expectation in

$$M_t^2 = M_s^2 + 2\int_s^t M_{u-}\,dM_u + N_t - N_s$$

and obtain

$$\begin{aligned}
\mathbb{E}[M_t^2|\mathcal{F}_s] &= M_s^2 + 2\mathbb{E}[\int_s^t M_{u-}\,dM_u|\mathcal{F}_s] + \mathbb{E}[N_t|\mathcal{F}_s] - N_s \\
&= M_s^2 + \mathbb{E}[M_t + \lambda t|\mathcal{F}_s] - N_s \\
&= M_s^2 + M_s + \lambda t - N_s \\
&= M_s^2 + \lambda(t-s).
\end{aligned}$$

Exercise 6.2.12 Integrating in (6.2.12) yields

$$F_t = F_s + \int_s^t \frac{\partial f}{\partial x}\, dW_t^1 + \int_s^t \frac{\partial f}{\partial y}\, dW_t^2.$$

One can check that $\mathbb{E}[F_t|\mathcal{F}_s] = F_s$.

Exercise 6.2.13 (a) $dF_t = 2W_t^1 dW_t^1 + 2W_t^2 dW_t^2 + 2dt$;

(b) $dF_t = \dfrac{2W_t^1 dW_t^1 + 2W_t^2 dW_t^2}{(W_t^1)^2 + (W_t^2)^2}$.

Exercise 6.2.14 Consider the function $f(x,y) = \sqrt{x^2+y^2}$. Since $\dfrac{\partial f}{\partial x} = \dfrac{x}{\sqrt{x^2+y^2}}$, $\dfrac{\partial f}{\partial y} = \dfrac{y}{\sqrt{x^2+y^2}}$, $\Delta f = \dfrac{1}{2\sqrt{x^2+y^2}}$, we get

$$dR_t = \frac{\partial f}{\partial x} dW_t^1 + \frac{\partial f}{\partial y} dW_t^2 + \Delta f\, dt = \frac{W_t^1}{R_t} dW_t^1 + \frac{W_t^2}{R_t} dW_t^2 + \frac{1}{2R_t} dt.$$

Chapter 7

Exercise 7.2.4 (a) Use integration formula with $g(x) = \tan^{-1}(x)$.

$$\int_0^T \frac{1}{1+W_t^2}\, dW_t = \int_0^T (\tan^{-1})'(W_t)\, dW_t = \tan^{-1} W_T + \frac{1}{2}\int_0^T \frac{2W_t}{(1+W_t)^2}\, dt.$$

(b) Use $\mathbb{E}\Big[\displaystyle\int_0^T \frac{1}{1+W_t^2}\, dW_t\Big] = 0$.

(c) Use Calculus to find minima and maxima of the function $\varphi(x) = \dfrac{x}{(1+x^2)^2}$.

Exercise 7.2.5 (a) Use integration by parts with $g(x) = e^x$ and get

$$\int_0^T e^{W_t}\, dW_t = e^{W_T} - 1 - \frac{1}{2}\int_0^T e^{W_t}\, dt.$$

(b) Applying the expectation we obtain

$$\mathbb{E}[e^{W_T}] = 1 + \frac{1}{2}\int_0^T \mathbb{E}[e^{W_t}]\, dt.$$

If let $\phi(T) = \mathbb{E}[e^{W_T}]$, then ϕ satisfies the integral equation

$$\phi(T) = 1 + \frac{1}{2}\int_0^T \phi(t)\, dt.$$

Differentiating yields the ODE $\phi'(T) = \frac{1}{2}\phi(T)$, with $\phi(0) = 1$. Solving yields $\phi(T) = e^{T/2}$.

Exercise 7.2.6 (*a*) Apply integration by parts with $g(x) = (x-1)e^x$ to get

$$\int_0^T W_t e^{W_t}\,dW_t = \int_0^T g'(W_t)\,dW_t = g(W_T) - g(0) - \frac{1}{2}\int_0^T g''(W_t)\,dt.$$

(*b*) Applying the expectation yields

$$\begin{aligned}\mathbb{E}[W_T e^{W_T}] &= \mathbb{E}[e^{W_T}] - 1 + \frac{1}{2}\int_0^T \Big[\mathbb{E}[e^{W_t}] + \mathbb{E}[W_t e^{W_t}]\Big]\,dt \\ &= e^{T/2} - 1 + \frac{1}{2}\int_0^T \Big(e^{t/2} + \mathbb{E}[W_t e^{W_t}]\Big)\,dt.\end{aligned}$$

Then $\phi(T) = \mathbb{E}[W_T e^{W_T}]$ satisfies the ODE $\phi'(T) - \phi(T) = e^{T/2}$ with $\phi(0) = 0$.

Exercise 7.2.7 (*a*) Use integration by parts with $g(x) = \ln(1 + x^2)$.
(*e*) Since $\ln(1 + T) \leq T$, the upper bound obtained in (*e*) is better than the one in (*d*), without contradicting it.

Exercise 7.3.3 By straightforward computation.

Exercise 7.3.4 By computation.

Exercise 7.3.15 (*a*) $\frac{e}{\sqrt{2}}\sin(\sqrt{2}W_1)$; (*b*) $\frac{1}{2}e^6\sin(2W_3)$; (*c*) $\frac{1}{\sqrt{2}}(e^{\sqrt{2}W_4-4} - 1)$.

Exercise 7.3.16 Apply Ito's formula to get

$$\begin{aligned}d\varphi(t, W_t) &= (\partial_t\varphi(t, W_t) + \frac{1}{2}\partial_x^2\varphi(t, W_t))dt + \partial_x\varphi(t, W_t)dW_t \\ &= G(t)dt + f(t, W_t)\,dW_t.\end{aligned}$$

Integrating between a and b yields

$$\varphi(t, W_t)|_a^b = \int_a^b G(t)\,dt + \int_a^b f(t, W_t)\,dW_t.$$

Chapter 8

Exercise 8.2.2 (*a*) $X_t = 1 + \sin t - \int_0^t \sin s\,dW_s$, $\mathbb{E}[X_t] = 1 + \sin t$, $Var[X_t] = \int_0^t (\sin s)^2\,ds = \frac{t}{2} - \frac{1}{4}\sin(2t)$;
(*b*) $X_t = e^t - 1 + \int_0^t \sqrt{s}\,dW_s$, $\mathbb{E}[X_t] = e^t - 1$, $Var[X_t] = \frac{t^2}{2}$;
(*c*) $X_t = 1 + \frac{1}{2}\ln(1+t^2) + \int_0^t s^{3/2}\,dW_s$, $\mathbb{E}[X_t] = 1 + \frac{1}{2}\ln(1+t^2)$, $Var(X_t) = \frac{t^4}{4}$.

Exercise 8.3.6 (a) $X_t = \frac{1}{3}W_t^3 - tW_t + e^t$; (b) $X_t = \frac{1}{3}W_t^3 - tW_t - \cos t$; (c) $X_t = e^{W_t - \frac{t}{2}} + \frac{t^3}{3}$; (d) $X_t = e^{t/2}\sin W_t + \frac{t^2}{2} + 1$.

Exercise 8.3.7 (a) $X_t = \frac{1}{4}W_t^4 + tW_t$; (b) $X_t = \frac{1}{2}W_t^2 + t^2W_t - \frac{t}{2}$; (c) $X_t = e^tW_t - \cos W_t + 1$; (d) $X_t = te^{W_t} + 2$.

Exercise 8.4.5 (a)

$$\begin{aligned} dX_t &= (2W_tdW_t + dt) + W_tdt + tdW_t \\ &= d(W_t^2) + d(tW_t) = d(tW_t + W_t^2) \end{aligned}$$

so $X_t = tW_t + W_t^2$.
(b) We have

$$\begin{aligned} dX_t &= (2t - \frac{1}{t^2}W_t)dt + \frac{1}{t}dW_t \\ &= 2tdt + d(\frac{1}{t}W_t) = d(t^2 + \frac{1}{t}W_t), \end{aligned}$$

so $X_t = t^2 + \frac{1}{t}W_t - 1 - W_1$.
(c) $dX_t = \frac{1}{2}e^{t/2}W_tdt + e^{t/2}dW_t = d(e^{t/2}W_t)$, so $X_t = e^{t/2}W_t$.
(d) We have

$$\begin{aligned} dX_t &= t(2W_tdW_t + dt) - tdt + W_t^2dt \\ &= td(W_t^2) + W_t^2dt - \frac{1}{2}d(t^2) \\ &= d\Big(tW_t^2 - \frac{t^2}{2}\Big), \end{aligned}$$

so $X_t = tW_t - \frac{t^2}{2}$.
(e) $dX_t = dt + d(\sqrt{t}W_t) = d(t + \sqrt{t}W_t)$, so $X_t = t + \sqrt{t}W_t - W_1$.

Exercise 8.5.4 (a) $X_t = X_0e^{4t} + \frac{1}{4}(1 - e^{4t}) + 2\int_0^t e^{4(t-s)}\,dW_s$;
(b) $X_t = X_0e^{3t} + \frac{2}{3}(1 - e^{3t}) + e^{3t}W_t$;
(c) $X_t = e^t(X_0 + 1 + \frac{1}{2}W_t^2 - \frac{t}{2}) - 1$;
(d) $X_t = X_0e^{4t} - \frac{t}{4} - \frac{1}{16}(1 - e^{4t}) + e^{4t}W_t$;
(e) $X_t = X_0e^{t/2} - 2t - 4 + 5e^{t/2} - e^t\cos W_t$;
(f) $X_t = X_0e^{-t} + e^{-t}W_t$.

Exercise 8.8.2 (a) The integrating factor is $\rho_t = e^{-\int_0^t \alpha\,dW_s + \frac{1}{2}\int_0^t \alpha^2 ds} = e^{-\alpha W_t + \frac{\alpha^2}{2}t}$, which transforms the equation in the exact form $d(\rho_t X_t) = 0$.

Then $\rho_t X_t = X_0$ and hence $X_t = X_0 e^{\alpha W_t - \frac{\alpha^2}{2}t}$.
(*b*) $\rho_t = e^{-\alpha W_t + \frac{\alpha^2}{2}t}$, $d(\rho_t X_t) = \rho_t X_t dt$, $dY_t = Y_t dt$, $Y_t = Y_0 e^t$, $\rho_t X_t = X_0 e^t$, $X_t = X_0 e^{(1-\frac{\alpha^2}{2})t + \alpha W_t}$.

Exercise 8.8.3 $\sigma t\, dA_t = dX_t - \sigma A_t\, dt$, $\mathbb{E}[A_t] = 0$, $Var(A_t) = \mathbb{E}[A_t^2] = \frac{1}{t^2}\int_0^t \mathbb{E}[X_s]^2\, ds = \frac{X_0^2}{t}$.

Exercise 8.10.1 Integrating yields $X_t = X_0 + \int_0^t (2X_s + e^{2s})\, ds + \int_0^t b\, dW_s$. Taking the expectation we get

$$\mathbb{E}[X_t] = X_0 + \int_0^t (2\mathbb{E}[X_s] + e^{2s})\, ds.$$

Differentiating we obtain $f'(t) = 2f(t) + e^{2t}$, where $f(t) = \mathbb{E}[X_t]$, with $f(0) = X_0$. Multiplying by the integrating factor e^{-2t} yields $(e^{-2t}f(t))' = 1$. Integrating yields $f(t) = e^{2t}(t + X_0)$.

Exercise 8.10.6 (*a*) Using product rule and Ito's formula, we get

$$d(W_t^2 e^{W_t}) = e^{W_t}(1 + 2W_t + \frac{1}{2}W_t^2)dt + e^{W_t}(2W_t + W_t^2)dW_t.$$

Integrating and taking expectations yields

$$\mathbb{E}[W_t^2 e^{W_t}] = \int_0^t \Big(\mathbb{E}[e^{W_s}] + 2\mathbb{E}[W_s e^{W_s}] + \frac{1}{2}\mathbb{E}[W_s^2 e^{W_s}]\Big)\, ds.$$

Since $\mathbb{E}[e^{W_s}] = e^{t/2}$, $\mathbb{E}[W_s e^{W_s}] = te^{t/2}$, if let $f(t) = \mathbb{E}[W_t^2 e^{W_t}]$, we get by differentiation

$$f'(t) = e^{t/2} + 2te^{t/2} + \frac{1}{2}f(t), \qquad f(0) = 0.$$

Multiplying by the integrating factor $e^{-t/2}$ yields $(f(t)e^{-t/2})' = 1 + 2t$. Integrating yields the solution $f(t) = t(1+t)e^{t/2}$.
(*b*) Similar method.

Exercise 8.10.8 (*a*) Using Exercise 3.1.17

$$\begin{aligned}
\mathbb{E}[W_t^4 - 3t^2|\mathcal{F}_t] &= \mathbb{E}[W_t^4|\mathcal{F}_t] - 3t^2 \\
&= 3(t-s)^2 + 6(t-s)W_s^2 + W_s^4 - 3t^2 \\
&= (W_s^4 - 3s^2) + 6s^2 - 6ts + 6(t-s)W_s^2 \neq W_s^4 - 3s^2.
\end{aligned}$$

Hence $W_t^4 - 3t^2$ is not a martingale.
(*b*) $\mathbb{E}[W_t^3|\mathcal{F}_s] = W_s^3 + 3(t-s)W_s \neq W_s^3$, and hence W_t^3 is not a martingale.

Exercise 8.10.10 (a) Similar method as in Example 8.10.9.
(b) Applying Ito's formula

$$d(\cos(\sigma W_t)) = -\sigma \sin(\sigma W_t)dW_t - \frac{1}{2}\sigma^2 \cos(\sigma W_t)dt.$$

Let $f(t) = \mathbb{E}[\cos(\sigma W_t)]$. Then $f'(t) = -\frac{\sigma^2}{2}f(t)$, $f(0) = 1$. The solution is $f(t) = e^{-\frac{\sigma^2}{2}t}$.
(c) Since $\sin(t+\sigma W_t) = \sin t \cos(\sigma W_t) + \cos t \sin(\sigma W_t)$, taking the expectation and using (a) and (b) yields

$$\mathbb{E}[\sin(t + \sigma W_t)] = \sin t\mathbb{E}[\cos(\sigma W_t)] = e^{-\frac{\sigma^2}{2}t}\sin t.$$

(d) Similarly starting from $\cos(t + \sigma W_t) = \cos t \cos(\sigma W_t) - \sin t \sin(\sigma W_t)$.

Exercise 8.10.11 From Exercise 8.10.10 (b) we have $\mathbb{E}[\cos(W_t)] = e^{-t/2}$. From the definition of expectation

$$\mathbb{E}[\cos(W_t)] = \int_{-\infty}^{\infty} \cos x \frac{1}{\sqrt{2\pi t}} e^{-\frac{x^2}{2t}}\, dx.$$

Then choose $t = 1/2$ and $t = 1$ to get (a) and (b), respectively.

Exercise 8.10.12 (a) Using a standard method involving Ito's formula we can get $E(W_t e^{bW_t}) = bte^{b^2t/2}$. Let $a = 1/(2t)$. We can write

$$\begin{aligned}
\int xe^{-ax^2+bx}\, dx &= \sqrt{2\pi t}\int xe^{bx}\frac{1}{\sqrt{2\pi t}}e^{-\frac{x^2}{2t}}\, dx \\
&= \sqrt{2\pi t}E(W_t e^{bW_t}) = \sqrt{2\pi t}bte^{b^2t/2} = \sqrt{\frac{\pi}{a}}\left(\frac{b}{2a}\right)e^{b^2/(4a)}.
\end{aligned}$$

The same method for (b) and (c).

Exercise 8.10.13 (a) We have

$$\begin{aligned}
\mathbb{E}[\cos(tW_t)] &= \mathbb{E}[\sum_{n\geq 0}(-1)^n \frac{W_t^{2n}t^{2n}}{(2n)!}] = \sum_{n\geq 0}(-1)^n \frac{\mathbb{E}[W_t^{2n}]t^{2n}}{(2n)!} \\
&= \sum_{n\geq 0}(-1)^n \frac{t^{2n}}{(2n)!}\frac{(2n)!t^n}{2^n n!} = \sum_{n\geq 0}(-1)^n \frac{t^{3n}}{2^n n!} \\
&= e^{-t^3/2}.
\end{aligned}$$

(b) Similar computation using $\mathbb{E}[W_t^{2n+1}] = 0$.

Chapter 9

Exercise 9.2.2 $A = \frac{1}{2}\Delta = \frac{1}{2}\sum_{k=1}^{n} \partial_k^2$.

Exercise 9.2.4 (*a*) We have

$$\begin{aligned} X_1(t) &= x_1^0 + W_1(t) \\ X_2(t) &= x_2^0 + \int_0^t X_1(s)\,dW_2(s) \\ &= x_2^0 + x_1^0 W_2(t) + \int_0^t W_1(s)\,dW_2(s). \end{aligned}$$

Exercise 9.5.2 (*c*) $\mathbb{E}[\tau]$ is maximum if $b - x_0 = x_0 - a$, i.e. when $x_0 = (a+b)/2$. The maximum value is $(b-a)^2/4$.

Exercise 9.6.1 Let $\tau_k = \min(k, \tau) \nearrow \tau$ and $k \to \infty$. Apply Dynkin's formula for τ_k to show that

$$\mathbb{E}[\tau_k] \le \frac{1}{n}(R^2 - |a|^2),$$

and take $k \to \infty$.

Exercise 9.6.3 x^0 and x^{2-n}.

Exercise 9.7.1 (*a*) We have $a(t,x) = x$, $c(t,x) = x$, $\varphi(s) = xe^{s-t}$ and $u(t,x) = \int_t^T xe^{s-t}\,ds = x(e^{T-t} - 1)$.
(*b*) $a(t,x) = tx$, $c(t,x) = -\ln x$, $\varphi(s) = xe^{(s^2-t^2)/2}$ and

$$u(t,x) = -\int_t^T \ln\left(xe^{(s^2-t^2)/2}\right) ds = -(T-t)\Big[\ln x + \frac{T}{6}(T+t) - \frac{t^2}{3}\Big].$$

Exercise 9.7.2 (*a*) $u(t,x) = x(T-t) + \frac{1}{2}(T-t)^2$.
(*b*) $u(t,x) = \frac{2}{3}e^x\left(e^{\frac{3}{2}(T-t)} - 1\right)$.
(*c*) We have $a(t,x) = \mu x$, $b(t,x) = \sigma x$, $c(t,x) = x$. The associated diffusion is $dX_s = \mu X_s ds + \sigma X_s dW_s$, $X_t = x$, which is the geometric Brownian motion

$$X_s = xe^{(\mu - \frac{1}{2}\sigma^2)(s-t) + \sigma(W_s - W_t)}, \qquad s \ge t.$$

The solution is

$$\begin{aligned} u(t,x) &= \mathbb{E}\Big[\int_t^T xe^{(\mu-\frac{1}{2}\sigma^2)(s-t)+\sigma(W_s-W_t)}\,ds\Big] \\ &= x\int_t^T e^{(\mu-\frac{1}{2}(s-t))(s-t)}\mathbb{E}[e^{\sigma(s-t)}]\,ds \\ &= x\int_t^T e^{(\mu-\frac{1}{2}(s-t))(s-t)}e^{\sigma^2(s-t)/2}\,ds \\ &= x\int_t^T e^{\mu(s-t)}\,ds = \frac{x}{\mu}\left(e^{\mu(T-t)} - 1\right). \end{aligned}$$

Chapter 10

Exercise 10.1.7 Apply Example 10.1.6 with $u = 1$.

Exercise 10.1.8 $X_t = \int_0^t h(s)\,dW_s \sim N(0, \int_0^t h^2(s)\,ds)$. Then e^{X_t} is log-normal with $\mathbb{E}[e^{X_t}] = e^{\frac{1}{2}Var(X_t)} = e^{\frac{1}{2}\int_0^t h(s)^2\,ds}$.

Exercise 10.1.9 (a) Using Exercise 10.1.8 we have

$$\begin{aligned}\mathbb{E}[M_t] &= \mathbb{E}[e^{-\int_0^t u(s)\,dW_s}e^{-\frac{1}{2}\int_0^t u(s)^2\,ds}]\\ &= e^{-\frac{1}{2}\int_0^t u(s)^2\,ds}\mathbb{E}[e^{-\int_0^t u(s)\,dW_s}] = e^{-\frac{1}{2}\int_0^t u(s)^2\,ds}e^{\frac{1}{2}\int_0^t u(s)^2\,ds} = 1.\end{aligned}$$

(b) Similar computation as (a).

Exercise 10.1.10 (a) Applying the product and Ito's formulas we get

$$d(e^{t/2}\cos W_t) = -e^{-t/2}\sin W_t\,dW_t.$$

Integrating yields

$$e^{t/2}\cos W_t = 1 - \int_0^t e^{-s/2}\sin W_s\,dW_s,$$

which is an Ito integral, and hence a martingale; (b) Similarly.

Exercise 10.1.12 Use that the function $f(x_1, x_2) = e^{x_1}\cos x_2$ satisfies $\Delta f = 0$.

Exercise 10.1.14 (a) $f(x) = x^2$; (b) $f(x) = x^3$; (c) $f(x) = x^n/(n(n-1))$; (d) $f(x) = e^{cx}$; (e) $f(x) = \sin(cx)$.

Exercise 10.3.10 Let $X_t = a\int_0^t e^{\frac{a^2 s}{2}}\,dB_s$. The quadratic variation is

$$\langle X, X\rangle_t = \int_0^t (dX_s)^2 = a^{a^2 t} - 1.$$

Then apply Theorem 10.3.1.

Exercise 10.3.11 (a) By direct computation;
(b) Let $X_t = \int_0^t e^{qu}\,dW_u$. The quadratic variation is

$$\langle X, X\rangle_t = \int_0^t e^{2qs}\,ds = \frac{e^{2qt}-1}{2q}.$$

Applying Theorem 10.3.1 yields the result.

Exercise 10.4.15 (*c*) We evaluate $\mathbb{E}^Q[X_t]$ in two ways. On one side $\mathbb{E}^Q[X_t] = 0$, because X_t is a Q-Brownian motion. On the other side, using Girsanov theorem

$$\begin{aligned}\mathbb{E}^Q[X_t] &= \mathbb{E}^P[X_t M_T] = \mathbb{E}^P\Big[\Big(\frac{\lambda t^2}{2} + W_t\Big)e^{-\lambda\int_0^T s\,dW_s}e^{-\frac{\lambda^2 T^3}{6}}\Big] \\ &= \frac{\lambda t^2}{2} + e^{-\frac{\lambda^2}{6}T^3}\mathbb{E}^P[W_t e^{-\lambda\int_0^t s\,dW_s}]\mathbb{E}^P[e^{-\lambda\int_t^T s\,dW_s}] \\ &= \frac{\lambda t^2}{2} + \mathbb{E}^P[W_t e^{-\lambda\int_0^t s\,dW_s}]e^{-\frac{\lambda t^3}{6}}.\end{aligned}$$

Equating to zero yields

$$\mathbb{E}^P[W_t e^{-\lambda\int_0^t s\,dW_s}] = -\frac{\lambda t^2}{2}e^{\frac{\lambda t^3}{6}}.$$

Chapter 11

Exercise 11.2.2 (*a*) Integrating yields

$$v_t = gt + \int_0^t v_s\,dW_s,$$

so $\mathbb{E}[v_t] = gt$.
(*b*) $a_t = g + \sigma v_t \mathcal{N}_t$, $\mathbb{E}[a_t] = g$.
(*c*) Multiply by the integrating factor $\rho_t = e^{-\sigma W_t + \frac{1}{2}\sigma^2 t}$ and obtain the exact equation

$$d(\rho_t v_t) = \rho_t g\,dt.$$

Integrating we get

$$v_t = g e^{\sigma W_t - \frac{1}{2}\sigma^2 t}\int_0^t e^{-\sigma W_s + \frac{1}{2}\sigma^2 s}\,ds.$$

Exercise 11.2.3 (*a*) Solving as a linear equation yields

$$v_t = e^{-2t} + 0.3e^{-2t}\int_0^t e^{2s}\,dW_s,$$

and hence v_t is normally distributed with mean $\mu = e^{-2t}$ and variance $\sigma^2 = \frac{0.09}{4}(1 - e^{-4t})$. In our case $\mu = 0.00247875$ and $\sigma^2 = 0.0224999$, $\sigma = 0.15$. Then

$$P(v_3 < 0.5) = P\Big(\frac{v_3 - \mu}{\sigma} < 0.650\Big) = 0.74.$$

Exercise 11.3.1 (*a*) Substitute $t = h$ in formula

$$N(t) = N(0)e^{-\lambda t} + \sigma e^{-\lambda t} \int_0^t e^{\lambda s}\, dW_s.$$

Then take the expectation in

$$N(0) + \sigma \int_0^h e^{\lambda s}\, dW_s = \frac{1}{2} e^{\lambda h} N(0)$$

and obtain $N(0) = \frac{1}{2}\mathbb{E}[e^{\lambda h}]N(0)$, which implies the desired result.
(*b*) Jensen's inequality for the random variable h becomes $\mathbb{E}[e^{\lambda h}] \geq e^{\mathbb{E}[\lambda h]}$. This can be written as $2 \geq e^{\lambda \mathbb{E}[h]}$.

Exercise 11.3.2 (*a*) Use that $N(t) = N(0)e^{-\lambda t}$. (*b*) $t = -(\ln 0.9)/\lambda$.

Exercise 11.4.1 (*a*) $Q_0(t) = c_1 e^{-t} + c_2 e^{-2t}$. (*c*) $Q_t = Q_t^p + Q_0(t)$.

Exercise 11.4.2 (*a*) Let $Z_t = (X_t, Y_t)^T$ and write the equation as $dZ_t = AZ_t + KdW_t$ and solve it as a linear equation. (*b*) Use substitutions $X_t = \theta_t$, $Y_t = \dot{\theta}_t$.

Exercise 11.5.1 (*a*) P_t is log-normally distributed, with

$$\begin{aligned} P(P_t \leq x) &= P\Big(P_0 e^{\int_0^t a(s)ds - \beta^2 t/2 + \beta B_t}\Big) \\ &= P\Big(B_t \leq \frac{1}{\beta}\ln\frac{x}{P_0} + \frac{\beta t}{2} - \frac{1}{\beta}\int_0^t a(s)ds\Big) \\ &= F_{B_t}\Big(\frac{1}{\beta}\ln\frac{x}{P_0} + \frac{\beta t}{2} - \frac{1}{\beta}\int_0^t a(s)ds\Big), \end{aligned}$$

where $F_{B_t} = \frac{1}{\sqrt{2\pi t}} e^{-\frac{u^2}{2t}}$.

Exercise 11.5.3 The noisy rate is $r_t = a(t) + \frac{dW_t}{dt}$, so $a(t) = t^2$. The expectation is given by $\mathbb{E}[P_T] = P_0 e^{\int_0^T a(s)\, ds} = P_0 e^{T^3/3}$. Then $T = (3\ln 2)^{1/3}$.

Exercise 11.7.1 $dC_t = a(C_0 - C_t)dt + bEdt + bdB_t + \sigma dW_t$.

Exercise 11.7.2 Use formula (11.7.37).

Bibliography

[1] B. Alshamary and O. Calin. *Stochastic optimization approach of car value depreciation.* Applied Stochastic Models in Business and Industry, vol.29, 3, 2012, pp.208-223.

[2] B. Alshamary and O. Calin. *Pricing a stochastic car value depreciation deal.* Applied Stochastic Models in Business and Industry, vol.30, 4, 2014, pp. 509-516.

[3] V.I. Arnold. *Ordinary Differential Equations.* MIT Press, Cambridge, MA, London, 1973.

[4] A. Bain and D. Crisan. *Fundamentals of Stochastic Filtering.* Springer, 2009.

[5] M. Baxter and A. Renie. *Financial Calculus.* Cambridge University Press, 1996.

[6] J. Bertoin. *Lévy Processes.* Cambridge University Press, 121, 1996.

[7] W.E. Boyce and C. DiPrima. *Elementary Differential Equations and Boundary Value Problems, 6th ed.* John Wiley and Sons, Inc., 1997.

[8] R. Brown. *A brief account of microscopical observations made in the months of June, July and August, 1827, on the particles contained in the pollen of plants; and on the general existence of active molecules in organic and inorganic bodies.* Phil. Mag. 4, pp. 161-173, 1828.

[9] Z. Brzezniak and T. Zastawniak. *Basic Stochastic Processes.* Springer, London, Berlin, Heidelberg, 1999.

[10] O. Calin, D.C. Chang, K. Furutani, and C. Iwasaki. *Heat Kernels for Elliptic and Sub-elliptic Operators.* Birkhauser, Applied and Numerical Harmonic Analysis, 2011.

[11] J. R. Cannon. *The One-dimensional Heat Equation.* Encyclopedia of Mathematics and Its Applications, 23, Cambridge University Press, 2008.

[12] K.L. Chung and R. Williams. *Introduction to Stochastic Integration.* Birkhäuser, 1990.

[13] W. D. D. Wackerly, Mendenhall, and R. L. Scheaffer. *Mathematical Statistics with Applications, 7th ed.* Brooks/Cole, 2008.

[14] J.L. Doob. *Stochastic Processes.* John Wiley and Sons, 1953.

[15] R. Durrett. *Stochastic Calculus.* CRC Press, 1996.

[16] E. B. Dynkin. *Markov Processes I, II.* Springer-Verlag, 1965.

[17] A. Einstein. *Uber die von der molekularkinetischen Theorie der Warme geforderte Bewegung von in ruhenden Flussigkeiten suspendierten Teilchen.* Annalen der Physik, 322 (8), pp. 549-560, 1905.

[18] I. Fényes. *Eine wahrscheinlichkeitstheoretische Begründung und Interpretation der Quantenmechanik.* Zeitschrift für Physik, 132, 1952, pp. 81-106.

[19] D. Freedman. *Brownian Motion and Diffusion.* Springer-Verlag, 1983.

[20] C. W. Gardiner. *Handbook of Stochastic Processes, 3nd ed.* Springer, 2004.

[21] P.R. Halmos. *Measure Theory.* Van Nostrand Company, Inc., 1950.

[22] T. Hida. *Brownian Motion.* Springer-Verlag, 1980.

[23] N. Ikeda and S. Watanabe. *Stochastic Differential Equations and Difussion Processes.* North-Holland (2nd ed., 1989.

[24] K. Ito. *Stochastic Integral.* Proc. Imp. Acad. Tokyo, 20, 1944, pp. 519-524.

[25] K. Ito. *On a stochastic integral equation.* Proc. Imp. Acad. Tokyo, 22, 1946, pp. 32-35.

[26] I. Karatzas and S. E. Shreve. *Brownian Motion and Stochastic Calculus, 2nd ed.* Springer-Verlag, 1991.

[27] F. Knight. *Essentials of Brownian Motion and Diffusion.* AMS, 1981, 1989.

[28] A. N. Kolmogorov. *Grundbegriffe der Wahrscheinlichkeitsrechnung.* Ergeb. Math. 2, 1933.

[29] N.V. Krylov and A.K.Zvonkin. *On strong solutions of stochastic differential equations.* Sel. Math. Sov. I, pp. 19-61, 1981.

[30] H.H. Kuo. *Introduction to Stochastic Integration.* Springer, Universitext, 2006.

[31] J. Lamperti. *Probability.* W. A. Benjamin, Inc., 1966.

[32] P. Langevin. *Sur la theorie du mouvement brownien.* C. R. Acad. Sci. (Paris) 146, pp.530-533, 1908.

[33] P. Lévy. *Processus Stochastiques et Mouvement Brownien.* Gauthier-Villars, Paris, 1948.

[34] P. Mörters and Y. Peres. *Brownian Motion.* Cambridge University Press, 2010.

[35] S. Neftici. *Mathematics of Financial Derivatives.* Academic Press, 1996.

[36] E. Nelson. *Dynamical Theories of Brownian Motion.* Princeton University Press, 2nd ed., 1967.

[37] B. Øksendal. *Stochastic Differential Equations, An Introduction with Applications, 6th ed.* Springer-Verlag Berlin Heidelberg New-York, 2003.

[38] J.W. Pitman. *One-dimensional Brownian motion and the three-dimensional Bessel process.* Adv. Appl. Probab. 7 pp. 511-526, 1975.

[39] F.A.A. Postali and P. Picchetti. *Geometric Brownian Motion and structural breaks in oil prices: A quantitative analysis.* Energy Economics, 28, pp. 506-522, 2006.

[40] P. Protter. *Stochastic Integration and Differential Equations.* 2nd ed. Springer-Verlag, 2004.

[41] D. Revuz and M. Yor. *Continuous Martingales and Brownian Motion, 3th ed.* Springer, 2005.

[42] S. O. Rice. *Mathematical Analysis of Random Noise.* Bell Syst. Tech.J., 23, 24 pp. 282-332, 46-156, 1944, 1945.

[43] S. M. Ross. *Stochastic Processes.* Second ed., John Wiley & Sons, Inc., 1996.

[44] W. Schottky. *Uber spontane Stromschwankungen in verschiedenen Elektrizitatsleitern.* Ann. Phys. 57, pp. 451-567, 1918.

[45] D. Sondermann. *Introduction to Stochastic Calculus for Finance.* Springer, Lecture Notes in Economics and Mathematical Systems, 2006.

[46] D.V. Widder. *The Heat Equation.* Academic Press, London, 1975.

[47] N. Wiener. *Differential Space.* J. Math. Phys. 58, 1923, pp.131-174.

[48] M. Yor. *On some exponential functionals of Brownian motion.* Adv. Appl. Prob., 24, pp. 509-531, 1992.

Index

Printed in the United States
By Bookmasters